Derivatization in Analytical Chemistry

Derivatization in Analytical Chemistry

Editor

Paraskevas D. Tzanavaras

MDPI • Basel • Beijing • Wuhan • Barcelona • Belgrade • Manchester • Tokyo • Cluj • Tianjin

Editor
Paraskevas D. Tzanavaras
Chemistry
Aristotle University
Thessaloniki
Greece

Editorial Office
MDPI
St. Alban-Anlage 66
4052 Basel, Switzerland

This is a reprint of articles from the Special Issue published online in the open access journal *Molecules* (ISSN 1420-3049) (available at: www.mdpi.com/journal/molecules/special_issues/derivatization_analytical_chemistry).

For citation purposes, cite each article independently as indicated on the article page online and as indicated below:

LastName, A.A.; LastName, B.B.; LastName, C.C. Article Title. *Journal Name* **Year**, *Volume Number*, Page Range.

ISBN 978-3-0365-4256-0 (Hbk)
ISBN 978-3-0365-4255-3 (PDF)

© 2022 by the authors. Articles in this book are Open Access and distributed under the Creative Commons Attribution (CC BY) license, which allows users to download, copy and build upon published articles, as long as the author and publisher are properly credited, which ensures maximum dissemination and a wider impact of our publications.

The book as a whole is distributed by MDPI under the terms and conditions of the Creative Commons license CC BY-NC-ND.

Contents

Preface to "Derivatization in Analytical Chemistry" vii

Apostolia Tsiasioti, Constantinos K. Zacharis and Paraskevas D. Tzanavaras
Single-Step Hydrolysis and Derivatization of Homocysteine Thiolactone Using Zone Fluidics: Simultaneous Analysis of Mixtures with Homocysteine Following Separation by Fluorosurfactant-Modified Gold Nanoparticles
Reprinted from: *Molecules* **2022**, *27*, 2040, doi:10.3390/molecules27072040 1

Ankhbayar Lkhagva and Hwan-Ching Tai
Dimethylcysteine (DiCys)/*o*-Phthalaldehyde Derivatization for Chiral Metabolite Analyses: Cross-Comparison of Six Chiral Thiols
Reprinted from: *Molecules* **2021**, *26*, 7416, doi:10.3390/molecules26247416 11

Dimitrios Tsikas
GC-MS Analysis of Biological Nitrate and Nitrite Using Pentafluorobenzyl Bromide in Aqueous Acetone: A Dual Role of Carbonate/Bicarbonate as an Enhancer and Inhibitor of Derivatization
Reprinted from: *Molecules* **2021**, *26*, 7003, doi:10.3390/molecules26227003 23

Petra Horká, Vladimír Vrkoslav, Jiří Kindl, Karolina Schwarzová-Pecková and Josef Cvačka
Structural Characterization of Unusual Fatty Acid Methyl Esters with Double and Triple Bonds Using HPLC/APCI-MS^2 with Acetonitrile In-Source Derivatization
Reprinted from: *Molecules* **2021**, *26*, 6468, doi:10.3390/molecules26216468 39

Kyong-Oh Shin and Kyungho Park
A Newly Developed HPLC-UV/Vis Method Using Chemical Derivatization with 2-Naphthalenethiol for Quantitation of Sulforaphane in Rat Plasma
Reprinted from: *Molecules* **2021**, *26*, 5473, doi:10.3390/molecules26185473 61

Carlos A. Valdez and Roald N. Leif
Analysis of Organophosphorus-Based Nerve Agent Degradation Products by Gas Chromatography-Mass Spectrometry (GC-MS): Current Derivatization Reactions in the Analytical Chemist's Toolbox
Reprinted from: *Molecules* **2021**, *26*, 4631, doi:10.3390/molecules26154631 71

Shusuke Uekusa, Mayu Onozato, Tatsuya Sakamoto, Maho Umino, Hideaki Ichiba and Kenji Okoshi et al.
Development of a Derivatization Reagent with a 2-Nitrophenylsulfonyl Moiety for UHPLC-HRMS/MS and Its Application to Detect Amino Acids Including Taurine
Reprinted from: *Molecules* **2021**, *26*, 3498, doi:10.3390/molecules26123498 89

Katarzyna Kurpet, Rafał Głowacki and Grażyna Chwatko
Simultaneous Determination of Human Serum Albumin and Low-Molecular-Weight Thiols after Derivatization with Monobromobimane
Reprinted from: *Molecules* **2021**, *26*, 3321, doi:10.3390/molecules26113321 101

Olga Begou, Kathrin Weber, Bibiana Beckmann and Dimitrios Tsikas
GC-MS Studies on Derivatization of Creatinine and Creatine by BSTFA and Their Measurement in Human Urine
Reprinted from: *Molecules* **2021**, *26*, 3206, doi:10.3390/molecules26113206 119

Svetlana Baskal, Alexander Bollenbach and Dimitrios Tsikas
GC-MS Discrimination of Citrulline from Ornithine and Homocitrulline from Lysine by Chemical Derivatization: Evidence of Formation of N^5-Carboxy-ornithine and N^6-Carboxy-lysine
Reprinted from: *Molecules* **2021**, *26*, 2301, doi:10.3390/molecules26082301 137

Eliise Tammekivi, Signe Vahur, Martin Vilbaste and Ivo Leito
Quantitative GC–MS Analysis of Artificially Aged Paints with Variable Pigment and Linseed Oil Ratios
Reprinted from: *Molecules* **2021**, *26*, 2218, doi:10.3390/molecules26082218 149

Svetlana Baskal, Alexander Bollenbach and Dimitrios Tsikas
Two-Step Derivatization of Amino Acids for Stable-Isotope Dilution GC–MS Analysis: Long-Term Stability of Methyl Ester-Pentafluoropropionic Derivatives in Toluene Extracts
Reprinted from: *Molecules* **2021**, *26*, 1726, doi:10.3390/molecules26061726 163

Yun Ai, Yan Ni Sun, Li Liu, Fang Yuan Yao, Yan Zhang and Feng Yi Guo et al.
Determination of Biogenic Amines in Different Parts of *Lycium barbarum* L. by HPLC with Precolumn Dansylation
Reprinted from: *Molecules* **2021**, *26*, 1046, doi:10.3390/molecules26041046 173

Ibrahim A. Darwish, Hany W. Darwish, Nasr Y. Khalil and Ahmed Y. A. Sayed
[-5]Experimental and Computational Evaluation of Chloranilic Acid as an Universal Chromogenic Reagent for the Development of a Novel 96-Microwell Spectrophotometric Assay for Tyrosine Kinase Inhibitors
Reprinted from: *Molecules* **2021**, *26*, 744, doi:10.3390/molecules26030744 185

Preface to "Derivatization in Analytical Chemistry"

Derivatization is one of the most widely used sample pretreatment techniques in Analytical Chemistry and Chemical Analysis. Reagent-based or reagent-less schemes offer improved detectability of target compounds, modification of the chromatographic properties and/or the stabilization of sensitive compounds until analysis. Either coupled with separation techniques or as a "stand alone" analytical procedure, derivatization offers endless possibilities in all aspects of analytical applications.

Paraskevas D. Tzanavaras
Editor

Article

Single-Step Hydrolysis and Derivatization of Homocysteine Thiolactone Using Zone Fluidics: Simultaneous Analysis of Mixtures with Homocysteine Following Separation by Fluorosurfactant-Modified Gold Nanoparticles

Apostolia Tsiasioti [1], Constantinos K. Zacharis [2] and Paraskevas D. Tzanavaras [1,*]

1. Laboratory of Analytical Chemistry, School of Chemistry, Faculty of Sciences, Aristotle University of Thessaloniki, 54124 Thessaloniki, Greece; atsiasioti@gmail.com
2. Laboratory of Pharmaceutical Analysis, Department of Pharmaceutical Technology, School of Pharmacy, Aristotle University of Thessaloniki, 54124 Thessaloniki, Greece; czacharis@pharm.auth.gr
* Correspondence: ptzanava@chem.auth.gr; Tel.: +30-23-1099-7721

Abstract: Herein, we report a new automated flow method based on zone fluidics for the simultaneous determination of homocysteine and homocysteine thiolactone using fluorimetric detection (λ_{ext} = 370 nm/λ_{em} = 480 nm). Homocysteine thiolactone is hydrolyzed on-line in alkaline medium (1 mol L^{-1} NaOH) to yield homocysteine, followed by reaction with o-phthalaldehyde in a single step. Derivatization is rapid without the need of elevated temperatures and stopped-flow steps, while specificity is achieved through a unique reaction mechanism in the absence of nucleophilic compounds. Mixtures of the analytes can be analyzed quantitatively after specific separation with fluorosurfactant-capped gold nanoparticles that are selectively aggregated by homocysteine, leaving the thiolactone analogue in solution. As low as 100 nmol L^{-1} of the analyte(s) can be quantified in aqueous solutions, while concentrations > 2 µmol L^{-1} can be analyzed in artificial and real urine matrix following 20-fold dilution. The percent recoveries ranged between 87 and 119%.

Keywords: homocysteine thiolactone; homocysteine; zone fluidics; o-phthalaldehyde; fluorosurfactant-modified gold nanoparticles

1. Introduction

Homocysteine thiolactone (HTL) is a well-known modifying factor of proteins, and its role in the pathogenesis of different diseases has started to be recognized [1–4]. HTL is a chemically reactive metabolite generated by methionyl-tRNA synthetase and cleared by the kidney [5]. There are numerous recent studies trying to elucidate the role of HTL in human health, including the oxidative status of liver and intestines [6], sperm function [7], blood vessel disfunction [8], and cardiovascular diseases [9].

HTL has, therefore, attracted the interest of analytical chemists and there are various methods in the literature reporting its determination in biological material, either alone [10–16] or in combination with HCY [17–19]. Due to the complexity of the biological matrices, the majority of the methods take advantage of the enhanced selectivity features of separation instrumental techniques, such as gas chromatography (GC) [11,13,17], liquid chromatography (HPLC) [15,16,18], and capillary electrophoresis [10,12,19,20]. Electrophoretic techniques offer low operational costs and high separation efficiency but generally low sensitivity. HTL/HCY can be detected directly using simple UV detection, but sensitivity enhancement to sub-micromolar levels requires preconcentration by either single-drop microextraction (SDME) [10,20], or by field-amplified sample stacking [10,12]. GC-MS is reported to be able to detect HTL/HCY selectively at micromolar levels with a derivatization/extraction step always being necessary to improve the volatility of the analytes [11,13,17]. HPLC is considered to be by far the most widely applied technique in bioanalysis and there are a couple of recent elegant reports on the analysis of HTL/HCY. For example, HTL/HCY were derivatized with *o*-phthalaldehyde on-column (the reagent was incorporated in the mobile phase), resulting in sharp peaks and fast elution. However, the stability of the reversed phase column under highly alkaline conditions (0.1 mol L^{-1} NaOH in the mobile phase) should always be of concern [13]. Alternatively, the analyte(s) can be determined by HPLC directly (UV at 240 nm) [16] or after post-column derivatization combined with cation exchange purification [15].

In our previous work, we have studied the selective reaction of HCY with *o*-phthalaldehyde (OPA) in highly alkaline medium under flow conditions using the concept of zone fluidics (ZF) [21]. Herein, we expand our work on investigating the potential of simultaneous determining of HCY and HTL based on the rapid alkaline hydrolysis of the latter under flow conditions [22]. Our goal is to achieve quantitative conversion of HTL to HCY and derivatization with OPA in a single run. Analysis of mixtures is accomplished by a simple (centrifugation-based) and elegant off-line step based on the different interactions of the analytes with fluorosurfactant (FSN)-modified gold nanoparticles (GNPs) [14]. To the best of our knowledge, this is the first automated flow assay for HTL reported in the literature.

2. Results and Discussion

2.1. Hydrolysis of HTL under Flow Conditions

HCY reacts with OPA under flow conditions and in highly alkaline medium (0.5 mol L^{-1} NaOH [21]) to form a highly fluorescent derivative in the absence of nucleophilic reagents. The chemical system is specific in the presence of cysteine and other common amino acids and highly selective against histidine, histamine, and glutathione. On the other hand, HTL can react with the derivatizing reagent only after cleavage of the thiolactone ring to yield HCY (Figure 1) [22].

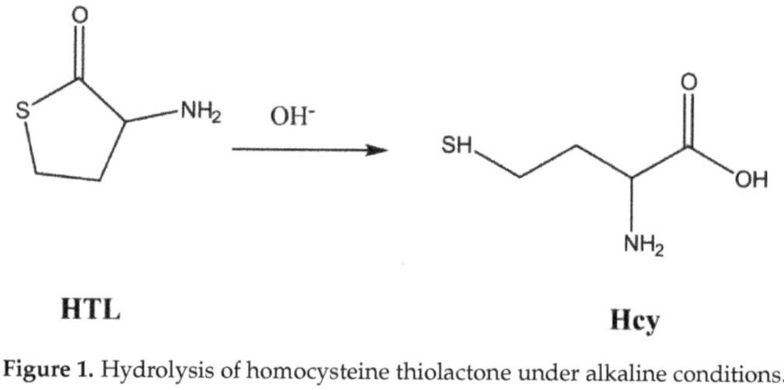

Figure 1. Hydrolysis of homocysteine thiolactone under alkaline conditions.

The potential of automating the hydrolysis and derivatization in a single step under zone fluidics was investigated using the setup described in Section 3.2. Equal amount concentrations of HCY and HTL (aqueous solutions of 0.75 µmol L^{-1} each) were processed sequentially using elevating concentrations of NaOH (0.5 to 2.0 mol L^{-1}). The experimental results are depicted in Figure 2 and clearly demonstrate the effective hydrolysis of HTL (97–101%) at all NaOH levels (at [NaOH] > 1 mol L^{-1}, the sensitivity decreased equally for both analytes). It is also worth mentioning that no heating of the reaction coil nor stopped-flow was necessary to improve the cleavage of the thiolactone ring, simplifying the on-line assay. Based on the findings in Figure 2, a concentration of NaOH of 1.0 mol L^{-1} was selected for further experiments.

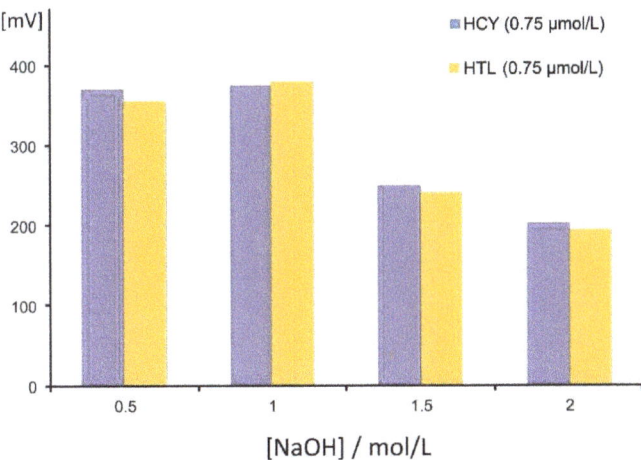

Figure 2. Effect of the concentration of NaOH on the hydrolysis of homocysteine thiolactone.

In a following series of experiments, the effective cleavage of the thiolactone ring under flow conditions was investigated at the entire practical linearity range in the artificial urine matrix (2 to 30 µmol L^{-1}). The experimental procedure involved the steps described in Section 3.4 under the "Analysis of HCY+HTL". The ratio of the slopes of the curves (29.3 (±0.8) for HTL and 29.8 (±0.6) for HCY) indicated 98.3% conversion within the whole concentration range.

2.2. Separation of HCY and HTL

Since HCY and HTL react in a rather identical way with OPA/NaOH under flow conditions, simultaneous analysis can be carried out only after a simple and yet effective separation step.

GNPs have been evolved as viable alternatives for both sample preparation and sensor development in bioanalysis [23–25]. Fluorosurfactant-capped GNPs (FSN-GNPs) have proven to offer enhanced specificity and, most importantly, stability under high salinity conditions [26]. FSN interacts with the nanoparticles through the hydrophilic end of the molecule, while the hydrophobic chains remain dispersed in the solution [27,28]. Small molecules, such as HCY, can penetrate the FSN layer and interact with the GNPs, causing aggregation. Larger molecules are repelled through strong hydrophobic interactions, offering unique selectivity properties.

On this basis, Huang and Cheng have reported quantitative removal of HCY (ca. 98%) using the FSN-GNP-based procedure described in Section 3.4 [14]. To verify their findings, artificial urine matrix spiked with HCY in the range of 2–30 µmol L^{-1} (final concentrations of 0.1–1.5 µmol L^{-1}) were processed either directly or following the separation step. The experimental results are depicted in Figure 3. Based on the ratios of the slopes, ca. 97.1% removal of HCY was achieved.

A second series of experiments confirmed the absence of interaction of the FSN-GNPs with HTL at two concentration levels, namely 5 and 10 µmol L^{-1}. Repetitive separation experiments resulted in satisfactory recoveries in the range of 95–108%, both in the absence and in the presence of HCY (20 µmol L^{-1}) (Figure 4).

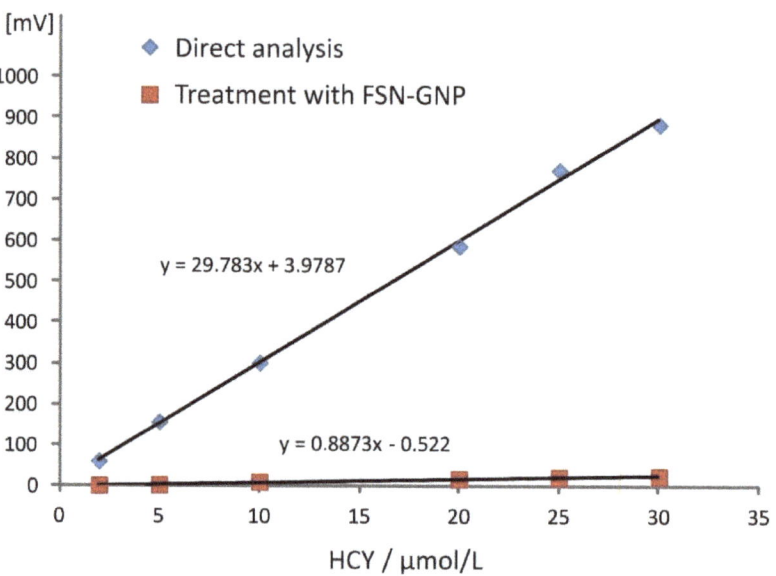

Figure 3. Study of the efficiency of the removal of homocysteine by the fluorosurfactant-capped gold nanoparticles.

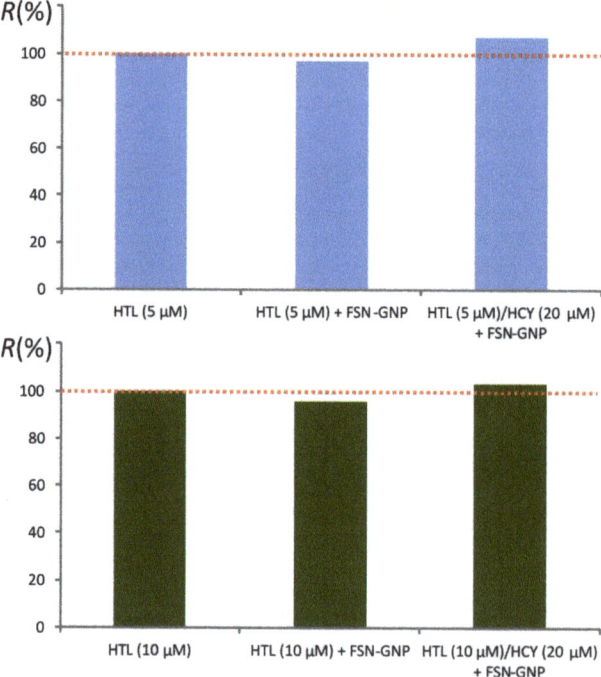

Figure 4. Recoveries of homocysteine thiolactone following treatment with fluorosurfactant-capped gold nanoparticles.

The repeatability of the separation protocol was evaluated by processing mixtures of HTL (10 µmol L^{-1}) and HCY (20 µmol L^{-1}) through six replicate experiments (EXP1–EXP6), followed by analysis by the developed ZF method. The RSD was quite satisfactory, being <5% (Figure S1).

2.3. Analytical Figures of Merit

Linearity for both analytes was evaluated in artificial urine matrix following the experimental procedure described in Section 3.4. The respective regression equations in the range of 2–30 µmol L^{-1} (corresponding to 0.1–1.5 µmol L^{-1} in aqueous solutions) were obtained in a cumulative way by incorporating data from independent analyses through six nonconsecutive working days (36 data points for each analyte):

$$F(HCY) = 30.78\ (\pm 0.91) \times [HCY] + 1.98\ (\pm 6.65),\ r = 0.991$$
$$F(HTL) = 29.12\ (\pm 0.63) \times [HTL] + 1.24\ (\pm 5.52),\ r = 0.998 \quad (1)$$

The LOD for both analytes was estimated to be 0.6 µmol L^{-1} (based on the standard deviation of the intercept rule) and the LLOQ = 2 µmol L^{-1} (lower level of the calibration graph with residuals of <±10%). Both values refer to the artificial urine matrix considering the 20-fold dilution.

The within-day precision was evaluated by repetitive injections (n = 8) of HCY and HTL in artificial urine matrix at a low (2 µmol L^{-1}) and a medium level (10 µmol L^{-1}). The RSD values were in the range of 1.1 and 2.4% in all cases. The between-days precision was validated by obtaining independent calibration curves for both analytes within six nonconsecutive working days. The respective RSD values of the slopes were 5.2% (HCY) and 5.9% (HTL).

2.4. Analysis of HCY and HTL in Artificial Urine

The feasibility of the proposed procedure was evaluated by analysis of various mixtures of HCY and HTL in artificial urine matrix. Based on our previous findings, 20-fold dilution of the matrix is adequate to avoid matrix effects (see Section 3.4) [21]. The experimental results are included in Table 1 (samples S1–S10). As indicated by the percent recoveries, the incorporation of the hydrolysis of HTL in alkaline medium and the derivatization reaction with OPA in a ZF configuration combined with an effective separation pretreatment step offers satisfactory accuracy (R = 89–116%).

Table 1. Determination of HCY/HTL in artificial urine (S1–S10) and in human urine (S11–S15).

Sample	HTL (µmol L^{-1})	Recovery (%)	HCY (µmol L^{-1})	Recovery (%)
S1	—	—	2	89 (±3)
S2	2	112 (±5)	—	—
S3	5	95 (±3)	5	109 (±5)
S4	5	109 (±5)	10	97 (±3)
S5	5	108 (±4)	20	110 (±4)
S6	—	—	10	91 (±4)
S7	10	115 (±5)	—	—
S8	10	112 (±4)	10	102 (±3)
S9	20	92 (±5)	5	116 (±5)
S10	20	90 (±2)	—	—
S11	5	87 (±5)	—	—
S12	5	89 (±4)	5	85 (±5)
S13	—	—	10	114 (±6)
S14	10	119 (±3)	5	107 (±4)
S15	10	102 (±4)	10	90 (±6)

Potential applicability has also been examined in pooled human urine, focusing on the evaluation of the matrix effect. Real urine samples (see Section 3.4) were spiked with the analytes and processed at two dilution factors, namely 1:10 and 1:20. The experimental results were quite similar to our findings for artificial urine [21]. Based on the ratios of the slopes, the matrix effect at 1:10 dilution was calculated to be -22.6%, whereas 1:20 dilution offered an acceptable matrix effect of -5.1%. As can be seen in Table 1 (S11–S15), the percent recoveries ranged between 87 and 119%.

3. Materials and Methods

3.1. Instrumentation

The zone fluidics (ZF) instrumentation consisted of the following parts: a Minipuls3 peristaltic pump (Gilson, Middleton, WI, USA), a micro-electrically actuated 10-port valve (Valco, Brockville, ON, Canada), and an RF-551 flow-through spectrofluorimetric detector operated at high sensitivity (Shimadzu, Kyoto, Japan). The flow connections were made of PTFE tubing (0.5 or 0.7 mm i.d.), except for the connection used in the peristaltic pump that was made of Tygon tubing. The reaction coil (RC, 100 cm × 0.5 mm i.d.) was tightly wrapped around a metallic rod (10 cm × 4.6 mm i.d.) and was thermostated at the desired temperature (±0.1 °C) using an HPLC column heater (Jones Chromatography).

The ZF system was operated through the LabVIEW®, a home-developed program (National Instruments, Austin, TX, USA). Data acquisition (as peak heights) was carried out through the Clarity® software (version 4.0.3, DataApex, Prague, Czech Republic).

3.2. Reagents and Materials

Homocysteine (HCY, Merck), homocysteine thiolactone (HTL, Sigma), o-phthalaldehyde (OPA, Fluka), NaOH (Sigma), and HCl (Sigma) were all of analytical grade. Doubly deionized water was produced by a B30 water purification system (Adrona SIA, Riga, Latvia).

The standard stock solutions of the analytes were prepared at a concentration level of 10 mmol L^{-1} by dissolving accurately weighted amounts in 50 mmol L^{-1} HCl and were kept at 4 °C. Working aqueous standards were prepared daily by serial dilutions in doubly deionized water in the range of 0.1–1.5 µmol L^{-1} for HCY and HTL (2–30 µmol L^{-1} in artificial urine). The OPA solution (c = 10 mmol L^{-1}) was prepared by dissolving the appropriate amount in 500 µL of MeOH + 9500 µL water and was stable in light-protected vials for 5 working days when stored at 4 °C. The NaOH solutions were prepared at the required concentration levels in water (0.5–2 mol L^{-1}).

FSN-modified GNPs were prepared following the procedure described by Huang and Cheng [14]. In brief, 54 µL of $HAuCl_4$ (10% m/v, Sigma) was added rapidly to an aqueous solution of sodium citrate (60 mL, 0.075% m/v) under vigorous boiling and continuous stirring. The resulting mixture was heated under reflux for an additional 15 min to obtain a deep-red-colored solution (λ_{max} = 520 nm). Modification of the GNPs with FSN was carried out by adding 240 µL of the surfactant (10% v/v, Sigma) in the GNPs (60 mL) at room temperature and under stirring, following by storage at 4 °C.

The composition of the artificial urine matrix can be seen in Table S1. All compounds were mixed in deionized water and the pH of the solution was adjusted to 6.0 by addition of 1.0 mol L^{-1} hydrochloric acid.

3.3. ZF Procedure

The ZF procedure for the determination of HCY and HTL consisted of the following steps (Figure 5) [21]: 50 µL of OPA (10 mmol L^{-1}), 50 µL of NaOH (1 mol L^{-1}), and 150 µL of standards were aspirated in this order in the holding coil through the respective ports of the multi-position valve. The flow was reversed and the zones passed through a 100-cm-long reaction coil at a flow rate of 0.6 mL min^{-1}, in which the HCY–OPA reaction product was formed (or hydrolysis and derivatization in the case of HTL). The derivatives were monitored fluorimetrically as peak heights at $\lambda_{ext}/\lambda_{em}$ = 370/480 nm.

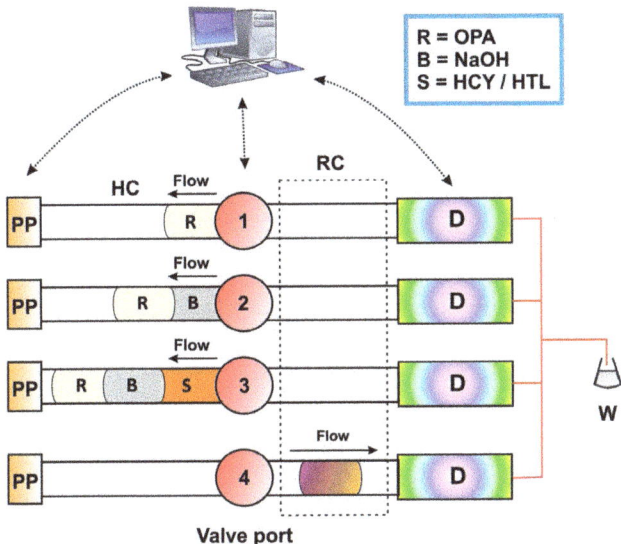

Figure 5. Schematic diagram of the zone fluidics setup: PP = peristaltic pump; HC = holding coil; RC = reaction coil; D = fluorimetric detector; W = waste.

HCY and HTL were determined in two successive runs; (i) in a first run, total HCY + HTL was determined without prior separation of the analytes, and (ii) HTL was quantified following separation using the FSN-GNP-based procedure in a second run. HCY was estimated by difference.

3.4. Preparation of Samples

Artificial urine matrix was prepared as described in Section 2.2 and aliquots were spiked with HCY, HTL, and their mixtures in the range of 2 to 30 µmol L^{-1}. In an analogous way, a pooled human urine sample from apparently healthy volunteers ($n = 6$, members of the lab) was also utilized.

Analysis of HCY+HTL: 50 µL of the spiked matrix (artificial or human urine) was diluted to 1000 µL with water and analyzed under the ZF conditions proposed above.

Analysis of HTL: 50 µL of the spiked matrix (artificial or human urine) was diluted to 500 µL with water, followed by the addition of 500 µL of FSN-GNP solution. The mixture was allowed to react for 20 min and the aggregated nanoparticles were separated by centrifugation (18,000 rpm, 20 min). The supernatant was analyzed directly by the ZF procedure.

4. Conclusions

The present report offers—to our opinion—some interesting features: (i) this is, to the best of our knowledge, the first method for the assay of HTL using automated flow methods; (ii) hydrolysis in alkaline medium and derivatization of HTL with OPA are carried out rapidly and quantitatively in a single step; (iii) due to the rapid and on-line character of the flow scheme, potential side reactions that are favored under alkaline batch conditions (e.g., formation of 2,5-diketopiperazine [29]) are avoided; (iv) there is no need for elevated temperatures and time-consuming stopped-flow mode; (v) high specificity is achieved by reacting with OPA in the absence of nucleophilic compounds; (vi) HTL and HCY can be quantified in their mixtures following a simple and yet efficient separation step based on the selectivity of FSN-capped GNPs; (vii) sub-micromolar levels can be analyzed in aqueous solutions and low micromolar levels in diluted artificial and human urine without matrix interferences; (viii) application to real human urine requires either the use of a matrix-matched calibration curve or at least 20-fold dilution. Further investigation is required in order to develop an analyte preconcentration scheme that will enable the quantification of the analytes in human urine at the nanomolar level.

Supplementary Materials: The following supporting information can be downloaded at: https://www.mdpi.com/article/10.3390/molecules27072040/s1, Figure S1: Repeatability of the removal of Homocysteine in the presence of Homocysteine thiolactone (six independent experiments, EXP1–EXP6); Table S1: Composition of the artificial urine matrix.

Author Contributions: Conceptualization, P.D.T.; methodology, P.D.T. and C.K.Z.; validation, A.T.; investigation, A.T.; data curation, P.D.T. and C.K.Z.; writing—original draft preparation, A.T.; writing—review and editing, P.D.T. All authors have read and agreed to the published version of the manuscript.

Funding: This research received no external funding.

Institutional Review Board Statement: Not applicable.

Informed Consent Statement: Not applicable.

Data Availability Statement: Not applicable.

Conflicts of Interest: The authors declare no conflict of interest.

Sample Availability: Samples of the compounds are not available from the authors.

References

1. Chubarov, A.S. Homocysteine Thiolactone: Biology and Chemistry. *Encyclopedia* **2021**, *1*, 445–459. [CrossRef]
2. Jakubwski, H. Copper, Heart Disease and Homocysteine Thilactone. *J. Intern. Med.* **2021**, *290*, 229–230. [CrossRef] [PubMed]
3. Paul, S.; Nandi, R.; Ghoshal, K.; Bhattacharyya, M.; Maiti, D.K. A Smart Sensor for Rapid Detection of Lethal Hydrazine in Human Blood and Drinking Water. *New J. Chem.* **2019**, *43*, 3303–3308. [CrossRef]
4. Gątarek, P.; Rosiak, A.; Borowczyk, K.; Głowacki, R.; Kałużna-Czaplińska, J. Higher Levels of Low Molecular Weight Sulfur Compounds and Homocysteine Thiolactone in the Urine of Autistic Children. *Molecules* **2020**, *25*, 973. [CrossRef]
5. Borowczyk, K.; Piechocka, J.; Głowacki, R.; Dhar, I.; Midtun, Ø.; Tell, G.S.; Ueland, P.M.; Nygård, O.; Jakubowski, H. Urinary excretion of homocysteine thiolactone and the risk of acute myocardial infarction in coronary artery disease patients: The WENBIT trial. *J. Intern. Med.* **2019**, *285*, 232–244. [CrossRef] [PubMed]
6. Stojanović, M.; Šćepanović, L.; Bosnić, O.; Mitrović, D.; Jozanov-Stankov, O.; Šćepanović, V.; Šćepanović, R.; Stojanović, T.; Ilić, S.; Djurić, D. Effects of Acute Administration of D,L-Homocysteine Thiolactone on the Antioxidative Status of Rat Intestine and Liver. *Acta Vet.* **2016**, *66*, 26–36. [CrossRef]
7. Aitken, R.J.; Flanagan, H.M.; Connaughton, H.; Whiting, S.; Hedges, A.; Baker, M.A. Involvement of homocysteine, homocysteine thiolactone, and paraoxonase type 1 (PON-1) in the etiology of defective human sperm function. *Andrology* **2016**, *4*, 345–360. [CrossRef]
8. Smith, R.M.; Kruzliak, P.; Adamcikova, Z.; Zulli, A. Role of Nox inhibitors plumbagin, ML090 and gp91ds-tat peptide on homocysteine thiolactone induced blood vessel dysfunction. *Clin. Exp. Pharmacol. Physiol.* **2015**, *42*, 860–864. [CrossRef]
9. Gu, W.; Lu, J.; Yang, G.; Dou, J.; Mu, Y.; Meng, J.; Pan, C. Plasma homocysteine thiolactone associated with risk of macrovasculopathy in Chinese patients with type 2 diabetes mellitus. *Adv. Ther.* **2008**, *25*, 914–924. [CrossRef]
10. Purgat, K.; Kośka, I.; Kubalczyk, P. The Use of Single Drop Microextraction and Field Amplified Sample Injection for CZE Determination of Homocysteine Thiolactone in Urine. *Molecules* **2021**, *26*, 5687. [CrossRef]
11. Piechocka, J.; Wrońska, M.; Chwatko, G.; Jakubowski, H.; Głowacki, R. Quantification of homocysteine thiolactone in human saliva and urine by gas chromatography-mass spectrometry. *J. Chromatogr. B* **2020**, *1149*, 122155. [CrossRef] [PubMed]
12. Furmaniak, P.; Kubalczyk, P.; Głowacki, R. Determination of homocysteine thiolactone in urine by field amplified sample injection and sweeping MEKC method with UV detection. *J. Chromatogr. B* **2014**, *961*, 36–41. [CrossRef] [PubMed]
13. Wrońska, M.; Chwatko, G.; Borowczyk, K.; Piechocka, J.; Kubalczyk, P.; Głowacki, R. Application of GC–MS technique for the determination of homocysteine thiolactone in human urine. *J. Chromatogr. B* **2018**, *1099*, 18–24. [CrossRef] [PubMed]
14. Huang, C.C.; Tseng, W.L. Role of fluorosurfactant-modified gold nanoparticles in selective detection of homocysteine thiolactone: Remover and sensor. *Anal. Chem.* **2008**, *80*, 6345–6350. [CrossRef] [PubMed]
15. Chwatko, G.; Jakubowski, H. The determination of homocysteine-thiolactone in human plasma. *Anal. Biochem.* **2005**, *337*, 271–277. [CrossRef]
16. Jakubowski, H. The determination of homocysteine-thiolactone in biological samples. *Anal. Biochem.* **2002**, *308*, 112–119. [CrossRef]
17. Piechocka, J.; Wieczorek, M.; Głowacki, R. Gas Chromatography–Mass Spectrometry Based Approach for the Determination of Methionine-Related Sulfur-Containing Compounds in Human Saliva. *Int. J. Mol. Sci.* **2020**, *21*, 9252. [CrossRef]

18. Głowacki, R.; Bald, E.; Jakubowski, H. An on-column derivatization method for the determination of homocysteine-thiolactone and protein N-linked homocysteine. *Amino Acids* **2011**, *41*, 187–194. [CrossRef]
19. Zinellu, A.; Sotgia, S.; Scanu, B.; Pisanu, E.; Sanna, M.; Sati, S.; Deiana, L.; Sengupta, S.; Carru, C. Determination of homocysteine thiolactone, reduced homocysteine, homocystine, homocysteine–cysteine mixed disulfide, cysteine and cystine in a reaction mixture by overimposed pressure/voltage capillary electrophoresis. *Talanta* **2010**, *82*, 1281–1285. [CrossRef]
20. Purgat, K.; Olejarz, P.; Kośka, I.; Głowacki, R.; Kubalczyk, P. Determination of homocysteine thiolactone in human urine by capillary zone electrophoresis and single drop microextraction. *Anal. Biochem.* **2020**, *596*, 113640. [CrossRef]
21. Tsiasioti, A.; Andreou, A.; Tzanavaras, P.D. Selective reaction of homocysteine with o-phthalaldehyde under flow conditions in highly alkaline medium: Fluorimetric determination using zone fluidics. *Luminescence* **2020**, *35*, 1402–1407. [CrossRef]
22. Jakubowski, H. Mechanism of the Condensation of Homocysteine Thiolactone with Aldehydes. *Chem.–Eur. J.* **2006**, *12*, 8039–8043. [CrossRef] [PubMed]
23. Bharadwaj, K.K.; Rabha, B.; Pati, S.; Sarkar, T.; Choudhury, B.K.; Barman, A.; Bhattacharjya, D.; Srivastava, A.; Baishya, D.; Edinur, H.A.; et al. Green Synthesis of Gold Nanoparticles Using Plant Extracts as Beneficial Prospect for Cancer Theranostics. *Molecules* **2021**, *26*, 6389. [CrossRef] [PubMed]
24. Rónavári, A.; Igaz, N.; Adamecz, D.I.; Szerencsés, B.; Molnar, C.; Kónya, Z.; Pfeiffer, I.; Kiricsi, M. Green Silver and Gold Nanoparticles: Biological Synthesis Approaches and Potentials for Biomedical Applications. *Molecules* **2021**, *26*, 844. [CrossRef] [PubMed]
25. Ait-Touchente, Z.; Falah, S.; Scavetta, E.; Chehimi, M.M.; Touzani, R.; Tonelli, D.; Taleb, A. Different Electrochemical Sensor Designs Based on Diazonium Salts and Gold Nanoparticles for Pico Molar Detection of Metals. *Molecules* **2020**, *25*, 3903. [CrossRef]
26. Lu, C.; Zu, Y.; Yam, V.W.W. Specific postcolumn detection method for HPLC assay of homocysteine based on aggregation of fluorosurfactant-capped gold nanoparticles. *Anal. Chem.* **2007**, *79*, 666–672. [CrossRef]
27. Li, F.; Zu, Y. Effect of Nonionic Fluorosurfactant on the Electrogenerated Chemiluminescence of the Tris(2,2′-bipyridine)ruthenium(II)/ Tri-n-propylamine System: Lower Oxidation Potential and Higher Emission Intensity. *Anal. Chem.* **2004**, *76*, 1768–1772. [CrossRef]
28. Lu, C.; Zu, Y. Specific detection of cysteine and homocysteine: Recognizing one-methylene difference using fluorosurfactant-capped gold nanoparticles. *Chem. Commun.* **2007**, *37*, 3871–3873. [CrossRef]
29. Chubarov, A.S.; Zakharova, O.D.; Koval, O.A.; Romaschenko, A.V.; Akulov, A.E.; Zavjalov, E.L.; Razumov, I.A.; Koptyug, I.V.; Knorre, D.G.; Godovikova, T.S. Design of protein homocystamides with enhanced tumor uptake properties for 19F magnetic resonance imaging. *Bioorg. Med. Chem.* **2015**, *23*, 6943–6954. [CrossRef]

Article

Dimethylcysteine (DiCys)/*o*-Phthalaldehyde Derivatization for Chiral Metabolite Analyses: Cross-Comparison of Six Chiral Thiols

Ankhbayar Lkhagva [1] and Hwan-Ching Tai [2],*

[1] Department of Chemistry, National University of Mongolia, Ulaanbaatar 14200, Mongolia; a_lkhagva@uncg.edu
[2] School of Public Health, Xiamen University, Xiamen 361102, China
* Correspondence: hctai@xmu.edu.cn; Tel.: +86-178-500-25-032

Abstract: Metabolomics profiling using liquid chromatography-mass spectrometry (LC-MS) has become an important tool in biomedical research. However, resolving enantiomers still represents a significant challenge in the metabolomics study of complex samples. Here, we introduced N,N-dimethyl-L-cysteine (dimethylcysteine, DiCys), a chiral thiol, for the *o*-phthalaldehyde (OPA) derivatization of enantiomeric amine metabolites. We took interest in DiCys because of its potential for multiplex isotope-tagged quantification. Here, we characterized the usefulness of DiCys in reversed-phase LC-MS analyses of chiral metabolites, compared against five commonly used chiral thiols: N-acetyl-L-cysteine (NAC); N-acetyl-D-penicillamine (NAP); isobutyryl-L-cysteine (IBLC); N-(*tert*-butoxycarbonyl)-L-cysteine methyl ester (NBC); and N-(*tert*-butylthiocarbamoyl)-L-cysteine ethyl ester (BTCC). DiCys and IBLC showed the best overall performance in terms of chiral separation, fluorescence intensity, and ionization efficiency. For chiral separation of amino acids, DiCys/OPA also outperformed Marfey's reagents: 1-fluoro-2-4-dinitrophenyl-5-L-valine amide (FDVA) and 1-fluoro-2-4-dinitrophenyl-5-L-alanine amide (FDAA). As proof of principle, we compared DiCys and IBLC for detecting chiral metabolites in aqueous extracts of rice. By LC–MS analyses, both methods detected twenty proteinogenic L-amino acids and seven D-amino acids (Ala, Arg, Lys, Phe, Ser, Tyr, and Val), but DiCys showed better analyte separation. We conclude that DiCys/OPA is an excellent amine-derivatization method for enantiomeric metabolite detection in LC-MS analyses.

Keywords: chiral metabolomics; rice water; d-amino acids; enantiomer separation; dimethyl labeling

---

1. Introduction

In the post-genomics era, metabolomics profiling has become an important tool in biomedical research [1–4]. For highly complex metabolomes, reversed-phase liquid chromatography–tandem MS (RP-LC-MS/MS) analyses is the standard tool for high-throughput discovery [5–9]. One of the fundamental limitations of RP-LC-MS is the lack of stereoselectivity, but many important metabolites are chiral molecules. Recently, chiral metabolomics has become an area of emerging interest [10–15].

Initial interests in chiral metabolomics began with D-amino acids, which are physiologically active substances in mammals [16,17]. In fact, D-serine, D-aspartate, D-alanine, and D-cysteine are found in many tissues and body fluids, and several D-amino acids are important neurotransmitters in the brain [18,19]. Enantiomeric amino acids and their derivatives may be useful biomarkers and novel drug candidates; their detection is important in pharmacological research, clinical analysis, agriculture, and food science [20–22]. Using isotope tagging, advanced MS instrumentation, and new MS data analysis schemes, it is possible to carry out non-targeted chiral metabolomics profiling and discover novel chiral biomarkers beyond just amino acids [23].

A classic reagent for the derivatization of amine metabolites is *o*-phthalaldehyde (OPA), widely utilized in commercial amino acid analyzers [24–27]. The chemical reaction with

OPA to form fluorescent isoindole derivatives requires a nucleophilic thiol. Coupling OPA to chiral thiols enables chiral separation via diastereomer formation. Chiral thiols tested for OPA derivatization included N-acetyl-L-cysteine (NAC) [28]; N-acetyl-D-penicillamine (NAP) [29]; isobutyryl-L-cysteine (IBLC) [30]; N-(*tert*-butoxycarbonyl)-L-cysteine methyl ester (NBC) [31]; N-(*tert*-butylthiocarbamoyl)-L-cysteine ethyl ester (BTCC) [32]; N-R-mandelyl-L-cysteine (NMC) [33,34]; and 2,3,4,6-tetra-*o*-acetyl-1-thio-β-D-glucopyranose (TATG) [34].

In advanced chiral metabolomics profiling, labeling with heavy isotopes is very important for quantification. To our knowledge, no study has introduced isotope labels via thiol/OPA derivatization. We are particularly interested in developing N,N-dimethyl-L-cysteine (DiCys) with OPA as a potential strategy for isotope tags in chiral metabolomics. DiCys can be easily synthesized in one step from L-cysteine by reductive amination (dimethyl labeling), using formaldehyde (CH_2O) and sodium cyanoborohydride ($NaBH_3CN$). The fact that CD_2O, $^{13}CH_2O$, $^{13}CD_2O$ and $NaBD_3CN$ are commercially available at relatively low costs means that +2, +4, +6, and +8 Da tags can be easily generated via dimethyl labeling [35,36]. Moreover, ^{13}C- and ^{15}N-cysteines are also commercially available, which means that up to 10-plex isotope labeling (+0 – +9 Da) is feasible (Figure S1, supplementary materials).

Due to the potential of DiCys/OPA as a versatile isotope-labeling method, this study sought to understand its performance in standard RP-LC-MS analyses of chiral metabolites. DiCys was evaluated against five commonly used chiral thiols: NAC, NAP, IBLC, NBC, and BTCC. The reaction mechanism of DiCys/OPA with amines is shown in Figure 1a, and the chemical structures of the other thiols are shown in Figure 1b. They were compared based on their fluorescence intensity, separation performance, stability, and ionization efficiency for amino acid enantiomers. DiCys/OPA was also compared against Marfey's reagents, which are commonly used for resolving chiral amino acids. Finally, we compared DiCys against IBLC in identifying D-amino acids in aqueous extracts of rice. Our data suggest that DiCys/OPA is an excellent derivatization method to resolve chiral amines in RP-LC-MS metabolomics profiling.

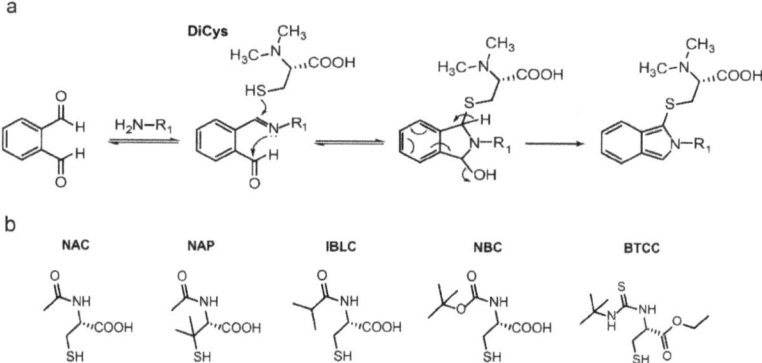

Figure 1. (a) Derivatization reaction of amino acids with DiCys/OPA. (b) Structures of chiral thiols: NAC, NAP, IBLC, NBC, and BTCC.

2. Results and Discussion

2.1. Stability and Fluorescence of DiCys Derivatives

Some of the most abundant amine-containing metabolites in biological samples are amino acids. L and D amino acid pairs are also among the most important enantiomeric metabolites in terms of biological functions. The charged carboxylate group makes it somewhat challenging to resolve all 20 proteinogenic amino acids by RP-HPLC. Therefore, we chose amino acids as model metabolites to study DiCys/OPA derivatization.

One of the reported disadvantages of OPA/thiol derivatization is the instability of the product [24,25]. Here, we evaluated the stability of OPA adducts with DiCys and five additional chiral thiols—NAC, NAP, IBLC, NBC, and BTCC. We monitored the fluorescence intensities of OPA/thiol-derivatized amino acids at 4 °C for 60 min (Figure S2 supplementary materials), and there was no visible sign of fluorophore breakdown, consistent with previous reports [37]. This should therefore be stable enough for routine LC-MS workflows.

We also quantified the fluorescence intensities of five L-amino acids derivatized with six chiral thiols (Figure 2a) after HPLC separation. Our results indicated that IBLC, NAC, and DiCys derivatives produced stronger fluorescence. In contrast, the NAP and NBC derivatives exhibited very low fluorescence intensities.

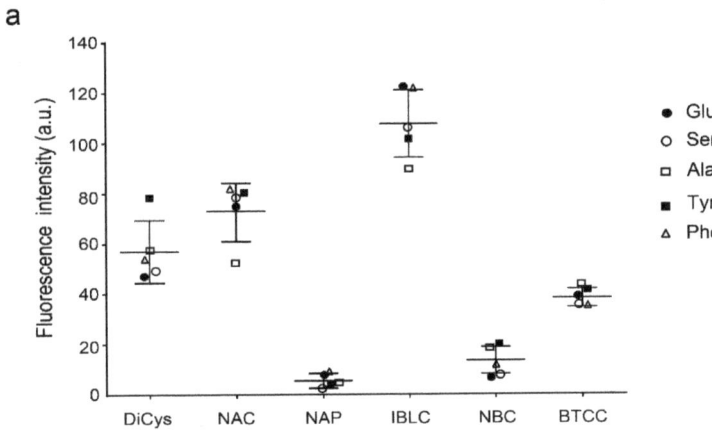

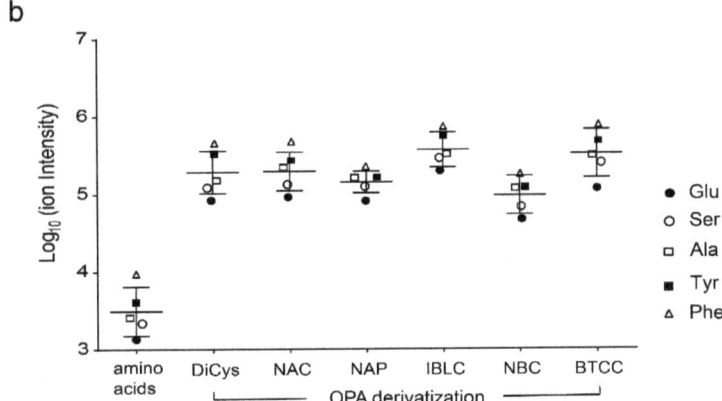

Figure 2. Fluorescence intensities (**a**) and ion intensities (**b**) of the OPA/thiol-derivatized L-amino acids (Glu, Ser, Ala, Tyr, and Phe) at equal concentrations. The center bar represents the mean, and the whiskers represent ±2 standard deviations.

2.2. Separation of Enantiomers

When Chernobrrovkin et al. compared NAC, NAP, IBLC, and NMC as chiral thiols for OPA derivatization [33], they found that NAC and NMC provided better chiral resolution than NAP and IBLC. However, the resolution factors may depend on the column; mobile phase composition; flow rate; and gradient [38,39]. We previously found that optimal elution condition for OPA adducts was around pH 8 instead of the typical acidic conditions [40]. Therefore, we conducted RP-LC separation at pH 8 to resolve five enantiomer pairs (Glu, Ser, Ala, Tyr, and Phe) (Figure 3). The best resolution was obtained with IBLC and DiCys, and the worst was with BTCC (Table S1 supplementary materials). Quantitative conversion to derivatized products for both enantiomers and the lack of racemization were confirmed by MS detection.

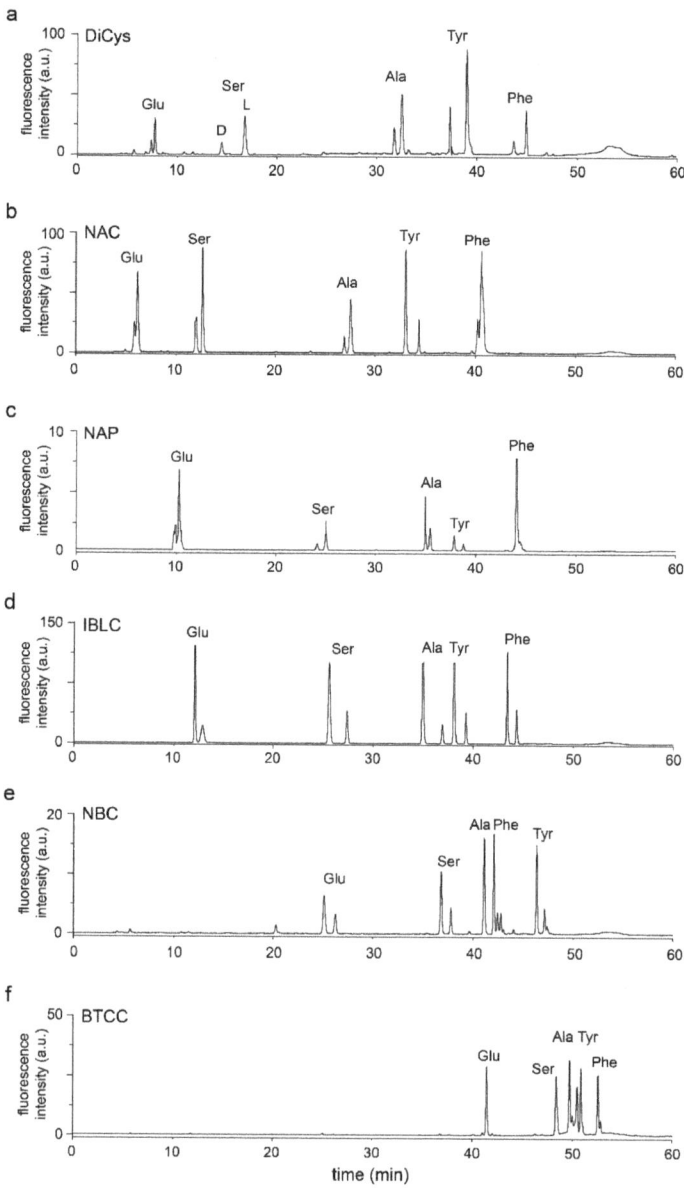

Figure 3. RP-HPLC analysis of amino acid enantiomers derivatized with OPA/thiol. The adducts of DiCys (**a**), NAC (**b**), NAP (**c**), IBLC (**d**), NBC (**e**), and BTCC (**f**) are detected by 340 nm excitation/450 nm emission. The ratio between L:D amino acids is 3:1.

2.3. Ionization Efficiency and MS/MS Properties

Amino acids exhibit low ionization efficiencies in ESI-MS experiments, and OPA derivatization may bring significant enhancements [40]. As shown in Figure 2b, all six thiol adducts have shown 25–100-fold higher ionization efficiencies over non-derivatized amino acids, making them generally useful for ESI-MS detection. In MS analyses, it is easy to identify derivatized amino acid enantiomers in the mass chromatogram based on double-peak detection via selected ion monitoring. To fragment OPA adducts requires relatively high collision energies: around 20 V [40]. The fragmentation patterns of DiCys/OPA adducts with seven amino acids are shown in Figure S3 supplementary materials, with a neutral loss of the thiol group in all cases.

2.4. Comparing DiCys/OPA against Marfey's Reagents

For the enantiomeric separation of amino acids, Marfey's reagent has been used widely [41–44]. This has led to the development of several Marfey variants, including: 1-fluoro-2-4-dinitrophenyl-5-L-alanine amide (FDAA); 1-fluoro-2-4-dinitrophenyl-5-L-Valina amide (FDVA); and the corresponding Phe, Ile, and Leu versions [44]. Here, we compared the most commonly used Marfey's variants, FDAA and FDVA, to the performance of DiCys/OPA. DiCys was the best of the three regarding chiral amino acid separation (Figure S4 supplementary materials). DiCys/OPA derivatives also have the advantage of being fluorogenic, while Marfey's derivatives are non-fluorescent. Therefore, we conclude that DiCys/OPA is highly suitable for resolving chiral analytes, better than popular methods such as NAC/OPA and Marfey's reagents.

2.5. Enantiomer Identification in Rice Water with DiCys/OPA

To test the usefulness of DiCys/OPA, we analyzed the aqueous extracts of edible rice, otherwise known as rice water. Rice water is the starchy water that remains after soaking or cooking rice, containing vitamins, amino acids, and minerals. It has been used traditionally in the treatment of skin and hair in Asian countries [45–47]. Little is known about the composition of amino acid enantiomers in rice water. Therefore, we separately applied DiCys/OPA and IBLC/OPA derivatization to rice water samples. Their RP-HPLC chromatograms are shown in Figure 4, and we observed almost twice as many visible fluorescent peaks with DiCys compared to IBLC. It shows that DiCys is suitable for separating a wide range of naturally occurring amine metabolites.

By MS and MS/MS detection, we could identify all twenty proteinogenic L-amino acids and seven D-amino acids (Ala, Arg, Lys, Phe, Ser, Tyr, and Val) in rice water samples using either DiCys or IBLC. Figure 5 shows the integrated ion intensities of individual amino acids. The retention time, precursor ion, and product ion information are listed in Tables S2 and S3 supplementary materials. The ratios between D/L amino acids are shown in Table 1. Interestingly, the highest D/L ratios were found for the two positively charged amino acids, Arg and Lys. Their physiological roles and gustatory effects remain undetermined.

3. Materials and Methods

3.1. Reagents

L and D amino acids (Glu, Ser, Ala, Tyr, Phe), L-Cys, OPA, ammonium bicarbonate, perchloric acid (ACS reagent, 70%), formaldehyde (37% w/w), dichloromethane, ninhydrin, fluorescamine, Ellman's reagent (DTNB), NAC, NAP, IBLC, NBC, and BTCC were purchased from Sigma-Aldrich (St. Louis, MO, USA). Methanol and acetonitrile (ACN) were purchased from Baker (Radnor, PA, USA). Boric acid and sodium tetraborate were purchased from Acros (Geel, Belgium). Sodium cyanoborohydride was purchased from Fluka (Buchs, Switzerland). FDAA and FDVA were purchased from Thermo Fisher (Waltham, MA, USA).

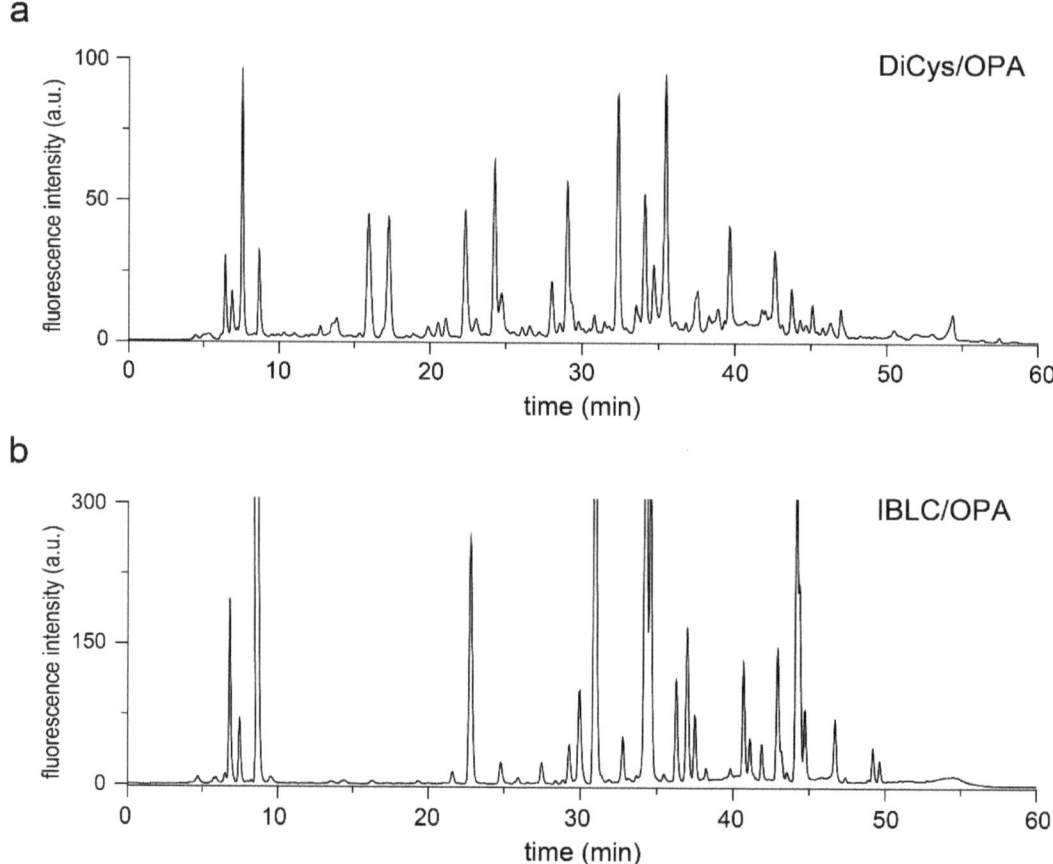

Figure 4. RP-HPLC analysis of free amines in rice water derivatized by DiCys/OPA (**a**) and IBLC/OPA (**b**), detected by 340 nm excitation/450 nm emission.

3.2. Synthesis of N,N-Dimethyl-L-Cysteine

A total of 100 mg of L-cysteine was dissolved in 10 mL dilute HCl (pH 2.5) and mixed with 8.25 mmol of sodium cyanoborohydride (NaBH$_3$CN) for 10 min at 4 °C. Then, 8.25 mmol of formaldehyde (37% w/w) was added, stirred for 30 min, and the reaction was monitored by ninhydrin staining on thin-layer chromatography plates. The DiCys product was purified via silica-gel column chromatography using MeOH/CH$_2$Cl$_2$. DiCys fraction was acidified to pH 2.5 by adding 0.1 N HCl and evaporated to dryness at 60 °C. DiCys powder was dissolved in deionized water and quantified using the Ellman assay. The reaction yield was 87%. HRMS (ESI/Q-TOF) m/z: M = C$_5$H$_{11}$NO$_2$S, calculated for [M + H]$^+$ = 150.0583, found 150.0589.

3.3. Rice Water Preparation

Sushi rice samples were purchased from a local grocery store in Taiwan. In total, 50 g of the dried rice was placed in 50 mL of deionized water. After shaking for 30 min, the solution was passed through filter paper. The rice water was lyophilized and dissolved in 250 µL of 0.01% perchloric acid and filtered twice through 0.22 µm nylon filters. Finally, we quantified total amines using a fluorescamine assay [48].

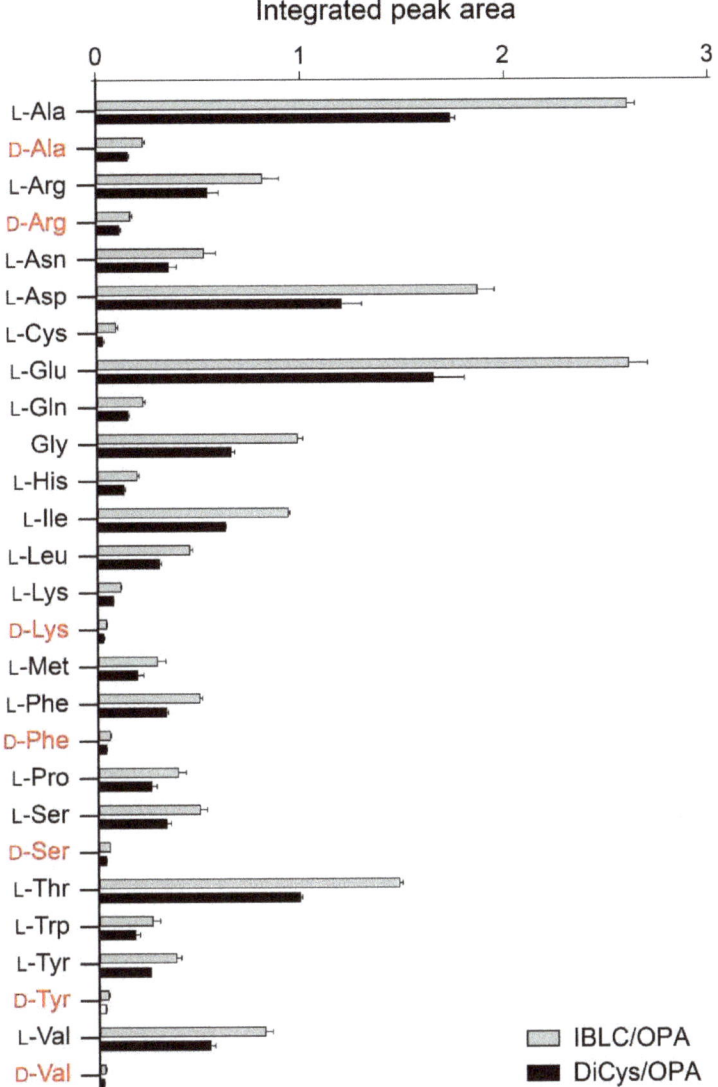

Figure 5. The amino acid contents of rice water measured by IBLC/OPA and DiCys/OPA derivatization. Bars indicate the integrated peak area of ion intensities from RP-LC-MS analyses. Error bars correspond to the standard error from three independent replicate experiments.

Table 1. The ratios of D-amino acid to L-amino acid in rice water.

Amino Acids	D/L Ratio
Ala	0.09
Arg	0.31
Lys	0.30
Phe	0.09
Ser	0.10
Tyr	0.12
Val	0.03

3.4. Derivatization Reactions

The following reagents were prepared freshly before use: L, D-glutamic acid; L, D-serine; L, D-alanine; L, D-arginine; D-valine; L, D-tyrosine; L, D-phenylalanine; and L, D-lysine. These reagents were used as amino acid standards and dissolved in 0.01% perchloric acid. The thiols (DiCys, NAC, NAP, IBLC, NBC, and BTCC) were dissolved in methanol to 150 mM. The OPA solution (50 mM) was prepared by dissolving 1.5 mg OPA in a mixture of 20 µL MeOH and 180 µL of 1 M borate buffer (pH 10.7); then, 5.03 mg of FDAA was dissolved in 500 µL ACN (37 mM), and 5.55 mg of FDVA in 500 µL ACN (37 mM).

After this, 20 µL of 50 mM OPA, 20 µL of 1 M borate buffer (pH 10.7), and 20 µL of 150 mM thiol were combined. We then added either 20 µL of 2.5 mM amino acid solution or rice water sample, and the mixture was vortexed and incubated at 4 °C for 2 min under dark conditions. The solution was diluted to a final volume of 200 µL with 50% ACN, and 20 µL of the mixture was injected into the HPLC.

A total of 20 µL of 37 mM Marfey's reagent (FDAA or FDVA) was mixed with 20 µL of 2.5 mM amino acid solution, 8 µL of 1 M NaHCO$_3$ (pH 8.0), and 31.5 µL of acetone. The mixture was incubated at 40 °C for 1 h, quenched by adding 6 µL of 2 M HCl, before 20 µL of the mixture was injected into the HPLC.

3.5. LC-MS Analysis

The Agilent 1260 HPLC system (Santa Clara, CA, USA) was equipped with an autosampler, a quaternary pump, a column oven, a UV-Vis absorbance detector, and a fluorescence detector. The Hydrosphere C18 column (250 × 4.6 mm, 5 µm bead diameter) used for separation was acquired from YMC (Kyoto, Japan). The aqueous mobile phase (A) consisted of 2 mM ammonium bicarbonate (pH 8.0), whilst mobile phase B contained 7% MeOH in ACN. Elution was performed at a flow rate of 1 mL/min at 40 °C using the following gradient program: 0–5 min, 10%; 5–10 min, 10–12%; 10–20 min, 12–22%; 20–30 min, 22–38%; 30–40 min, 38–60%; 40–47 min, 60–83%; 47–50 min, 83–100%; 50–54 min, 100%; 54–57 min, 100–10%; 57–60 min, 10%. The HPLC was connected to Bruker micrOTOF-QII (Bremen, Germany) operated in positive mode. Full MS spectra were recorded from (m/z) 100 to 600. ESI source parameters were nebulizer gas (nitrogen) at 0.3 bar, drying gas (nitrogen) at 4 L/min, and 180 °C.

4. Conclusions

We systematically evaluated the suitability of six chiral thiols (DiCys, NAC, NAP, IBLC, NBC, and BTCC) for OPA-assisted separation of amino acid enantiomers. The best separation efficiencies in RP-HPLC were found with DiCys and IBLC. For fluorescence detection, IBLC, NAC, and DiCys gave stronger signals while NAP only gave very weak signals. All six reagents enhanced ionization efficiencies by 25–100 fold, useful for MS detection. Previously, IBLC has been a popular reagent for resolving chiral amino acids [30,49], and our data supported its usefulness. More importantly, our study was the first to introduce DiCys/OPA for enantiomeric separation, and its performance was comparable to IBLC in our tests. DiCys also outperformed Marfey's reagents FDAA and FDVA, which were specially developed for chiral separation purposes. We conclude that DiCys is a highly versatile reagent for resolving enantiomeric amines in chiral metabolomics experiments.

The greatest advantage of DiCys is its potential for multiplex heavy-isotope labeling. Using well-established chemistries [35], heavy isotope versions from +1 Da to +9 Da may be easily synthesized in one step using commercially available reagents. This may provide 10-plex labeling at an affordable cost for high-throughput metabolomics experiments. Combined with fluorogenic detection and excellent chiral separation, DiCys is one of the most versatile amine derivatization reagents currently available.

In real-world metabolomics profiling of rice water, DiCys provided better separation of amine metabolites than IBLC. Both allowed us to detect twenty proteinogenic L-amino acids and identify seven D-amino acids—Ala, Arg, Lys, Phe, Ser, Tyr, and Val. These D-amino acids are primarily associated with sweetness for humans [50,51], suggesting that D-amino acids may be important for the gustatory taste. Moreover, D-amino acids synthesized by gut microbiomes may affect our immune systems [52,53]. How D-amino acids in rice diet may affect our gut microbiome–immune axis will warrant further investigation.

Supplementary Materials: The following are available online: Figures S1–S4 and Tables S1–S3.

Author Contributions: A.L. conducted the experiments. A.L. and H.-C.T. designed the experiments and wrote the manuscript. All authors have read and agreed to the published version of the manuscript.

Funding: This research received no external funding.

Institutional Review Board Statement: Not applicable.

Informed Consent Statement: Not applicable.

Data Availability Statement: Not applicable.

Acknowledgments: We thank Yun-Shiuan Leung and the Mass Spectrometry Core Facility at Department of Chemistry, National Taiwan University for LC–MS support.

Conflicts of Interest: The authors declare no conflict of interest.

Sample Availability: Not applicable.

References

1. Weckwerth, W. Metabolomics: An integral technique in systems biology. *Bioanalysis* **2010**, *2*, 829–836. [CrossRef] [PubMed]
2. Sevin, D.C.; Kuehne, A.; Zamboni, N.; Sauer, U. Biological insights through nontargeted metabolomics. *Curr. Opin. Biotechnol.* **2015**, *34*, 1–8. [CrossRef] [PubMed]
3. Beger, R.D.; Dunn, W.; Schmidt, M.A.; Gross, S.S.; Kirwan, J.A.; Cascante, M.; Brennan, L.; Wishart, D.S.; Oresic, M.; Hankemeier, T.; et al. Metabolomics enables precision medicine: "A White Paper, Community Perspective". *Metabolomics* **2016**, *12*, 149. [CrossRef] [PubMed]
4. Johnson, C.H.; Ivanisevic, J.; Siuzdak, G. Metabolomics: Beyond biomarkers and towards mechanisms. *Nat. Rev. Mol. Cell Biol.* **2016**, *17*, 451–459. [CrossRef]
5. Fuhrer, T.; Zamboni, N. High-throughput discovery metabolomics. *Curr. Opin. Biotechnol.* **2015**, *31*, 73–78. [CrossRef]
6. Zampieri, M.; Sekar, K.; Zamboni, N.; Sauer, U. Frontiers of high-throughput metabolomics. *Curr. Opin. Chem. Biol.* **2017**, *36*, 15–23. [CrossRef]
7. Huan, T.; Wu, Y.; Tang, C.; Lin, G.; Li, L. DnsID in MyCompoundID for rapid identification of dansylated amine- and phenol-containing metabolites in LC-MS-based metabolomics. *Anal. Chem.* **2015**, *87*, 9838–9845. [CrossRef]
8. Zhou, R.; Guo, K.; Li, L. 5-Diethylamino-naphthalene-1-sulfonyl chloride (DensCl): A novel triplex isotope labeling reagent for quantitative metabolome analysis by liquid chromatography mass spectrometry. *Anal. Chem.* **2013**, *85*, 11532–11539. [CrossRef]
9. SG, B.G.; Gowda, D.; Liang, C.; Li, Y.; Kawakami, K.; Fukiya, S.; Yokota, A.; Chiba, H.; Hui, S.P. Chemical Labeling Assisted Detection and Identification of Short Chain Fatty Acid Esters of Hydroxy Fatty Acid in Rat Colon and Cecum Contents. *Metabolites* **2020**, *10*, 398.
10. Toyo'oka, T. Derivatization-based High-throughput Bioanalysis by LC-MS. *Anal. Sci.* **2017**, *33*, 555–564.
11. Koga, R.; Yoshida, H.; Nohta, H.; Hamase, K. Multi-dimensional hplc analysis of metabolic related chiral amino acids-method development and biological/clinical applications. *Chromatography* **2019**, *40*, 1–8. [CrossRef]
12. Pandey, R.; Collins, M.; Lu, X.; Sweeney, S.R.; Chiou, J.; Lodi, A.; Tiziani, S. Novel Strategy for Untargeted Chiral Metabolomics using Liquid Chromatography-High Resolution Tandem Mass Spectrometry. *Anal. Chem.* **2021**, *93*, 5805–5814. [CrossRef]
13. Takayama, T.; Mochizuki, T.; Todoroki, K.; Min, J.Z.; Mizuno, H.; Inoue, K.; Akatsu, H.; Noge, I.; Toyo'oka, T. A novel approach for LC-MS/MS-based chiral metabolomics fingerprinting and chiral metabolomics extraction using a pair of enantiomers of chiral derivatization reagents. *Anal. Chim. Acta* **2015**, *898*, 73–84. [CrossRef]
14. Mochizuki, T.; Todoroki, K.; Inoue, K.; Min, J.Z.; Toyo'oka, T. Isotopic variants of light and heavy L-pyroglutamic acid succinimidyl esters as the derivatization reagents for DL-amino acid chiral metabolomics identification by liquid chromatography and electrospray ionization mass spectrometry. *Anal. Chim. Acta* **2014**, *811*, 51–59. [CrossRef]
15. Takayama, T.; Kuwabara, T.; Maeda, T.; Noge, I.; Kitagawa, Y.; Inoue, K.; Todoroki, K.; Min, J.Z.; Toyo'oka, T. Profiling of chiral and achiral carboxylic acid metabolomics: Synthesis and evaluation of triazine-type chiral derivatization reagents for carboxylic acids by LC-ESI-MS/MS and the application to saliva of healthy volunteers and diabetic patients. *Anal. Bioanal. Chem.* **2015**, *407*, 1003–1014. [CrossRef] [PubMed]
16. Hamase, K.; Morikawa, A.; Etoh, S.; Tojo, Y.; Miyoshi, Y.; Zaitsu, K. Analysis of small amounts of D-amino acids and the study of their physiological functions in mammals. *Anal. Sci.* **2009**, *25*, 961–968. [CrossRef] [PubMed]

17. Miyoshi, Y.; Koga, R.; Oyama, T.; Han, H.; Ueno, K.; Masuyama, K.; Itoh, Y.; Hamase, K. HPLC analysis of naturally occurring free d-amino acids in mammals. *J. Pharm. Biomed. Anal.* **2012**, *69*, 42–49. [CrossRef]
18. Bastings, J.J.; van Eijk, H.M.; Olde Damink, S.W.; Rensen, S.S. d-amino Acids in Health and Disease: A Focus on Cancer. *Nutrients* **2019**, *11*, 2205. [CrossRef] [PubMed]
19. Kumar, A.P.; Jin, D.; Lee, Y.-I. Recent development on spectroscopic methods for chiral analysis of enantiomeric compounds. *Appl. Spectrosc. Rev.* **2009**, *44*, 267–316. [CrossRef]
20. Fanali, C.; D'Orazio, G.; Gentili, A.; Fanali, S. Analysis of enantiomers in products of food interest. *Molecules* **2019**, *24*, 1119. [CrossRef]
21. Hui, M.; Cheung, S.-W.; Chin, M.-L.; Chu, K.-C.; Chan, R.C.-Y.; Cheng, A.F.-B. Development and application of a rapid diagnostic method for invasive Candidiasis by the detection of D-/L-arabinitol using gas chromatography/mass spectrometry. *Diagn. Microbiol. Infect. Dis.* **2004**, *49*, 117–123. [CrossRef] [PubMed]
22. Nguyen, L.A.; He, H.; Pham-Huy, C. Chiral drugs: An overview. *Int. J. Biomed. Sci.* **2006**, *2*, 85–100. [PubMed]
23. Takayama, T.; Mizuno, H.; Toyo'oka, T.; Akatsu, H.; Inoue, K.; Todoroki, K. Isotope Corrected Chiral and Achiral Nontargeted Metabolomics: An Approach for High Accuracy and Precision Metabolomics Based on Derivatization and Its Application to Cerebrospinal Fluid of Patients with Alzheimer's Disease. *Anal. Chem.* **2019**, *91*, 4396–4404. [CrossRef]
24. Guranda, D.T.; Kudryavtsev, P.A.; Khimiuk, A.Y.; Švedas, V.K. Efficient enantiomeric analysis of primary amines and amino alcohols by high-performance liquid chromatography with precolumn derivatization using novel chiral SH-reagent N-(R)-mandelyl-(S)-cysteine. *J. Chromatogr. A* **2005**, *1095*, 89–93. [CrossRef] [PubMed]
25. Molnár-Perl, I. HPLC of amino acids as o-phthalaldehyde derivatives. *J. Chromatogr. Libr.* **2005**, *70*, 163–198.
26. Molnár-Perl, I. Advancement in the derivatizations of the amino groups with the o-phthaldehyde-thiol and with the 9-fluorenylmethyloxycarbonyl chloride reagents. *J. Chromatogr. B* **2011**, *879*, 1241–1269. [CrossRef]
27. Gowda, S.G.B.; Nakahashi, A.; Yamane, K.; Nakahashi, S.; Murai, Y.; Siddegowda, A.K.C.; Hammam, M.A.S.; Monde, K. Facile Chemoselective Strategy toward Capturing Sphingoid Bases by a Unique Glutaraldehyde-Functionalized Resin. *ACS Omega* **2018**, *3*, 753–759. [CrossRef]
28. Yokoyama, T.; Tokuda, M.; Amano, M.; Mikami, K. Simultaneous determination of primary and secondary D- and L-amino acids by reversed-phase high-performance liquid chromatography using pre-column derivatization with two-step labelling method. *Biosci. Biotechnol. Biochem.* **2017**, *81*, 1681–1686. [CrossRef]
29. Brückner, H.; Wittner, R.; Godel, H. Amino acid analysis by derivatization with o-phthaldialdehyde and chiral thiols. In *Amino Acids*; Lubec, G., Rosenthal, G.A., Eds.; Springer: Dordrecht, The Netherlands, 1990; pp. 143–151.
30. Fitznar, H.P.; Lobbes, J.M.; Kattner, G. Determination of enantiomeric amino acids with high-performance liquid chromatography and pre-column derivatisation with o-phthaldialdehyde and N-isobutyrylcysteine in seawater and fossil samples (mollusks). *J. Chromatogr. A* **1999**, *832*, 123–132. [CrossRef]
31. Brückner, H.; Zivny, S. High-performance liquid chromatographic resolution of (R, S)-α-alkyl-α-amino acids as diastereomeric derivatives. *Amino Acids* **1993**, *4*, 157–167. [CrossRef]
32. Nimura, N.; Fujiwara, T.; Watanabe, A.; Sekine, M.; Furuchi, T.; Yohda, M.; Yamagishi, A.; Oshima, T.; Homma, H. A novel chiral thiol reagent for automated precolumn derivatization and high-performance liquid chromatographic enantioseparation of amino acids and its application to the aspartate racemase assay. *Anal. Biochem.* **2003**, *315*, 262–269. [CrossRef]
33. Chernobrovkin, M.; Shapovalova, E.; Guranda, D.; Kudryavtsev, P.; Švedas, V.; Shpigun, O. Chiral high-performance liquid chromatography analysis of α-amino acid mixtures using a novel SH reagent—N-R-mandelyl-L-cysteine and traditional enantiomeric thiols for precolumn derivatization. *J. Chromatogr. A* **2007**, *1175*, 89–95. [CrossRef]
34. Lóki, K.; Varga-Visi, É.; Albert, C.; Csapó, J. Separation and determination of the tryptophan enantiomers. *Acta Univ. Sapientiae* **2008**, *1*, 61–71.
35. Hsu, J.L.; Chen, S.H. Stable isotope dimethyl labelling for quantitative proteomics and beyond. *Philos. Trans. A Math. Phys. Eng. Sci.* **2016**, *374*, 20150364. [CrossRef] [PubMed]
36. Kovanich, D.; Cappadona, S.; Raijmakers, R.; Mohammed, S.; Scholten, A.; Heck, A.J. Applications of stable isotope dimethyl labeling in quantitative proteomics. *Anal. Bioanal. Chem.* **2012**, *404*, 991–1009. [CrossRef] [PubMed]
37. Chen, R.F.; Scott, C.; Trepman, E. Fluorescence properties of o-phthaldialdehyde derivatives of amino acids. *Biochim. Biophys. Acta Protein Struct.* **1979**, *576*, 440–455. [CrossRef]
38. Gurram, I.; Kavitha, M.; Nagabhushnam, M.; Bonthagara, B.; Reddy, D.N. Overview of validation, basic concepts and analytical method process validation. *Indian J. Pharm. Sci.* **2017**, *4*, 1665–1680.
39. Sabir, A.; Moloy, M.; Bhasin, P. HPLC Method Development and Validation: A Review. *Int. Res. J. Pharm.* **2015**, *4*, 39–46. [CrossRef]
40. Lkhagva, A.; Shen, C.-C.; Leung, Y.-S.; Tai, H.-C. Comparative study of five different amine-derivatization methods for metabolite analyses by liquid chromatography-tandem mass spectrometry. *J. Chromatogr. A* **2020**, *1610*, 460536. [CrossRef]
41. Ayon, N.J.; Sharma, A.D.; Gutheil, W.G. LC-MS/MS-based separation and quantification of Marfey's reagent derivatized proteinogenic amino acid DL-stereoisomers. *J. Am. Soc. Mass Spectrom.* **2018**, *30*, 448–458. [CrossRef]
42. Bhushan, R.; Brückner, H. Marfey's reagent for chiral amino acid analysis: A review. *Amino Acids* **2004**, *27*, 231–247. [CrossRef] [PubMed]

43. Bruckner, H.; Gah, C. High-performance liquid chromatographic separation of DL-amino acids derivatized with chiral variants of Sanger's reagent. *J. Chromatogr. A* **1991**, *555*, 81–95. [CrossRef]
44. Marfey, P. Determination of D-amino acids. II. Use of a bifunctional reagent, 1, 5-difluoro-2, 4-dinitrobenzene. *Carlsberg Res. Commun.* **1984**, *49*, 591–596. [CrossRef]
45. Inamasu, S.; Ikuyama, R.; Fujisaki, Y.; Sugimoto, K.I. The effect of rinse water obtained from the washing of rice (YU-SU-RU) as a hair treatment. *J. Soc. Cosmet. Chemists Japan* **2010**, *44*, 29–32. [CrossRef]
46. Kalman, D.S. Amino acid composition of an organic brown rice protein concentrate and isolate compared to soy and whey concentrates and isolates. *Foods* **2014**, *3*, 394–402. [CrossRef]
47. Wong, H. Rice water in treatment of infantile gastroenteritis. *Lancet* **1981**, *2*, 102–103. [CrossRef]
48. Chung, L. A fluorescamine assay for membrane protein and peptide samples with non-amino-containing lipids. *Anal. Biochem.* **1997**, *248*, 195–201. [CrossRef]
49. Müller, C.; Fonseca, J.R.; Rock, T.M.; Krauss-Etschmann, S.; Schmitt-Kopplin, P. Enantioseparation and selective detection of D-amino acids by ultra-high-performance liquid chromatography/mass spectrometry in analysis of complex biological samples. *J. Chromatogr. A* **2014**, *1324*, 109–114. [CrossRef]
50. Kawai, M.; Sekine-Hayakawa, Y.; Okiyama, A.; Ninomiya, Y. Gustatory sensation of L-and D-amino acids in humans. *Amino Acids* **2012**, *43*, 2349–2358. [CrossRef]
51. Delompré, T.; Guichard, E.; Briand, L.; Salles, C. Taste Perception of Nutrients Found in Nutritional Supplements: A Review. *Nutrients* **2019**, *11*, 2050. [CrossRef]
52. Sasabe, J.; Miyoshi, Y.; Rakoff-Nahoum, S.; Zhang, T.; Mita, M.; Davis, B.M.; Hamase, K.; Waldor, M.K. Interplay between microbial D-amino acids and host d-amino acid oxidase modifies murine mucosal defence and gut microbiota. *Nat. Microbiol.* **2016**, *1*, 16125. [CrossRef] [PubMed]
53. Sasabe, J.; Suzuki, M. Emerging Role of D-Amino Acid Metabolism in the Innate Defense. *Front. Microbiol.* **2018**, *9*, 933. [CrossRef] [PubMed]

Article

GC-MS Analysis of Biological Nitrate and Nitrite Using Pentafluorobenzyl Bromide in Aqueous Acetone: A Dual Role of Carbonate/Bicarbonate as an Enhancer and Inhibitor of Derivatization

Dimitrios Tsikas

Core Unit Proteomics, Institute of Toxicology, Hannover Medical School, 30625 Hannover, Germany; Tsikas.dimitros@mh-hannover.de

Citation: Tsikas, D. GC-MS Analysis of Biological Nitrate and Nitrite Using Pentafluorobenzyl Bromide in Aqueous Acetone: A Dual Role of Carbonate/Bicarbonate as an Enhancer and Inhibitor of Derivatization. *Molecules* **2021**, *26*, 7003. https://doi.org/10.3390/molecules26227003

Academic Editor: Paraskevas D. Tzanavaras

Received: 1 November 2021
Accepted: 18 November 2021
Published: 19 November 2021

Publisher's Note: MDPI stays neutral with regard to jurisdictional claims in published maps and institutional affiliations.

Copyright: © 2021 by the author. Licensee MDPI, Basel, Switzerland. This article is an open access article distributed under the terms and conditions of the Creative Commons Attribution (CC BY) license (https://creativecommons.org/licenses/by/4.0/).

Abstract: Carbon dioxide (CO_2) and carbonates, which are widely distributed in nature, are constituents of inorganic and organic matter and are essential in vegetable and animal organisms. CO_2 is the principal greenhouse gas in the atmosphere. In human blood, CO_2/HCO_3^- is an important buffering system. Inorganic nitrate (ONO_2^-) and nitrite (ONO^-) are major metabolites and abundant reservoirs of nitric oxide (NO), an endogenous multifunctional signaling molecule. Carbonic anhydrase (CA) is involved in the reabsorption of nitrite and nitrate from the primary urine. The measurement of nitrate and nitrite in biological samples is of particular importance. The derivatization of nitrate and nitrite in biological samples alongside their ^{15}N-labeled analogs, which serve as internal standards, is a prerequisite for their analysis by gas chromatography–mass spectrometry (GC-MS). A suitable derivatization reagent is pentafluorobenzyl bromide (PFB-Br). Nitrate and nitrite are converted in aqueous acetone to PFB-ONO_2 and PFB-NO_2, respectively. PFB-Br is also useful for the GC-MS analysis of carbonate/bicarbonate. This is of particular importance in conditions of pharmacological CA inhibition, for instance by acetazolamide, which is accompanied by elevated concomitant excretion of nitrate, nitrite and bicarbonate, as well as by urine alkalization. We performed a series of experiments with exogenous bicarbonate ($NaHCO_3$) added to human urine samples (range, 0 to 100 mM), as well as with endogenous bicarbonate resulting from the inhibition of CA activity in healthy subjects before and after ingestion of pharmacological acetazolamide. Our results indicate that bicarbonate enhances the derivatization of nitrate with PFB-Br. In contrast, bicarbonate decreases the derivatization of nitrite with PFB-Br. Bicarbonate is not a catalyst, but it enhances PFB-ONO_2 formation and inhibits PFB-NO_2 formation in a concentration-dependent manner. The effects of bicarbonate are likely to result from its reaction with PFB-Br to generate PFB-OCOOH. Nitrate reacts with concomitantly produced PFB-OCOOH to form PFB-ONO_2 in addition to the direct reaction of nitrate with PFB-Br. By contrast, nitrite does not react with PFB-OCOOH to form PFB-NO_2. Sample acidification by small volumes of 20 wt.% aqueous acetic acid abolishes the effects of exogenous and endogenous bicarbonate on nitrite measurement.

Keywords: acetazolamide; carbonic anhydrase; derivatization; enhancement; GC-MS; inhibition; pentafluorobenzyl bromide

1. Introduction

Pentafluorobenzyl bromide (2,3,4,5,6-pentafluorobenzyl bromide, PFB-Br) is a useful derivatization reagent for different classes of organic substances including carboxylic acids and amines, as well as inorganic anions, including nitrate and nitrite [1]. Derivatization with PFB-Br can be performed in water-free organic solvents such as acetonitrile, as well as in aqueous systems in the presence of water-miscible organic solvents such as acetone. Use of an acetone-aqueous sample in a volume ratio of 4:1 enables derivatization in homogenous phase [1]. Reactions with PFB-Br are nucleophilic substitutions of bromide by a nucleophile, which can be water, halogenides such as chloride, and other inorganic ions. The derivatization time is dependent upon the nucleophilicity and other factors, such as the stability of the PFB derivatives. The reaction products are lipophilic and extractable into organic solvents such as toluene and are best suitable for gas chromatography–mass spectrometry (GC-MS) analysis. Due to the fluorine atoms in PFB derivatives, their GC-MS analysis in the negative-ion chemical ionization mode revealed the highest sensitivity.

We previously developed a GC-MS method for the simultaneous quantitative measurement of nitrite and nitrate in different biological samples including human plasma, urine and saliva [2]. The reaction of PFB-Br with nitrite in aqueous acetone leads to the formation of the nitro derivative (PFB-NO$_2$), yet not of the expected nitrous acid ester (PFB-ONO) (Scheme 1, reaction A). The reaction of PFB-Br with nitrate generates the nitric acid ester (PFB-ONO$_2$) (Scheme 1, reaction B). Kinetic investigations showed that nitrite reacts with PFB-Br more rapidly and to a higher extent than nitrate, even at room temperature [2]. Yet, PFB-NO$_2$ seems to be readily susceptible to hydrolysis. As a compromise, we measured nitrite and nitrate simultaneously using GC-MS after derivatization with PFB-Br at 50 °C for 60 min. This procedure enhances the yield of the derivatization for nitrate and decreases the yield for nitrite due to hydrolysis. Yet, the use of the stable isotope analogs [^{15}N]nitrite and [^{15}N]nitrate guarantees highly accurate quantitative measurements in biological samples [2].

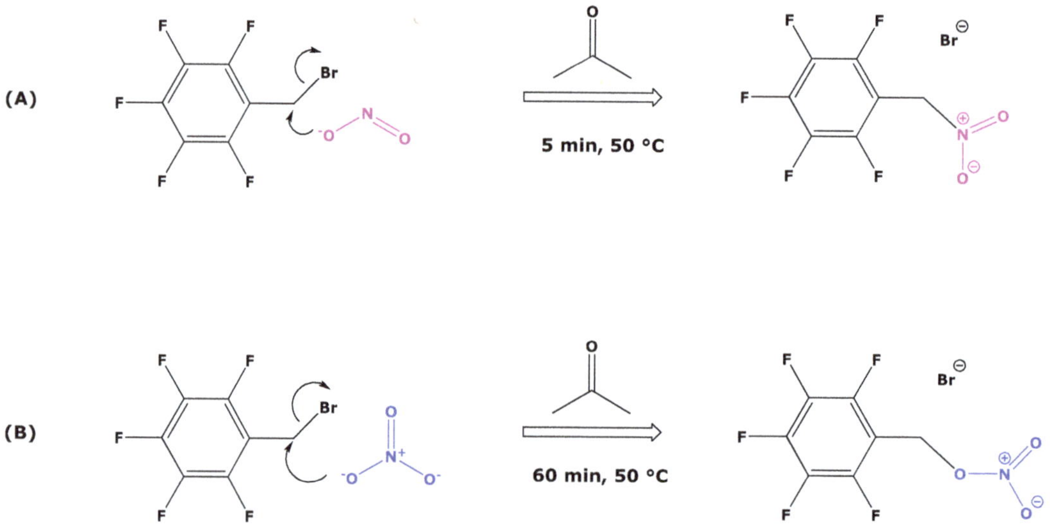

Scheme 1. Derivatization of (**A**) nitrite and (**B**) nitrate in aqueous acetone with pentafluorobenzyl bromide at 50 °C. The optimum derivatization time is 5 min for nitrite and 60 min for nitrate.

Carbon dioxide (CO$_2$) and carbonates are widely distributed in nature. In human blood, CO$_2$/HCO$_3^-$ is an important buffering system. In human urine, carbonate and bocrabonate are physiologically excreted [3,4]. Like nitrite and nitrate, we found that carbonate can react with PFB-Br under experimental conditions similar to those of nitrite and nitrate [5]. We observed the formation of the expected PFB-OCOOH derivative, albeit in low yield [5]. This behavior resembles in part that of nitrite derivatization with PFB-Br, which did not form any isolable PFB-ONO but exclusively (PFB-NO$_2$).

In previous work, we observed an interaction of carbonate/bicarbonate with the analysis of nitrite in urine samples of subjects who took acetazolamide [6], a clinical drug [7]. Acetazolamide inhibits carboanhydrase (CA) activity in the kidneys, and because of this, it increases the excretion of bicarbonate and pH of the urine. Urine acidification of carbonate-containing urine samples with acetic acid (20 wt.%) did not influence the derivatization of nitrate but seemed to increase the yield of PFB-NO$_2$ [6]. In the present work, we investigated the derivatization of nitrite and nitrate in human urine and plasma under various conditions, aiming to reveal potential mechanisms and solutions for the derivatization of nitrite and nitrate in the presence of high carbonate/bicarbonate concentrations.

2. Materials and Methods

2.1. Chemicals and Materials

2,3,4,5,6-Pentafluorobenzyl bromide (PFB-Br), sodium nitrite (purity 99.99+%), sodium [^{15}N]nitrite and sodium [^{15}N]nitrate (declared as 99 atom% at ^{15}N each) were obtained from Sigma-Aldrich (Steinheim, Germany). Toluene was purchased from Baker (Deventer, The Netherlands). Sodium bicarbonate and carbonate, acetone and glacial acetic acid were from Merck (Darmstadt, Germany). ^{2}H-Labelled creatinine ([*methylo*-^{2}H$_3$]creatinine, >99 atom% ^{2}H) was obtained from Aldrich. PFB-Br is corrosive and an eye irritant. Inhalation and contact with skin and eyes should be avoided. All work should be and was performed in a well-ventilated fume hood. Separate stock solutions of salts were prepared by dissolving accurately weighed amounts of commercially available unlabeled and stable-isotope-labeled salts in deionized water. Stock solutions were diluted with deionized water as appropriate.

Glassware for GC-MS (1.5-mL autosampler vials and 0.2-mL microvials) including the fused-silica capillary column Optima 17 (15 m × 0.25 mm I.D., 0.25-micrometer film thickness) were purchased from Macherey-Nagel (Düren, Germany).

2.2. Derivatization Procedures for Nitrite and Nitrate and GC-MS Analyses

In general, the following derivatization procedure was used. Deviations are reported in the individual experiments. A total of 100-µL aliquots of a sample were added to 400-µL aliquots of acetone and 10-µL aliquots of PFB-Br, and the samples were heated at 50 °C for 5 min or 60 min. After derivatization, acetone was removed under a stream of nitrogen, and analytes were extracted by vortexing with toluene (1 mL). Nitrite and nitrate were measured simultaneously in 100 µL urine specimens by a previously reported, fully validated GC-MS method immediately after acidification by using a 20 wt.% acetic acid solution and derivatization by PFB-Br, as described elsewhere [2]. ^{15}N-Labelled nitrite (final concentration, 4 µM or 1 µM) and ^{15}N-labeled nitrate (final concentration, 400 µM or 100 µM) were used as internal standards for urinary nitrite and nitrate, respectively. To investigate the effect of the CO_2/Na_2CO_3/$NaHCO_3$ system on nitrite and nitrate analysis, samples were derivatized before and after acidification by 20 wt.% acetic acid to reach a final pH value of about 4.5 in order to remove CO_2 from urine samples, as described elsewhere [2]. Urinary excretion of nitrite and nitrate was corrected for urinary creatinine excretion. Creatinine-corrected excretion rates are reported as µmol of nitrite or nitrate per mmol of creatinine.

Aliquots (1 µL) of the toluene extracts were injected into the GC-MS apparatus (model DSQ from ThermoFisher; Dreieich, Germany) in the splitless mode. Quantification was performed by selected-ion monitoring (SIM) of mass-to-charge (m/z) of m/z 46 for [^{14}N]nitrite, m/z 47 for [^{15}N]nitrite, m/z 62 for [^{14}N]nitrate and m/z 63 for [^{15}N]nitrate using a dwell time of 50 ms for each ion. The measured peak area (PA) values of unlabeled and labeled nitrite and nitrate and the peak area ratio (PAR) of unlabeled to labeled nitrite or nitrate were used in calculations.

2.3. Analyses in Urine Samples Collected in Previous Studies

In the present study, we used urine samples collected in a previously reported study [6], which had been performed as follows in brief. Six apparently healthy volunteers (2 females, aged 25 and 44 years; 4 males, aged 24–49 years) had participated in the study. In the morning (8 a.m.), the volunteers were orally given one to two tablets acetazolamide (Acemit® 250 mg, medphano/Berlin, Germany) corresponding to a dose of about 5 mg/kg bodyweight. First, volunteers emptied their bladder and collected the first urine specimen (time—2 h) followed by two collections at time—1 h and time 0 h. Immediately after collection of the 0 h urine sample, acetazolamide was taken by a glass of drinking water. Four urine samples were collected in polypropylene tubes in 30 min intervals and another four urine samples in 60 min intervals subsequently. Immediately after each collection, tubes were closed and put on ice. Urine samples were collected at several time points before and after acetazolamide ingestion, portioned in 1 mL and 10 mL aliquots and stored either at +5 °C for pH and carbonate measurement on the same day, or at −20 °C until analysis for nitrite, nitrate and creatinine on next day.

In some analyses, spot urine was collected by the author without any medication and used in some in the vitro studies on the effects of bicarbonate on nitrite and nitrate analyses.

2.4. Statistical Analysis

Results are expressed as mean with standard error of the mean, or as mean with standard deviation, as specified in the respective experiments. Differences between neighbor values were analyzed with paired or unpaired t tests as appropriate. p values ≤ 0.05 were considered statistically significantly different. Calculations and graphs were performed using GraphPad Prism version 7.0 (San Diego, CA, USA).

3. Results

3.1. Effect of Exogenous Bicarbonate and Hydroxide on the Derivatization of [^{15}N]nitrate, [^{15}N]nitrite and Endogenous Nitrate and Nitrite in Human Urine

A healthy volunteer provided a morning urine. The freshly collected urine sample was spiked with 400 μM [^{15}N]nitrate and 4 μM [^{15}N]nitrite. This pooled urine sample was divided into two equal fractions that were spiked with freshly prepared 100 mM NaHCO$_3$ or 100 mM NaOH. Thereafter, each two 100 μL aliquots of the urine samples were derivatized with PFB-Br at 50 °C for varying times (0, 5, 10, 20, 30, 45 and 60 min) without prior acidification, as well as after acidification with 20 wt.% acetic acid. All nitrite and nitrate species were measured simultaneously by GC-MS using SIM of m/z 46 for [^{14}N]nitrite, m/z 47 for [^{15}N]nitrite, m/z 62 for [^{14}N]nitrate and m/z 63 for [^{15}N]nitrate. The main results of this experiment are shown in Figure 1.

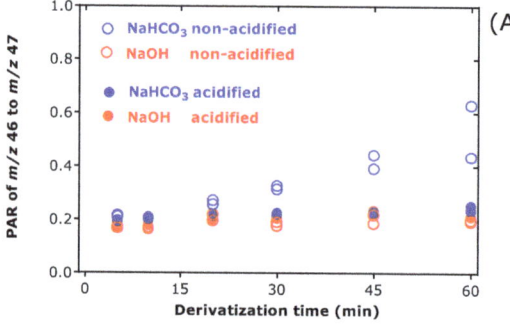
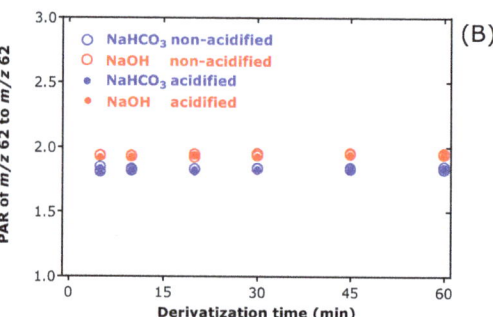

Figure 1. Effects of exogenous NaHCO$_3$ and NaOH on nitrite and nitrate analysis by GC-MS after derivatization with PFB-Br for the indicated times at 50 °C in a pooled human urine sample treated with 100 mM NaHCO$_3$ or with 100 mM NaOH. The samples were derivatized before and after acidification with 20 wt.% acetic acid. The concentrations of the internal standards in the urine samples were 400 μM [^{15}N]nitrate and 4 μM [^{15}N]nitrite. Data are shown from two independent experiments. All nitrite and nitrate species were measured simultaneously by GC-MS using selected-ion monitoring of (**A**) m/z 46 for [^{14}N]nitrite and m/z 47 for [^{15}N]nitrite, and of (**B**) m/z 62 for [^{14}N]nitrate and m/z 63 for [^{15}N]nitrate.

The PAR of m/z 46 to m/z 47 for nitrite and the PAR of m/z 62 to m/z 63 for nitrate behaved differently in the two urine samples. On the other hand, the PAR values behaved very similarly in the acidified urine samples that originally contained 100 mM NaHCO$_3$ or 100 mM NaOH. The greatest differences between the NaHCO$_3$- and NaOH-treated urine samples were observed for nitrite. The PAR of m/z 46 to m/z 47 increased after 20 min of derivatization only in the non-acidified, $NaHCO_3$-treated urine sample. This finding strongly indicates that it is not the alkalinity, but the $CO_2/Na_2CO_3/NaHCO_3$ system of the urine that interferes with the analysis of nitrite in human urine. This interference can be eliminated by acidification of the urine sample, most likely by instantaneous conversion of Na_2CO_3 and $NaHCO_3$ to highly volatile CO_2 [8].

3.2. Effect of Exogenous Bicarbonate on the Derivatization of [^{15}N]nitrate, [^{15}N]nitrite and Endogenous Nitrate and Nitrite in Human Urine

A pooled human urine sample was spiked with [^{15}N]nitrate and [^{15}N]nitrite at final concentrations of 100 µM and 1 µM, respectively. Each two 100 µL aliquots of this sample were spiked with an aqueous solution of $NaHCO_3$ to reach final added concentrations of 0, 20, 40, 60, 80 and 100 mM. After immediate derivatization with PFB-Br for 60 min, the derivatives were extracted with toluene (1 mL) and 1 µL aliquots thereof were analyzed by GC-MS in the SIM mode. Figure 2A shows that the peak area of [^{15}N]nitrate increases linearly with increasing concentration of added $NaHCO_3$, whereas the peak area of [^{15}N]nitrite decreases at added $NaHCO_3$ concentrations of 60, 80 and 100 mM. Figure 2B shows that the peak area of endogenous nitrate (i.e., [^{14}N]nitrate) increases linearly with increasing concentration of added $NaHCO_3$, whereas the peak area of endogenous nitrite (i.e., [^{14}N]nitrite) increases at added $NaHCO_3$ concentrations up to 60 mM with a tendency to slightly decrease at 80 mM and 100 mM $NaHCO_3$. Figure 2C shows the PAR of m/z 46 to m/z 47 for nitrite and the PAR of m/z 62 to m/z 63 for nitrate. The PAR of m/z 62 to m/z 63 is constant, i.e., independent of the added $NaHCO_3$ concentration. The mean PAR m/z 62 to m/z 63 was 7.579 (RSD, 0.8%) corresponding to a concentration of 758 µM for endogenous nitrate in the urine sample. Previously, we found that the derivatization of nitrate with PFB-Br is incomplete even for 60 min at 50 °C [2]. The results of Figure 2 suggest that the formation of the PFB-O$^{15}NO_2$ and PFB-O$^{14}NO_2$ increases to the same extent in dependency on the $NaHCO_3$ concentration in the urine. Thus, nitrate can be reliably measured in human urine in the presence of high bicarbonate concentrations, as they may occur upon acetazolamide administration [6]. On the other hand, the results of Figure 2 indicate that $NaHCO_3$, at concentration of 60, 80 and 100 mM, decreases the formation of PFB-$^{15}NO_2$ and PFB-$^{14}NO_2$. This may eventually lead to overestimation and thus to inaccurate measurement of nitrite in urine samples that contain high concentrations of bicarbonate.

3.3. Effects of Exogenous and Endogenous Bicarbonate, Acidification and Derivatization Time on the Derivatization of Nitrate and Nitrite in Human Urine

Two pooled human urine samples, i.e., Urine X and Urine Y, were spiked with [^{15}N]nitrate and [^{15}N]nitrite at the final added concentrations of 400 µM and 4 µM, respectively. Urine X was obtained from a volunteer who orally received acetazolamide. Urine Y was collected by another volunteer who did not receive any drug. The concentrations of bicarbonate, nitrate and nitrite were unknown in both urine samples. Urine Y was freshly spiked with 100 mM $NaHCO_3$. Aliquots (100 µL) of the urines were derivatized with PFB-Br at 50 °C for different incubation times (range, 0 to 60 min) without and with acidification of the samples and analyzed by GC-MS as described above. The GC-MS chromatograms from these analyses are shown in Figure 3. The results of this experiment are illustrated in Figure 4.

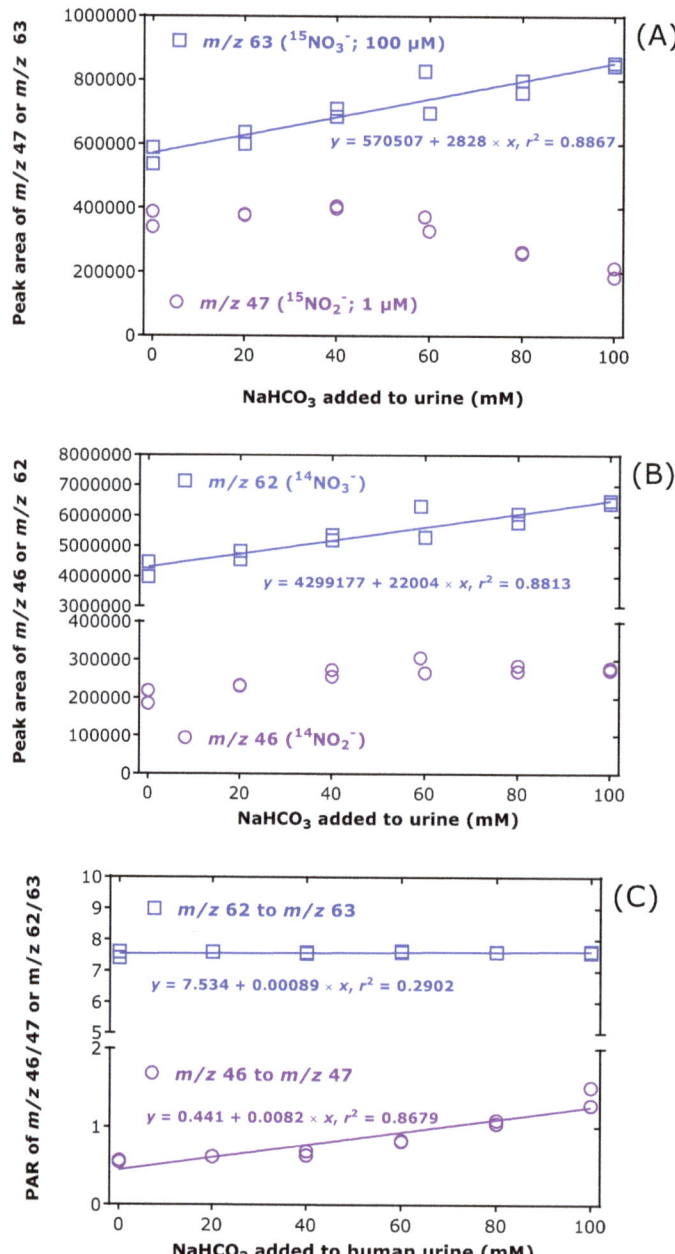

Figure 2. Plots of the peak area of (**A**) m/z 63 for [^{15}N]nitrate and m/z 47 for [^{15}N]nitrite, of (**B**) m/z 62 for [^{14}N]nitrate and m/z 46 for [^{14}N]nitrite in a human urine sample spiked with the indicated bicarbonate concentrations, and of (**C**) the peak area ratio (PAR) of m/z 62 to m/z 63 for nitrate and of m/z 46 to m/z 47 for nitrite. The added concentrations were 100 µM for [^{15}N]nitrate and 1 µM for [^{15}N]nitrite. Data are shown from two independent experiments. The derivatization time with PFB-Br was each 60 min.

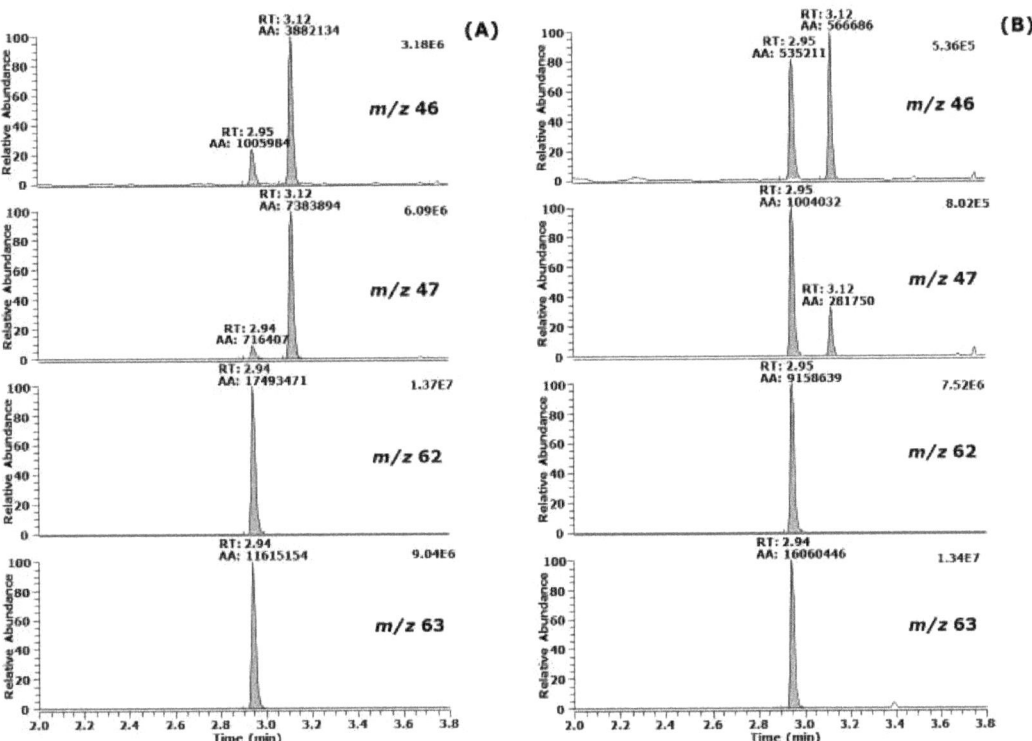

Figure 3. Representative partial GC-MS chromatograms from the simultaneous analysis of nitrite and nitrate in human urine samples without external addition of NaHCO$_3$ (**A**) and with addition of 100 mM NaHCO$_3$ (**B**). Derivatization with PFB-Br for 60 min at 50 °C in aqueous acetone was performed. The concentrations of the internal standards in the urine samples were 400 μM [^{15}N]nitrate and 4 μM [^{15}N]nitrite. GC-MS analysis was performed using selected-ion monitoring of m/z 46 for [^{14}N]nitrite, m/z 47 for [^{15}N]nitrite, m/z 62 for [^{14}N]nitrate and m/z 63 for [^{15}N]nitrate. The retention time (RT) was 2.95 min for nitrate and 3.12 min for nitrite. Note the decrease of the intensity and the peak area values of m/z 46 for [^{14}N]nitrite and m/z 47 for [^{15}N]nitrite in presence of bicarbonate (**B**). AA means peak area calculated in the automated mode. The ThermoFisher quadrupole GC-MS apparatus model DSQ in the negative-ion chemical ionization mode was used.

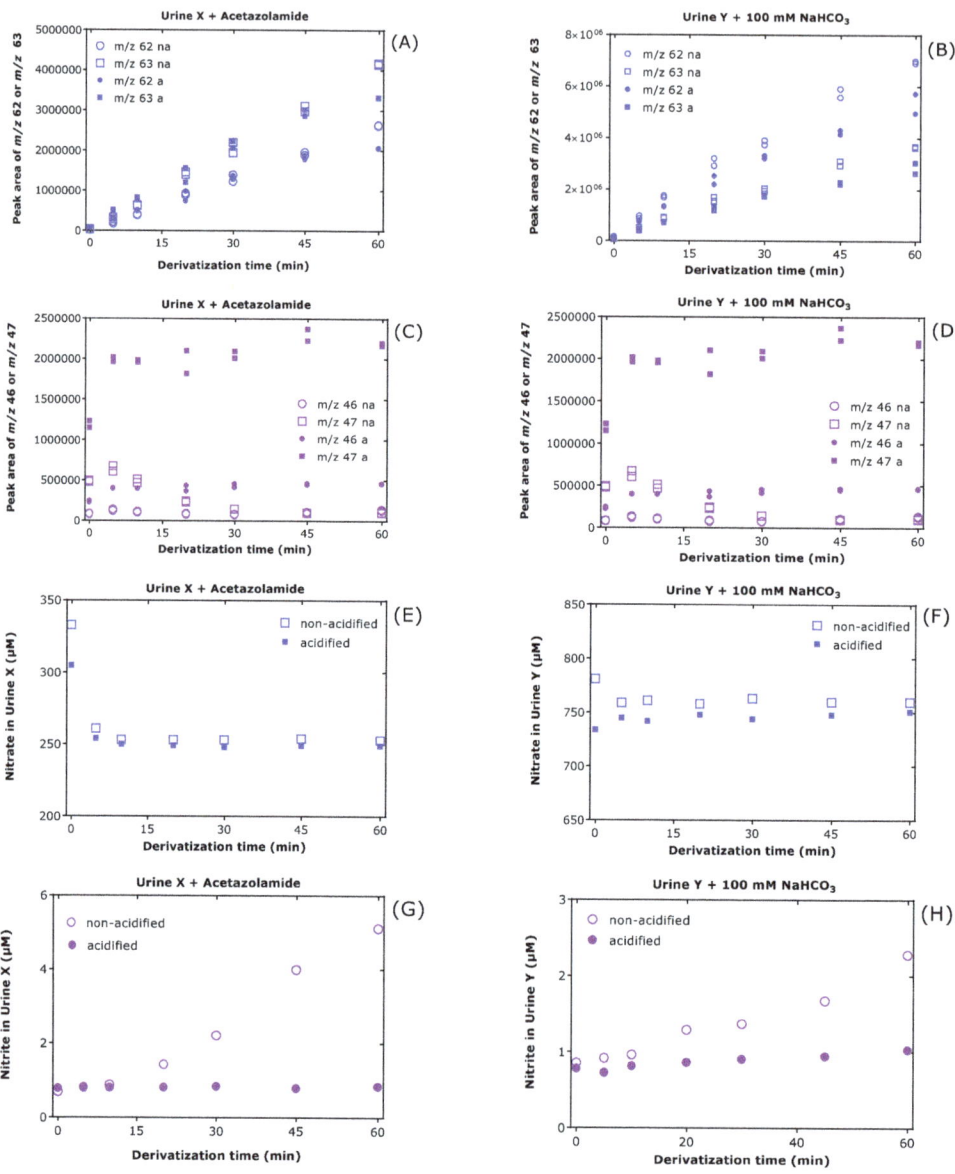

Figure 4. Effects of exogenous bicarbonate and of the ingestion of acetazolamide on nitrate and nitrite analysis by GC-MS after derivatization with PFB-Br for the indicated times at 50 °C in two different human urine samples. The samples were derivatized before and after acidification with 20 wt.% acetic acid. The concentrations of the internal standards in the urine samples were 400 µM [^{15}N]nitrate and 4 µM [^{15}N]nitrite. Urine X (**A,C,E,G**) was collected upon ingestion of a 500-milligram tablet acetazolamide from a previous study [6]. Urine Y (**B,D,F,H**) was freshly collected by a healthy volunteer (author of the present article) and spiked with 100 mM NaHCO$_3$. Data are shown from two independent experiments. All nitrite and nitrate species were measured simultaneously by GC-MS using selected-ion monitoring of m/z 46 for [^{14}N]nitrite, m/z 47 for [^{15}N]nitrite, (**B**) m/z 62 for [^{14}N]nitrate and m/z 63 for [^{15}N]nitrate. Left panels, Urine X; right panels, Urine Y. Acidified, close symbols; non-acidified, open symbols.

The effects of the derivatization time and acidification of the urine samples were qualitatively closely comparable. The peak area values of m/z 62 and m/z 63 increased with derivatization time and were lower in the acidified samples of Urine X and Urine Y (Figure 4A,B). The peak area values of m/z 46 and m/z 47 changed with derivatization time and were higher in the acidified samples of Urine X and Urine Y (Figure 4C,D). The highest peak area values were obtained at the derivatization time of 5 min. With the exception of the derivatization time of 0 min (urine sample was treated with PFB-Br and extracted immediately), the concentration of endogenous nitrate was independent of the derivatization time, but it was constantly lower in the acidified urine samples (Figure 4E,F). The concentration of endogenous nitrite was independent of the derivatization time of the acidified samples, but it increased constantly with the derivatization time larger than 10 min (Figure 4G,H).

3.4. Effect of Exogenous Bicarbonate on the Derivatization of [^{15}N]nitrite and Endogenous Nitrite in Human Plasma

A pooled human plasma sample was spiked with [^{15}N]nitrite at a final concentration of 1 µM. Each two 100-µL aliquots of this sample were spiked with an aqueous solution of NaHCO$_3$ to reach final added concentrations of 0, 10, 20, 40, 60, 80 and 100 mM. After derivatization with PFB-Br for 5 min, the derivatives were extracted with toluene and analyzed by GC-MS in the SIM mode. The results of these measurements are shown in Figure 5.

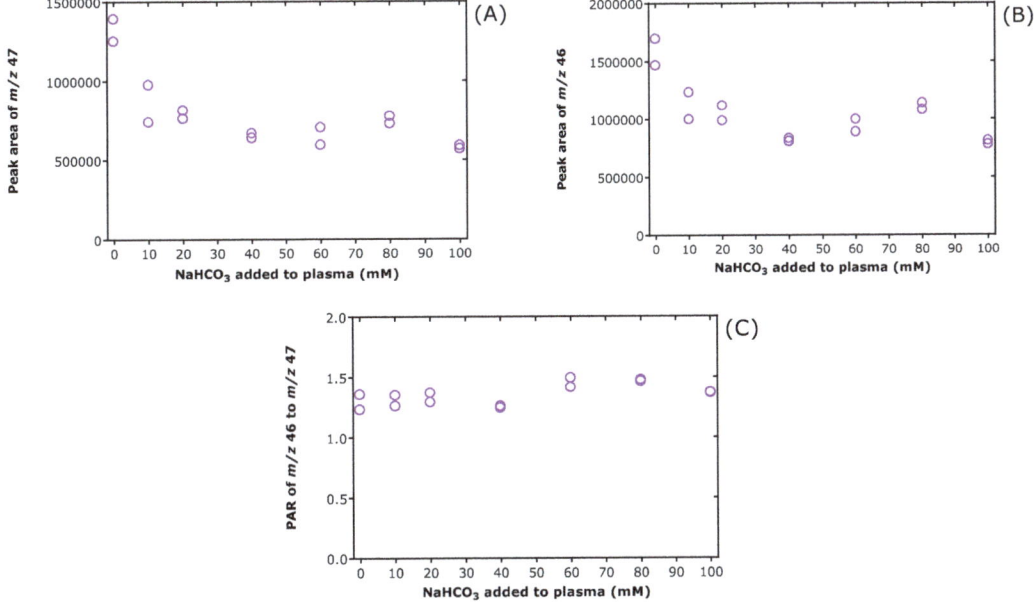

Figure 5. Plots of the peak area of (**A**) m/z 47 for exogenous nitrite (i.e., [^{15}N]nitrite, 1 µM), of (**B**) m/z 46 for endogenous nitrite (i.e., [^{14}N]nitrite) in a human plasma sample spiked with the indicated bicarbonate concentrations, and of (**C**) the resulting peak area ratio (PAR) of m/z 46 to m/z 47 for nitrite. Data are shown from two independent experiments. The derivatization time with PFB-Br was 5 min each. No [^{15}N]nitrate was added to the plasma sample.

The peak area of m/z 47 for the internal standard [^{15}N]nitrite and of m/z 46 for the endogenous nitrite ([^{14}N]nitrite) decreased considerably at added NaHCO$_3$ concentrations of 10 mM, 20 mM and 40 mM (Figure 5A,B). However, the PAR of m/z 46 to m/z 47 was largely independent of the NaHCO$_3$ concentration (Figure 5C). The mean PAR was

1.357 (6.43%), indicating a concentration of 1.36 µM for endogenous nitrite in the plasma sample. These results suggest that the exogenous $NaHCO_3$ concentration equally inhibits the formation of PFB-$^{15}NO_2$ and PFB-$^{14}NO_2$ by maximally 50% under these conditions. Thus, the endogenous $CO_2/Na_2CO_3/NaHCO_3$ system, which amounts to about 20 mM in total of human blood, is likely to inhibit the derivatization of nitrite with PFB-Br, yet without affecting analytical reliability in terms of accuracy. Unlike urinary nitrite, sample acidification is not pressingly needed in the quantitative analysis of nitrite in human plasma using PFB-Br derivatization.

4. Discussion

The derivatization of organic and inorganic anions such as nitrite, nitrate, chloride and carbonate is an indispensable analytical procedure in gas chromatography-based techniques for the vast majority of natural compounds. Pentafluorobenzyl bromide (PFB-Br) is a highly versatile derivatization reagent because of its favorable physicochemical properties with respect to both chromatography and detection due to its strongly electron-capturing F atoms [1]. The latter property leads to unbeatable amol-sensitivity in GC-MS-based approaches operating in the chemical ionization mode for numerous analytes in virtually all kinds of biological samples. As endogenous metabolites of NO and of environmental NO_x species produced by human and natural activities, nitrite and nitrate are of general interest. Many different methods have been reported for the analysis of nitrite and nitrate in the last two centuries [9]. We found that endogenous nitrite and nitrate can be simultaneously derivatized with PFB-Br in numerous biological fluids and tissues in their acetonic solutions and suspensions [2]. Nitrite and nitrate are converted by PFB-Br to PFB-NO_2 and PFB-ONO_2, respectively, virtually without the need of any catalyst. Nitrite and nitrate can be simultaneously quantitated by GC-MS as PFB-NO_2 and PFB-ONO_2 by using commercially available salts of [^{15}N]nitrite and [^{15}N]nitrate as internal standards, without problems arising from the need to chemically or enzymatically reduce nitrate to nitrite prior to derivatization.

PFB-Br is also suitable for the analysis of carboxylic groups-containing compounds in water-free organic solvents. Yet, the derivatization of fatty acids and their metabolites with PFB-Br requires the use of an organic base as a catalyst. Captopril is a carboxylic drug, and its derivatization with PFB-Br to its PFB ester has been reported to be catalyzed by carbonate [10]. Carbonate is often used in analytical derivatization, for instance, that of dimethylamine with pentafluorobenzoyl chloride, notably by means of extractive pentafluorobenzoylation [11,12]. However, the underlying mechanism of the catalytic action of carbonate is not yet fully understood. Previously, we had no indication that the derivatization of nitrite and nitrate with PFB-Br required carbonate/bicarbonate as a catalyst. We found that PFB-Br is suitable for the derivatization of carbonate/bicarbonate under the derivatization conditions of nitrite and nitrate [5]. In aqueous acetone, carbonate was found to form many reaction products. Two major carbonate derivatives were identified as $CH_3COCH_2-C(OH)(OPFB)_2$ and $CH_3COCH=C(OPFB)_2$, suggesting unique acetone-involving reactions. Two minor carbonate derivatives were PFB-OCOOH and $O=CO_2-(PFB)_2$. The GC-MS spectra of the PFB-$O^{12}COOH$ and PFB-$O^{13}COOH$ derivatives are shown in Figure 6 [5]. To the best of our knowledge, benzyl and pentafluorobenzyl esters of carbonate/bicarbonate have not been reported elsewhere, nor are they commercially available. PFB-OCOOH is presumably labile in aqueous solutions, but isolable by solvent extraction with toluene and apparently stable therein for GC-MS analysis [5]. It is proposed that PFB-OCOOH and $O=CO_2-(PFB)_2$ are formed by the reaction of carbonate with one and two PFB-Br molecules, respectively, yet without the incorporation of acetone in these derivatives (Scheme 2).

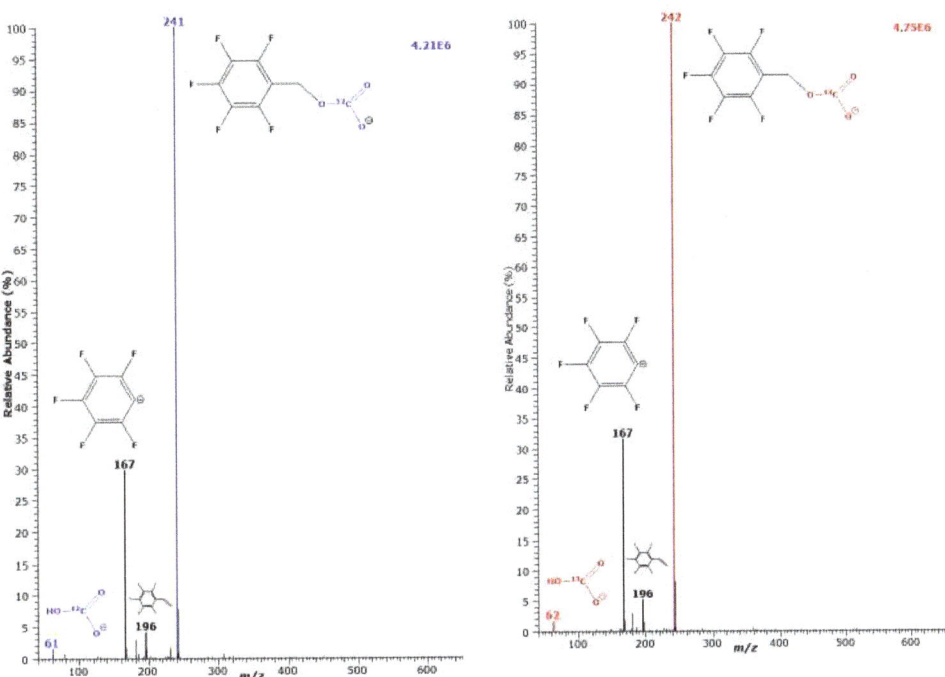

Figure 6. Negative-ion chemical ionization GC-MS spectra of PFB-O^{12}COOH and PFB-O^{13}COOH, two minor derivatization products from the separate reaction of ^{12}C- and ^{13}C-carbonate in aqueous acetone with pentafluorobenzyl bromide (PFB-Br) for 60 min at 50 °C. PFB-O^{12}COOH (M, 242.1) and PFB-O^{13}COOH (M, 243.1) have the same retention time of 6.97 min. Inserts indicate the structure of the proposed anions of ^{12}C-carbonate (left, blue) and ^{13}C-carbonate (right, red). The anions m/z 167 and m/z 196 are derived from the derivatization reagent PFB-Br. These mass spectra have been previously reported as supplementary information in Ref. [5].

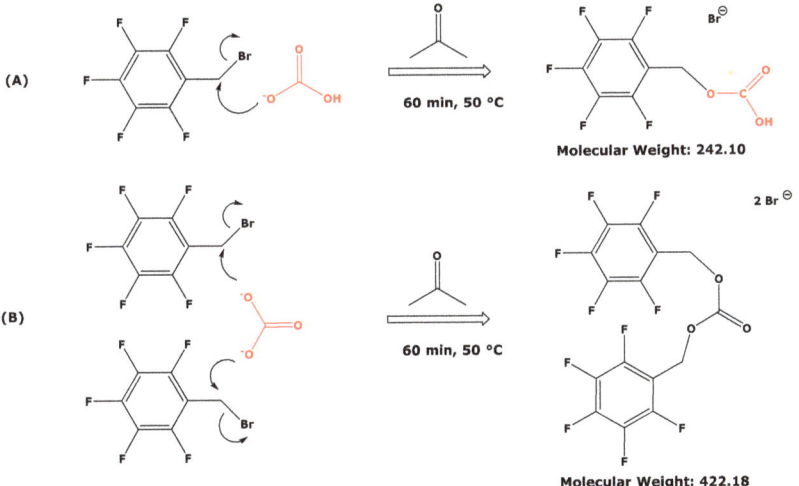

Scheme 2. Minor derivatization products from the reaction of carbonate in aqueous acetone with (**A**) one and (**B**) two pentafluorobenzyl bromide molecules at 50 °C. PFB-OCOOH (M, 241.1); PFB-O(CO)-O-PFB (M. 422.2).

In experiments with acetazolamide, a diuretic drug that massively enhances the excretion of carbonate/bicarbonate in the urine, we observed that the derivatization of nitrite with PFB-Br in urine samples of humans who ingested acetazolamide was decreased, suggesting a strong inhibitory effect of the conversion of nitrite into PFB-NO$_2$ [6]. This effect was abolished by acidifying the urine samples with aqueous acetic acid to pH values around 4.5 [6]. It is known that the derivatization of nitrite and halides with PFB-Br in its solutions in acetone, acetonitrile or ethanol are dependent on the pH value of the derivatization mixture [13]. It is also known that nitrite can react with CO_2 [14]. Thus, the CO_2/Na_2CO_3/$NaHCO_3$ system can interfere with the PFB-Br derivatization of nitrite and nitrate in biological samples by several different mechanisms. In the present work, we addressed such potential mechanisms.

In order to gain more mechanistic information in this study, we used [^{15}N]nitrite and [^{15}N]nitrate, human urine and plasma samples spiked with NaHCO$_3$ in relevant physiological and pharmacological concentration ranges. We also used urine samples from volunteers who took acetazolamide at pharmacological doses (around 5 mg/kg bodyweight).

The results of the present and previous studies suggest that carbonate does not catalyze the derivatization of nitrate to PFB-ONO$_2$ with PFB-Br in homogenous phase in aqueous acetone. Carbonate/bicarbonate-containing biological samples, notably human urine, are by nature alkaline. However, it is not the alkalinity itself but the presence of carbonate/bicarbonate at high concentrations that is responsible in part for the massive impairment of nitrite derivatization with PFB-Br. The concentration of carbonate/bicarbonate in human plasma is of the order of 20 mM, that of nitrate in the range of 20 to 100 µM and that of nitrite of the order of 1 µM. The concentration of carbonate/bicarbonate in human urine is usually below 10 mM, that of nitrate of the order of 1000 µM and that of nitrite of the order of 10 µM. In the case of acetazolamide ingestion, blood carbonate/bicarbonate and blood pH only slightly change. However, in urine, carbonate/bicarbonate can reach concentrations up to about 100 mM upon acetazolamide intake, whereas nitrate and nitrite concentrations in urine instead decrease due to the diuretic effects of the drug. Thus, in such urine samples, carbonate/bicarbonate are present in a high molar excess of nitrate and especially of nitrite. As an example, we considered a urine sample that contains 10 mM carbonate/bicarbonate, 1000 µM nitrate and 10 µM nitrite. Under regular derivatization conditions (100 µL urine, 10 µL PFB-Br equivalent to 70 µmol), PFB-Br is present in a very high molar excess in carbonate/bicarbonate (69:1 µmol), nitrate (690:1) and nitrite (6900:1). In addition, human urine contains many other organic and inorganic substances at mM concentrations, such as creatinine [15] and chloride [16], which can react with PFB-Br under the same derivatization conditions. Thus, nitrate and nitrite compete with many other species for PFB-Br. High increases of the concentration of competing nucleophiles with small changes of nitrate and nitrite concentrations, for instance carbonate/bicarbonate, would decrease the molar ratio of PFB-Br to nitrate and nitrite. As the yield of derivatization reactions with PFB-Br also depends upon the PFB-Br concentration [13], high increases of competitive analytes would decrease the derivatization yield of PFB-NO$_2$ and PFB-ONO$_2$. Our studies show that increasing carbonate/bicarbonate concentrations decrease the yield of PFB-NO$_2$, but they increase the yield of PFB-ONO$_2$. One may therefore conclude that competition alone cannot explain the opposite effects of carbonate/bicarbonate on nitrate and nitrite derivatization with PFB-Br.

A more convincing assumption could be that intermediate derivatives of carbonate/bicarbonate with PFB-Br, notably PFB-OCOOH and PFB-O(CO)-O-PFB, interact with nitrate and nitrite to produce diametrically opposed effects on the conversion of nitrate to PFB-ONO$_2$ and of nitrite to PFB-NO$_2$ (Scheme 3). The nucleophilic attack of nitrate on the benzyl groups of PFB-OCOOH and PFB-O(CO)-O-PFB would then increase the formation of PFB-ONO$_2$, thereby releasing carbonate. As PFB-ONO has not been detected thus far, it is possible that nitrite attacks the benzyl group of PFB-Br exclusively with its N atom to produce to PFB-NO$_2$. This reaction occurs more rapidly and abundantly than the reaction of nitrate with PFB-Br to generate PFB-ONO$_2$ [2]. That carbonate/bicarbonate inhibited the formation of PFB-NO$_2$ suggests that nitrite cannot react with PFB-OCOOH and PFB-O(CO)-O-PFB to form PFB-NO$_2$ and to release carbonate. Yet, it is also possible that carbonate reacts with PFB-NO$_2$ and PFB-ONO$_2$ to generate PFB-OCOOH, thereby releasing nitrite and nitrate, respectively (Scheme 3). Thus, the reaction of carbonate/bicarbonate with PFB-NO$_2$ is considered irreversible and decreases the concentration of PFB-NO$_2$ during the derivatization. On the other hand, the reaction of carbonate/bicarbonate with PFB-ONO$_2$ is considered reversible and eventually increases the concentration of PFB-ONO$_2$ during the derivatization, in addition to the direct reaction of nitrate with PFB-Br.

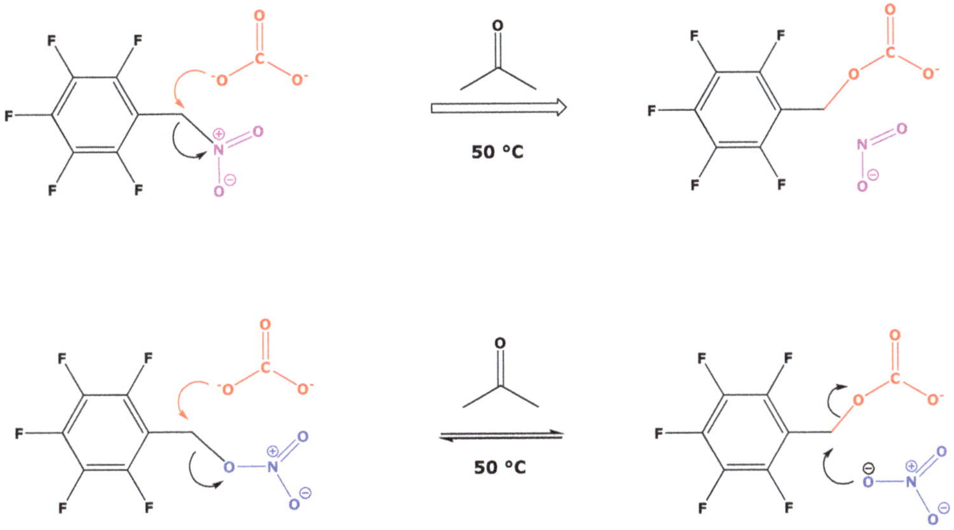

Scheme 3. Proposed reactions for the nucleophilic substitution by carbonate from (**A**) PFB-NO$_2$ and (**B**) PFB-ONO$_2$ to release nitrite and nitrate, respectively, during the derivatization in aqueous acetone with pentafluorobenzyl bromide at 50 °C in the presence of carbonate. Reaction (**A**) is considered irreversible, reaction B is assumed reversible.

In theory, carbonate/bicarbonate may contribute to nitrate in the case of chromatographic co-elution of PFB-OCOOH and PFB-ONO$_2$. This is because the ^{13}C isotope of carbonate forms PFB-O^{13}COOH, which ionizes to form m/z 62, albeit to a very low extent of about 2% (Figure 6). Such a contribution is considered very low because of the natural abundance of ^{13}C of 1.1%. Nevertheless, the contribution of carbonate/bicarbonate to nitrate may be higher in the case that ^{13}C-carbonate is used as the internal standard for endogenous carbonate/bicarbonate.

5. Conclusions

The derivatization of inorganic anions such as nitrite, nitrate and carbonate/bicarbonate in biological and environmental samples with PFB-Br using acetone as the organic solvent is performed in the homogenous phase and makes their GC-MS analysis possible. Although quantitative analysis is realized by using stable isotope labeled analogs, such as ^{15}N-nitrite, ^{15}N-nitrate and ^{13}C-carbonate, interferences may occur due to their ubiquitous occurrence in the form of their contaminants. Another potential source of interference with the analysis of nitrite and nitrate in human urine may be the occurrence of carbonate/bicarbonate at manifold higher concentrations than under normal conditions. We identified such a condition when humans ingested pharmacological doses of the diuretic acetazolamide, an inhibitor of human CA II and CA IV. Acetazolamide potently inhibits renal CA activity and, in this way, massively increases the excretion of carbonate/bicarbonate in the urine and alkalizes the urine to pH values of about 8. Our studies indicate that carbonate/bicarbonate exerts diametrically opposed effects on the analysis of nitrite and nitrate by GC-MS when derivatized with PFB-Br. Carbonate/bicarbonate increases the formation of PFB-ONO_2 but decreases the formation of PFB-NO_2. These effects are not due to the concurrent alkalization of the urine by drugs such as acetazolamide. Rather, carbonate/bicarbonate increases the formation of PFB-ONO_2 by enhancing the reaction of nitrate with an intermediate and isolable reaction product of carbonate/bicarbonate with PFB-Br, i.e., PFB-OCOOH. Unlike in other derivatization reactions such as with pentafluorobenzoyl chloride, carbonate/bicarbonate does not act as a catalyst in the derivatization of nitrate and nitrite with PFB-Br. On the other hand, carbonate/bicarbonate decreases the formation of PFB-NO_2 most likely due the inability of nitrite to attack the PFB-OCOOH via its N atom. Such effects are much less pronounced in human plasma, in part because of the lower carbonate/bicarbonate concentration and in part due to the higher buffer capacity of the plasma compared to urine. A very simple and effective solution of the negative effect of high carbonate/bicarbonate concentrations on nitrite measurement in urine as PFB-NO_2 is mild acidification by adding small volumes of 20 wt.% acetic acid to the urine.

Funding: This research received no external funding.

Institutional Review Board Statement: Ethical review and approval were waived for the present work, due to the use of human urine samples originally collected in a previous study cited in this work (i.e., Ref. [6]).

Informed Consent Statement: Not applicable.

Data Availability Statement: Not applicable.

Acknowledgments: The technical assistance of Anja Mitschke is gratefully acknowledged.

Conflicts of Interest: The author declares no conflict of interest.

Sample Availability: Samples of the compounds are not available from the authors.

References

1. Tsikas, D. Pentafluorobenzyl bromide-A versatile derivatization agent in chromatography and mass spectrometry: I. Analysis of inorganic anions and organophosphates. *J. Chromatogr. B Analyt. Technol. Biomed. Life Sci.* **2017**, *1043*, 187–201. [CrossRef] [PubMed]
2. Tsikas, D. Simultaneous derivatization and quantification of the nitric oxide metabolites nitrite and nitrate in biological fluids by gas chromatography/mass spectrometry. *Anal. Chem.* **2000**, *72*, 4064–4072. [CrossRef] [PubMed]
3. Arabaci, B.; Gulcin, I.; Alwasel, S. Capsaicin: A potent inhibitor of carbonic anhydrase isoenzymes. *Molecules* **2014**, *19*, 10103–10114. [CrossRef] [PubMed]
4. Topal, F.; Gulcin, I.; Dastan, A.; Guney, M. Novel eugenol derivatives: Potent acetylcholinesterase and carbonic anhydrase inhibitors. *Int. J. Biol. Macromol.* **2017**, *94*, 845–851. [CrossRef] [PubMed]
5. Tsikas, D.; Chobanyan-Jürgens, K. Quantification of carbonate by gas chromatography-mass spectrometry. *Anal. Chem.* **2010**, *82*, 7897–7905. [CrossRef] [PubMed]
6. Chobanyan-Jürgens, K.; Schwarz, A.; Böhmer, A.; Beckmann, B.; Gutzki, F.M.; Michaelsen, J.T.; Stichtenoth, D.O.; Tsikas, D. Renal carbonic anhydrases are involved in the reabsorption of endogenous nitrite. *Nitric Oxide* **2012**, *26*, 126–131. [CrossRef] [PubMed]
7. Supuran, C.T. Emerging role of carbonic anhydrase inhibitors. *Clin. Sci.* **2021**, *28*, 1233–1249. [CrossRef] [PubMed]
8. Boztaş, M.; Çetinkaya, Y.; Topal, M.; Gülçin, İ.; Menzek, A.; Şahin, E.; Tanc, M.; Supuran, C.T. Synthesis and carbonic anhydrase isoenzymes I, II, IX, and XII inhibitory effects of dimethoxybromophenol derivatives incorporating cyclopropane moieties. *J. Med. Chem.* **2015**, *58*, 640–650. [CrossRef] [PubMed]
9. Tsikas, D. Analysis of nitrite and nitrate in biological fluids by assays based on the Griess reaction: Appraisal of the Griess reaction in the L-arginine/nitric oxide area of research. *J. Chromatogr. B Analyt. Technol. Biomed. Life Sci.* **2007**, *851*, 51–70. [CrossRef]
10. Liu, Y.C.; Wu, H.L.; Kou, H.S.; Chen, S.H.; Wu, S.M. Derivatization-gas chromatographic determination of captopril. *Anal. Lett.* **1995**, *28*, 1465–1481. [CrossRef]
11. Tsikas, D.; Thum, T.; Becker, T.; Pham, V.V.; Chobanyan, K.; Mitschke, A.; Beckmann, B.; Gutzki, F.M.; Bauersachs, J.; Stichtenoth, D.O. Accurate quantification of dimethylamine (DMA) in human urine by gas chromatography-mass spectrometry as pentafluorobenzamide derivative: Evaluation of the relationship between DMA and its precursor asymmetric dimethylarginine (ADMA) in health and disease. *J. Chromatogr. B Analyt. Technol. Biomed. Life Sci.* **2007**, *851*, 229–239. [CrossRef] [PubMed]
12. Chobanyan, K.; Mitschke, A.; Gutzki, F.M.; Stichtenoth, D.O.; Tsikas, D. Accurate quantification of dimethylamine (DMA) in human plasma and serum by GC-MS and GC-tandem MS as pentafluorobenzamide derivative in the positive-ion chemical ionization mode. *J. Chromatogr. B Analyt. Technol. Biomed. Life Sci.* **2007**, *51*, 240–249. [CrossRef] [PubMed]

13. Wu, H.L.; Chen, S.H.; Lin, S.J.; Hwang, W.R.; Funazo, K.; Tanaka, M.; Shono, T. Gas chromatographic determination of inorganic anions as pentafluorobenzyl derivatives. *J. Chromatogr. A* **1983**, *269*, 183–190.
14. Tsikas, D.; Böhmer, A.; Gros, G.; Endeward, V. Evidence of the chemical reaction of ^{18}O-labeled nitrite with CO_2 in aqueous buffer of neutral pH and the formation of ^{18}OCO by isotope ratio mass spectrometry. *Nitric Oxide* **2016**, *55*, 25–35. [CrossRef] [PubMed]
15. Tsikas, D.; Wolf, A.; Mitschke, A.; Gutzki, F.M.; Will, W.; Bader, M. GC-MS determination of creatinine in human biological fluids as pentafluorobenzyl derivative in clinical studies and biomonitoring: Inter-laboratory comparison in urine with Jaffe, HPLC and enzymatic assays. *J. Chromatogr. B Analyt. Technol. Biomed. Life Sci.* **2010**, *878*, 2582–2592. [CrossRef] [PubMed]
16. Tsikas, D.; Fauler, J.; Frölich, J.C. Determination of chloride in biological fluids as pentafluorobenzylchloride by reversed-phase high-performance liquid chromatography and UV detection. *Chromatographia* **1992**, *33*, 317–320. [CrossRef]

Article

Structural Characterization of Unusual Fatty Acid Methyl Esters with Double and Triple Bonds Using HPLC/APCI-MS² with Acetonitrile In-Source Derivatization

Petra Horká [1,2], Vladimír Vrkoslav [1], Jiří Kindl [1], Karolina Schwarzová-Pecková [2] and Josef Cvačka [1,2,*]

[1] Institute of Organic Chemistry and Biochemistry of the Czech Academy of Sciences, Flemingovo Náměstí 542/2, 166 00 Prague 6, Czech Republic; peta.machajda@gmail.com (P.H.); vladimir.vrkoslav@uochb.cas.cz (V.V.); kindlj@ftz.czu.cz (J.K.)

[2] Department of Analytical Chemistry, Faculty of Science, Charles University, Hlavova 2030/8, 128 43 Prague 2, Czech Republic; karolina.schwarzova@natur.cuni.cz

* Correspondence: josef.cvacka@uochb.cas.cz; Tel.: +420-220-183-303

Abstract: Double and triple bonds have significant effects on the biological activities of lipids. Determining multiple bond positions in their molecules by mass spectrometry usually requires chemical derivatization. This work presents an HPLC/MS method for pinpointing the double and triple bonds in fatty acids. Fatty acid methyl esters were separated by reversed-phase HPLC with an acetonitrile mobile phase. In the APCI source, acetonitrile formed reactive species, which added to double and triple bonds to form $[M + C_3H_5N]^{+\bullet}$ ions. Their collisional activation in an ion trap provided fragments helpful in localizing the multiple bond positions. This approach was applied to fatty acids with isolated, cumulated, and conjugated double bonds and triple bonds. The fatty acids were isolated from the fat body of early-nesting bumblebee *Bombus pratorum* and seeds or seed oils of *Punicum granatum*, *Marrubium vulgare*, and *Santalum album*. Using the method, the presence of the known fatty acids was confirmed, and new ones were discovered.

Keywords: acetonitrile-related adducts; acetylenic lipids; double and triple bond localization; in-source derivatization; mass spectrometry

1. Introduction

The localization of double and triple bonds (DBs and TBs) is a key step in the structural characterization of fatty acids (FAs). The biological functions of lipids are often linked to the specific arrangement of multiple bonds in their FA chains. Lipids with unusually arranged double bonds and triple bonds are of interest because of their function in living organisms and their unique biological effects and potential use in medicine [1,2]. Mass spectrometry is useful for pinpointing the double bonds in FAs and their methyl esters (FAMEs), especially when combined with chromatography. The early methods were based on the electron ionization of derivatized lipids. Derivatization, either at the double bond site or at the carboxylic group, is required because of the bond migration along the aliphatic chains during electron ionization. Numerous FA derivatives, including pyrrolidides [3], 4,4-dimethyloxazoline (DMOX) [4], or dimethyl disulfide (DMDS) derivatives [5], have found their use in GC/MS. Later, HPLC/MS-based methods began to be developed. Unlike GC/MS, these methods also make it possible to analyze less volatile and non-volatile FAs and their derivatives. A number of methods have been proposed for localizing double bonds using electrospray ionization, including Paternò–Büchi photochemical derivatizations [6–8], epoxidation in low-temperature plasma [9,10] and negative-ion paper-spray ionization [11], post-column epoxidation and peroxidation [12], charge-switch derivatization with *N*-(4-aminomethylphenyl) pyridinium (AMPP) [13,14], or ozone-induced dissociation (OzID) [15,16], and combining charge-switch derivatization with OzID [17].

Besides electrospray ionization, atmospheric-pressure chemical ionization (APCI) can also be applied for localizing double bonds in HPLC/MS [18–23]. The methods rely on acetonitrile-related reactive species formed in the ion sources. The use of even-electron (1-methyleneimino)-1-ethenylium as a reagent for derivatizing double bonds was initially developed for chemical ionization [24–27] and later applied in APCI-MS [18]. Using helium as a nebulizing gas, $C_3H_4N^+$ adducts ($[M + 54]^+$) of triacylglycerols (TGs) were formed, and their CID spectra indicated the positions of the original double bonds [18]. Later, we showed that APCI sources operated under conventional conditions with nitrogen nebulizing gas yield odd-electron $C_3H_5N^{+\bullet}$ adducts ($[M + 55]^{+\bullet}$) [19]. The collision activation of the adducts induced cleavages of C–C bonds next to the original double bond, leading to pairs of diagnostic fragments indicating the double bond position. The advantage of this approach lies in its simplicity: the only requirement for an HPLC/APCI-MS^2 method is the presence of acetonitrile in the mobile phase. The method has been applied for the structure elucidation of various unsaturated lipids, including FAMEs [20,28], hydroxy-FAMEs [23], wax esters [19], diol diesters [22], or TGs [21].

To date, only a few methods for determining the position of triple bonds in lipids have been published [27,29–32]. Triple bonds in FAs can be pinpointed after DMOX derivatization using GC/MS [31]. While a conjugated system of double bonds manifests itself by a series of fragments differing by 12 Da, triple bond-related fragments differ by 10 Da. It allows for the structural characterization of conjugated ene–yne acids. Still, the fragmentation of conjugated yne–yne or yne–yne–ene bonds is more complex, and the spectra are difficult to interpret [30]. Using this approach, many acetylenic lipids have been identified in plants [29,30,32]. The position of a triple bond can also be determined using acetonitrile chemical ionization based on (1-methyleneimino)-1-ethenylium adducts formation [27]. To the best of our knowledge, no method for localizing triple bonds using HPLC/MS has appeared in the literature so far.

Double bond positions in FAs reflect specificities of desaturases involved in their biosynthesis. Most monounsaturated FAs have a double bond in 9-position. Other positions are also relatively common, for instance, 7-position in algae, 5- and 10-positions in bacteria, or 6-position in plants [33]. Double bonds in polyunsaturated FAs are typically spaced by one methylene group (methylene interrupted). FAs with double bonds separated by two or more methylene units are found, for instance, in marine sponges *Microciona prolifera* (FA 26:2n-17,21 and FA 26:3n-7,17,21) [34,35], *Dysidea fragilis* (FA 25:3n-8,16,20; FA 25:3n-6,16,20; FA 24:3n-7,15,19 and FA 24:2n-7,17) [36], or *Hymeniacidon sanguinea* (e.g., FA 28:2n-9,19,23; FA 26:2n-17,21; FA 26:3n-7,17,21; FA 24:2n-15,19 and FA 24:3n-7,15,19) [37]. More than twenty different FAs with double bonds separated by two or more methylene units were identified in the gonads of limpets *Cellana grata* [38], *Collisella dorsuosa* [38], and *Cellana toreum* [39,40]. Unusual FAs with 24, 26, and 28 carbon atoms were found in TGs isolated from the fat body of early-nesting bumblebee *Bombus pratorum*. FA 26:2n-7c,17c occupied one, two, or all three positions in the TGs [41].

FAs with a conjugated system of double bonds are mostly represented by conjugated linoleic acids (CLAs) and conjugated linolenic acids (CLnAs), which are collective terms for the positional and geometric isomers of octadecadienoic and octadecatrienoic acids, respectively. CLAs exist naturally at higher concentrations in animal products, such as milk fat, cheese, and ruminant meat [42,43]. Two double bonds in CLAs are primarily in positions 9 and 11, or 10 and 12, and each of the double bonds can be either *cis* or *trans* [44]. CLAs are important for human nutrition. For instance, *cis*-9,*trans*-11 and *trans*-10,*cis*-12 isomers reduce carcinogenesis and atherosclerosis, increase bone and muscle mass, and exhibit antidiabetic effects [42,45]. CLnAs are found in plant seed oils, including oils from *Vernicia fordii* (α and β-eleostearic acid; FA 18:3n-5t,7t,9c and FA 18:3n-5t,7t,9t) [46], *Catalpa bignonoides* (catalpic acid; FA 18:3n-5c,7t,9t) [47], *Jacaranda mimosifolia* (jacaric acid; FA 18:3n-6c,8t,10c) [48], or *Calendula officinalis* (alfa-calendic acid; FA 18:3n-6c,8t,10t and beta-calendic acid; FA 18:3n-6t,8t,10t [49]. A rich source of CLnA is pomegranate (*Punicum granatum*) seed oil (PSO). It contains punicic acid (FA 18:3n-5*c*,7*t*,9*c*), α-eleostearic acid (18:3n-5*t*,7*t*,9*c*), β-eleostearic acid (18:3n-5*t*,7*t*,9*t*), and catalpic acid (18:3n-5*c*,7*t*,9*t*) [50–54]. The structures of several other CLnAs in PSO remain to be clarified [51,55]. CLnAs are known for their antioxidant, anti-inflammatory, anti-atherosclerotic, antitumor, and serum lipid-lowering activities. They help fight against cancers, obesity, diabetes, and heart diseases [53,56,57].

Lipids with two cumulated double bonds (allenic lipids) are found in Lamiaceae family plants; elsewhere in nature, they are rare [58,59]. The first known C18 allenic FA, laballenic acid (FA 18:2n-12,13), was isolated from *Leonotis nepetaefolia* seed oil [60] and later reported also from other Lamiaceae species [61]. Lamenallenic acid (FA 18:3n-2*t*,12,13) was discovered in *Lamium purpureum* seed oil [62]. Phlomic acid (FA 20:2n-12,13) was found in several species of *Phlomis* genus (Lamiaceae) [61]. Seeds of *Marrubium vulgare* contain laballenic acid and phlomic acid [59]. Allenic lipids are known for their anticancer, anti-inflammatory, antiviral, and antibacterial activities [1].

FAs with triple bonds (acetylenic FAs) are relatively widely distributed in nature. They are found in plants, fungi, microorganisms, and invertebrates [58,63–65]. FAs and other acetylenic lipids in plants serve as chemical protection against microorganisms. They are toxic to bacteria, viruses, and insects [2,66–69]. Many acetylenic lipids exhibit fungicidal, phototoxic, antitumor, and other properties [1], which render them potentially useful in medicine. The chain length and triple bond positions affect their fungicidal properties [2,68]. The structures and cytotoxic activities of acetylenic lipids were reviewed recently [70]. Some plant FAs contain triple and double bonds conjugated, e.g., pyrulic acid (FA 17:2n-7,9^{TB}), ximenynic (also termed santalbic) acid (FA 18:2n-7*t*,9^{TB}), or heisteric acid (FA 18:3n-7*t*,9^{TB},11*c*) from *Heisteria silvanii* seed oil [32]. *Ximenia americana* contains FA 18:1n-13^{TB} and FA 18:4n-2,4,8,6^{TB} [2,66]. Santalbic acid (FA 18:2n-7*t*,9^{TB}), identified for the first time in *Santalum album* [58,63], is one of the few acetylenic FAs occurring at higher levels in plants. It is found in the seed oils of the Santalaceae, Olacaceae, and Opiliaceae families, where it can reach up to 95% of the total FAs [71,72]. Other biologically active acetylenic acids are crepenynic acid (FA 18:2n-6^{TB},9*c*), tariric acid (FA 18:1n-12^{TB}), stearolic acid (FA 18:1n-9^{TB}), or nonadec-6-ynoic acid (FA 19:1n-12 ^{TB}) [67,73–75]. FAs with a triple bond can also be found in water mosses [40,76–79].

This work deals with the localization of double and triple bonds in FAMEs. The conversion of lipids or lipid mixtures to FAMEs is frequently used in lipidomics workflows because the GC or LC analysis of FAMEs provides quick and valuable information on the fatty acyl chains. Here, FAME standards and FAMEs obtained by the transesterification of the TGs from biological samples were analyzed by HPLC/APCI-MS/MS using an acetonitrile mobile phase. Isolated, cumulated, and conjugated double bonds and triple bonds were localized using the fragmentation of $[M + C_3H_5N]^{+\bullet}$ adducts generated in the ion source. To the best of our knowledge, the localization of triple bonds in FAMEs by RP-HPLC with MS detection is reported here for the first time.

2. Results and Discussion

The chromatographic separation of FAMEs was achieved on the Develosil RP-Aqueous C30 column using isocratic elution with acetonitrile. The mobile phase in the APCI source formed reactive species, which added to double and triple bonds. The adducts were isolated and activated in the ion trap to generate ions bearing information on the original double or triple bond position. The diagnostic ions formed by the cleavages of adjacent C–C bonds were marked α if they carried the ester moiety or ω if they contained the terminal-carbon end without the ester group. The diagnostic peaks corresponding to cleavages before the first and after the last unsaturated bond in polyunsaturated FAMEs tended to be more abundant than the others. This phenomenon was used for deducing the arrangement of the double and triple bonds in polyunsaturated chains. A parameter named "multiple bond region" (MBR) was calculated and tabulated for various theoretically possible arrangements of multiple bonds (Table 1). The MBR value was calculated using theoretical *m/z* values of the adduct (precursor) and α and ω fragments corresponding to cleavages before the first and after the last unsaturated bond as follows:

$$\text{MBR} = m/z\,(\alpha) + m/z\,(\omega) - m/z\left([M+55]^{+\bullet}\right) \quad (1)$$

Table 1. Multiple bond region (MBR) values for common arrangements of double bonds (DBs) and triple bonds (TBs) in polyunsaturated chains.

MBR	Arrangement of Multiple Bonds
79	One triple bond –C≡C–
81	One double bond –CH=CH–
93	Two cumulated double bonds –CH=C=CH–
103	Two conjugated triple bonds –C≡C–C≡C–
105	One double bond and one triple bond, conjugated –CH=CH–C≡C–
107	Two conjugated double bonds –CH=CH–CH=CH–
119	One double bond and one triple bond, methylene-interrupted –CH=CH–CH$_2$–C≡CH–
121	Two methylene-interrupted double bonds –CH=CH–CH$_2$–CH=CH–
133	Three conjugated double bonds –CH=CH–CH=CH–CH=CH–
161	Three methylene-interrupted double bonds –CH=CH–CH$_2$–CH=CH–CH$_2$–CH=CH–
14n + 107	Two double bonds interrupted by several methylenes (–CH$_2$–)$_n$

The experimental MBR values calculated for the adduct and the most abundant α and ω fragments in the spectra were then compared to theoretical MBRs. For instance, the MS/MS spectrum of [M + 55]$^{+\bullet}$ adduct of unknown FA at *m/z* 347.0 provided the most abundant α and ω peaks at *m/z* 290.2 and *m/z* 190.2, respectively. The calculated MBR value (290 + 190 − 347 = 133) suggested FAME with three conjugated double bonds (Table 1). Diagnostic ions were accompanied by less abundant satellite peaks differing from α and ω ions by 14 or 15 Da. These fragments representing cleavages at more distant C–C bonds were important for distinguishing double and triple bonds. The elemental composition of the major fragments in the spectra of FAME standards was confirmed by Orbitrap high-resolution data (Supplementary Materials Table S1).

2.1. Mass Spectra of Standards with Conjugated Double Bonds

The system with two conjugated double bonds was investigated using standards of FAME 18:2n-7t,9t (Mangold's acid methyl ester) and FAME 18:2n-7c,9c (ricinenic acid methyl ester). The fragments in the MS/MS spectrum for FAME 18:2n-7t,9t (Figure 1) were rationalized as follows: α n-7 peak at *m/z* 264.1, α n-9 peak at *m/z* 238.2, ω n-7 peak at *m/z* 166.1, and ω n-9 peak at *m/z* 192.1. The MBR value calculated from the two most intense fragments in the spectrum (i.e., *m/z* 192.1 and *m/z* 264.1) was 107. Despite the presence of satellite fragments differing by 14 Da from the diagnostics peaks, the spectrum provided clear evidence of two conjugated double bonds in the n-7 and n-9 positions. The spectrum of FAME 18:2n-7c,9c having the opposite geometry on both double bonds looked similar (Figure S1), which confirmed the negligible effect of double bond geometry on the adduct fragmentation documented earlier [19].

The MS/MS spectrum of punicic acid methyl ester with three conjugated double bonds (FAME 18:3n-5c,7t,9c) is shown in Figure 2. The major fragments in the spectrum were formed by cleavages before and after the series of double bonds. They were easily distinguishable from the other ions. The most abundant fragments α n-5 at *m/z* 290.2 and ω n-9 at *m/z* 190.2 delimited the group of conjugated double bonds and corresponded to an MBR value of 133. The fragments formed by the cleavages between conjugated double bonds α n-7 (*m/z* 264.3), α n-9 (*m/z* 238.2), ω n-7 (*m/z* 164.2), and ω n-5 (*m/z* 138.2) were of low intensities but discernable in the spectrum. The same diagnostic fragments and MBR value could theoretically be expected for a FAME with two cumulated double bonds separated by one methylene group from the third double bond. Such an arrangement of double bonds would be, however, clearly distinguishable because the system of cumulated double bonds manifests itself by abundant α + 1 Da ion (Section 2.3.3). Such an ion (*m/z* 251 or *m/z* 291 in this case) is not present in the spectrum. Therefore, the spectrum in Figure 2 can be unambiguously interpreted as FAME 18:3n-5,7,9.

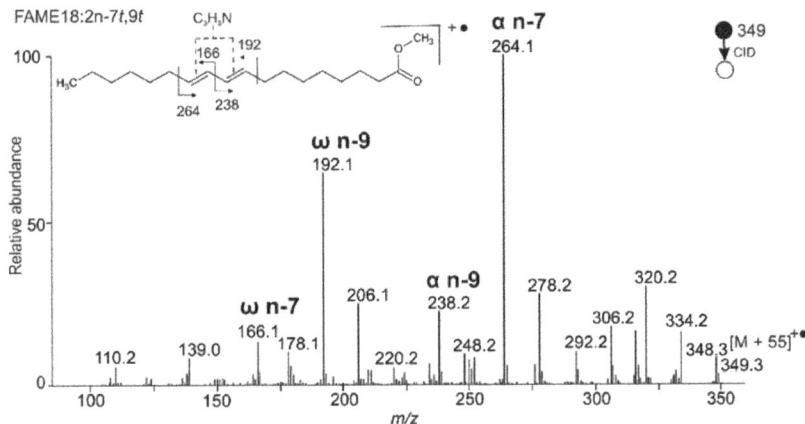

Figure 1. APCI MS/MS CID spectrum of [M + 55]$^{+\bullet}$ adduct of Mangold's acid methyl ester (FAME 18:2n-7*t*,9*t*); MBR = 264 + 192 − 349 = 107.

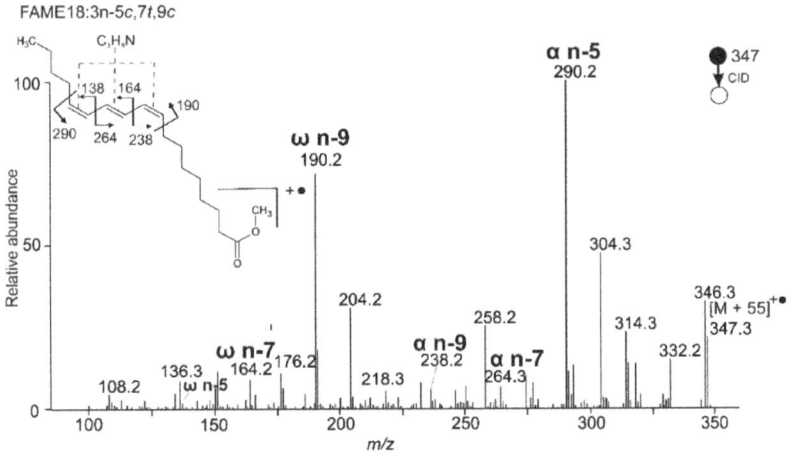

Figure 2. APCI MS/MS CID spectrum of [M + 55]$^{+\bullet}$ adduct of punicic acid methyl ester (FAME 18:3n-5*c*,7*t*,9*c*); MBR = 290 + 190 − 347 = 133.

2.2. Mass Spectra of Standards with a Triple Bond

Figure 3 shows the MS/MS spectrum of FAME 18:1n-9^{TB} (stearolic acid methyl ester) [M + 55]$^{+\bullet}$ adduct. The abundant fragments *m/z* 236.2 (α n-9^{TB}) and *m/z* 192.2 (ω n-9^{TB}) clearly indicated a triple bond in the n-9 position. Unlike FAMEs with double bonds, the satellite fragments differed by +15 Da from α^{TB} and ω^{TB} (*m/z* 207.1 and *m/z* 251.1, respectively). The intensities of the diagnostic fragments and their +15 Da satellites were similar, allowing us to recognize these peaks in the spectrum easily. Such a pattern distinctly indicated a triple bond. Satellite fragments differing by +14 Da, typical for double bonds, were present at significantly lower intensities.

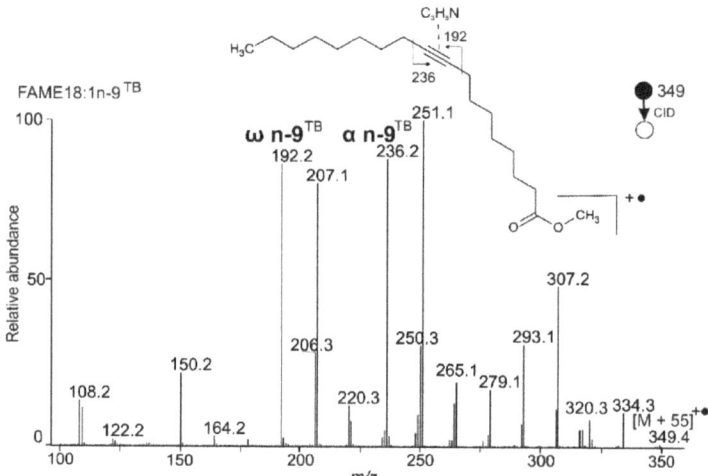

Figure 3. APCI MS/MS CID spectrum of [M + 55]$^{+\bullet}$ adduct of stearolic acid methyl ester (FAME 18:1n-9^{TB}); MBR = 236 + 192 − 349 = 79.

The satellite fragment ions made it also possible to characterize FAMEs with a combination of double and triple bonds. For instance, crepenynic acid methyl ester with one double bond and one triple bond (FAME 18:2n-6^{TB},9c) provided a spectrum with the most abundant peak at m/z 150.1 (Figure 4). This signal is a diagnostic fragment for triple bond (ω n-6^{TB}) because its satellite appears at a 15 Da higher m/z value (m/z 165.0). Analogously, the m/z 276.1 with its satellite at m/z 291.1 is the triple bond diagnostic peak (α n-6^{TB}). Fragment m/z 190.1 indicates a double bond (ω n-9) because its satellite peak appears at m/z 204.1.

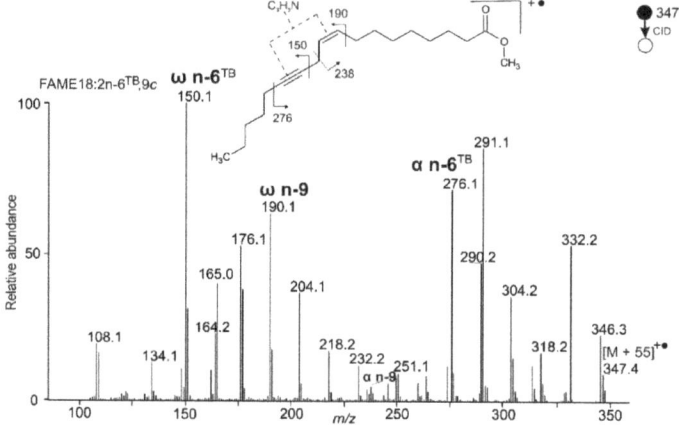

Figure 4. APCI MS/MS CID spectrum of the [M + 55]$^{+\bullet}$ adduct of crepenynic acid methyl ester (FAME 18:2n-6^{TB},9c); MBR = 276 + 190 − 347 = 119.

2.3. Analysis of Natural Samples

The fragmentation of FAME standards with various arrangements of double and triple bonds helped us characterize the FAMEs isolated from biological samples. The identification procedure was initiated by deducing the number of carbons and level of unsaturation from the m/z values of the protonated FAMEs. The second step examined the

MS/MS spectra of [M + 55]$^{+\bullet}$ ions to identify the diagnostic fragments and their satellites. The m/z values of the diagnostic fragments, MBR values, and the mass difference between the diagnostic fragments and satellites were used to deduce the positions of double and triple bonds. Finally, the retention times were checked for their consistency with the expected elution order of FAMEs [80,81].

2.3.1. FAMEs from the Fat Body of *Bombus pratorum*

The early-nesting bumblebee *Bombus pratorum* is widespread in Europe. It is one of the earliest bumblebee species to emerge from hibernation each year. The fat body of *B. pratorum* males contains TGs with long, diunsaturated fatty acyls, which are structurally related to its marking pheromone [41].

The chromatogram of *B. pratorum* FAMEs is shown in Figure 5. The MS/MS spectra of diunsaturated FAMEs (Figure 6) provided abundant and recognizable α and ω fragments interpreted as FAME 24:2n-7,17, FAME 25:2n-7,17, and 26:2n-7,17. The double bond positions were in excellent agreement with previous work, where the positions of the double bonds were established using dimethyl disulfide derivatization [41].

Altogether, nine saturated, fourteen monounsaturated, five diunsaturated, and one triunsaturated FAMEs were detected (Table 2). Nine of them (FAME 17:1n-7; FAME 17:0; FAME18:2n-3,6; FAME 18:1n-8; FAME 19:1n-7; FAME 19:0; FAME 22:1n-7; FAME 23:0; FAME 25:2n-7,17) are reported here for *B. pratorum* for the first time. To the best of our knowledge, FA 25:2n-7,17 has not been mentioned in the literature so far. FAMEs 25:2 are very rare in nature; the only known source of such acids are marine sponges producing different isomers [37,82–84]. FA 25:2n-7,17 likely serves as a precursor for tetracosadiene, a minor component (0.02 to 0.3%; I. Valterová 2021, personal communication, 18 April) of *B. pratorum* males' secretion. Bumblebee males use the secretion to mark their patrolling routes [85].

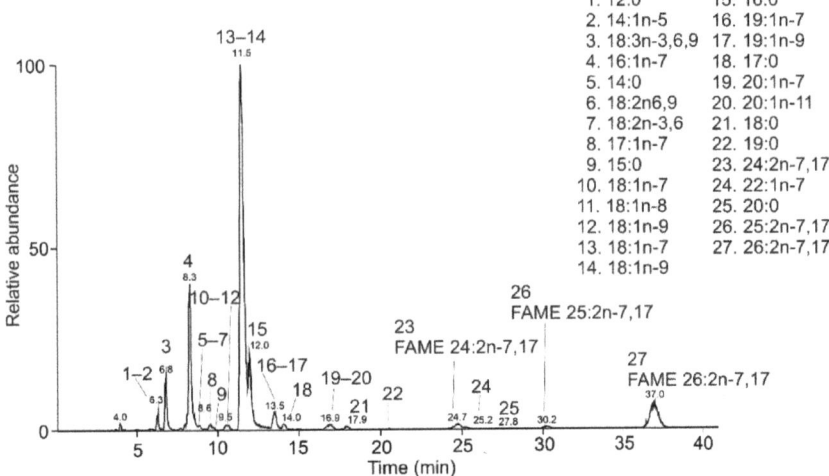

Figure 5. HPLC/APCI-MS base-peak chromatogram of FAMEs from the fat body of *Bombus pratorum* and the list of identified species.

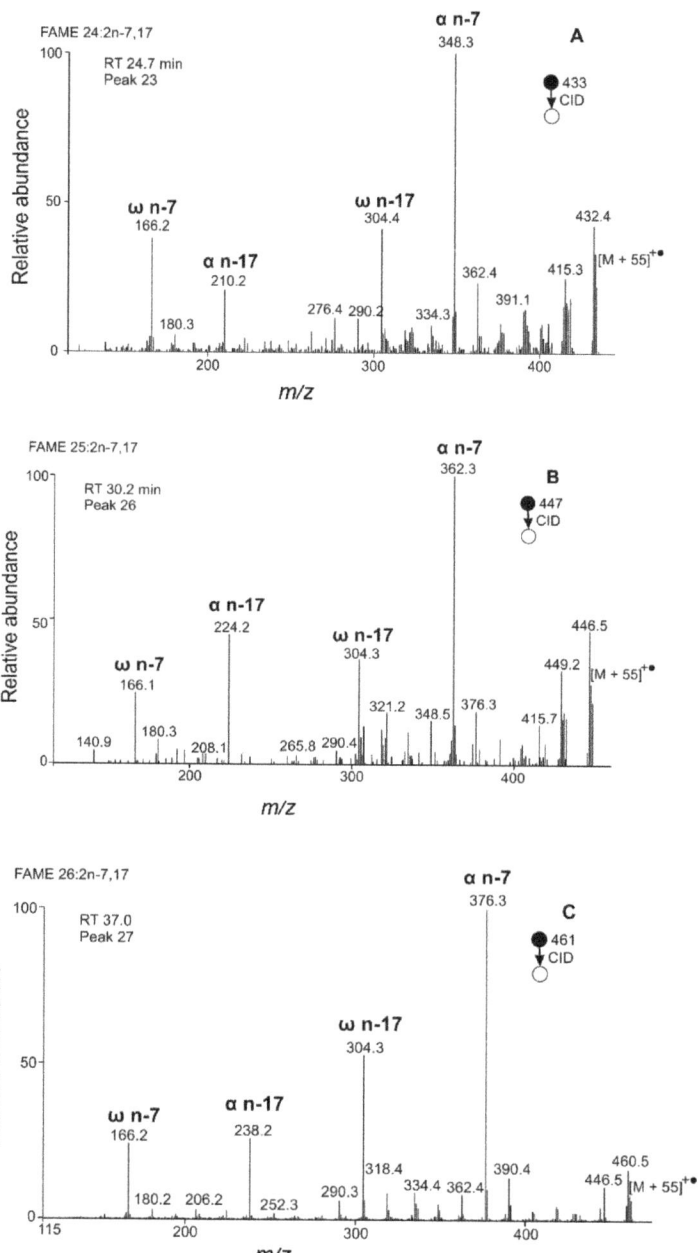

Figure 6. APCI MS/MS spectra of the [M + 55]$^{+\bullet}$ adducts of FAME from *B. pratorum* interpreted as FAME 24:2n-7,17 (**A**), 25:2 n-7,17 (**B**), and 26:2n-7,17 (**C**).

Table 2. FAMEs identified in TG fraction of *B. pratorum* fat body lipids.

FAME	t_R (min)	Rel. Peak Area (%)	Literature Data (%) *
12:0	6.2	1.3	3.7 ± 4.2
14:1n-5	6.5	<0.1	0.2 ± 0.3
18:3n-3,6,9	6.8	4.3	2.7 ± 0.3
16:1n-7	8.3	13.7	7.7 ± 1.9
16:1n-5	-	-	0.8 ± 0.4
14:0	8.4	1.7	5.4 ± 2.1
18:2n-6,9	8.5	0.7	0.8 ± 0.6
18:2n-3,6	8.5	0.3	-
17:1n-7	9.4	0.4	-
15:0	9.7	0.2	<0.1
18:1n-7	10.5	0.6	-
18:1n-8	10.6	<0.1	-
18:1n-9	10.6	0.3	-
18:1n-7	11.5	35.2	17.3 ± 3.6
18:1n-9	11.5	19.3	35.1 ± 3.2
16:0	12.0	5.5	18.5 ± 2.4
19:1n-9	13.5	2.6	-
19:1n-7	13.5	0.6	-
17:0	14.0	0.5	-
20:1n-7	16.9	0.8	<0.1
20:1n-11	16.9	0.6	<0.1
18:0	17.9	0.1	0.7 ± 0.3
19:0	21.4	<0.1	-
24:2n-7,17	24.7	1.4	-
22:1n-7	25.2	0.3	-
20:0	28.0	<0.1	0.2 ± 0.1
22:0	-	-	0.3 ± 0.1
25:2n-7,17	30.2	0.5	-
26:2n7,17	37.0	9.2	5.0 ± 2.4
23:0	38.5	<0.1	-
24:1n-15	39.3	<0.1	<0.1
24:0	-	-	<0.1
26:1n-17	-	-	0.2 ± 0.1

* Mean ± SD values of relative peak area values obtained by integrating GC/MS peaks; data for five bumblebee individuals. From ref. [41].

2.3.2. FAMEs from Pomegranate Seed Oil

Pomegranate (*Punicic granatum*) seed oil (PSO) is a rich source of FAs with conjugated double bonds. Cold-pressed PSO was transesterified, and the resulting mixture was analyzed by HPLC/MS. Many isomeric species with similar retention times tended to coelute. Still, the partial separation of the peaks allowed us to identify most of these lipids (Figure 7).

All the abundant peaks corresponded to CLnAs. The highest intensity exhibited an isomer with t_R 11.2 min, which was interpreted as FAME 18:3n-5,7,9. Its MS/MS spectrum (Figure 8A) showed abundant diagnostic peaks m/z 190.2 and m/z 290.1, corresponding to an MBR value of 133. The spectrum closely matched the punicic acid methyl ester shown in Figure 2. Interestingly, two less abundant isomers with the same diagnostic fragments were detected at t_R 12.0 min and t_R 14.8 min (Supplementary Materials Figure S2). These species were isomers with the same double bond positions but different double bond geometries. The geometrical isomers of punicic acid, namely FAME 18:3n-5t,7t,9c (β-eleostearic acid); FAME 18:3n-5c,7t,9t (α-eleostearic acid), and FAME 18:3n-5t,7t,9t (catalpic acid) were detected in pomegranate seed oil previously [51,86–89]. As the elution of the FAs in reversed-phase systems proceeds from *cis* to *trans* isomers [20,90], the later eluting isomers likely contained a higher number of *trans* double bonds. The MS/MS spectra of FAMEs with three conjugated double bonds in different positions are shown in Figure 8B–F. In all of them, the MBR value was 133, and the diagnostic fragments allowed us to interpret them as

FAME 18:3n-4,6,8 (Figure 8B), FAME 18:3n-3,5,7 (Figure 8C), FAME 18:3n-2,4,6 (Figure 8D), FAME 18:3n-8,10,12 (Figure 8E), and FAME 18:3n-9,11,13 (Figure 8F). The retention times of the latter two CLnAs were close to each other, which resulted in mixed spectra. Overall, ten CLnAs, one methylene-interrupted (18:2n-6,9), and two monounsaturated (18:1n-9 and 20:1n-9) FAMEs were identified in the PSO (Table 3). The results were in good agreement with previous analyses of PSO by silver-ion HPLC [51]. HPLC-based approaches to CLnAs analysis offer a higher number of isomers detected than GC [55,88,91,92]. We found four new CLnAs in the PSO, which, to the best of our knowledge, have not yet been described in the literature: two geometric isomers of 18:3n-2,4,6 (Figures 8D and S2), FAME 18:3n-8,10,12 (Figure 8E), and FAME 18:3n-9,11,13 (Figure 8F). They are characterized by the double bonds closer to the terminal carbon end (FAME 18:3n-2,4,6) or methyl ester group (FAME 18:3n-8,10,12 and FAME 18:3n-9,11,13).

Table 3. FAMEs identified in TG fraction of pomegranate seed oil.

FAME	t_R (min)	Rel. Peak Area (%)	References
18:3n-5,7,9	11.2	36.0	[51,86,87,89,93–95]
18:3n-5,7,9	12.0	24.2	[51,86,87,89]
18:3n-4,6,8	12.3	2.5	[51]
18:3n-3,5,7	12.9	1.7	[51]
18:2n-6,9	13.6	1.5	[86,87,89,93–95]
18:3n-2,4,6	13.7	1.0	-
18:3n-5,7,9	14.8	9.4	[51,86,87,89]
18:3n-4,6,8	15.4	10.1	[51]
18:3n-8,10,12	15.6	3.3	-
18:3n-9,11,13	15.8	2.7	-
18:3n-2,4,6	16.6	2.3	-
18:1n-9	19.5	4.5	[86,87,89,93–95]
20:1n-9	30.1	0.8	[86,89]
14:0	-	-	[87,93]
16:0	-	-	[86,87,89,93–95]
18:0	-	-	[86,87,89,93–95]
18:3n-3,6,9	-	-	[95]
20:0	-	-	[86,87,89,95]
22:0	-	-	[93]
24:0	-	-	[86,87]
24:1	-	-	[86]

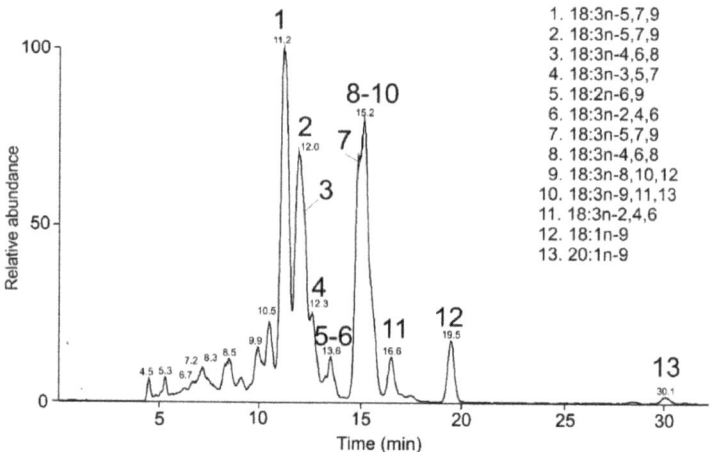

Figure 7. HPLC/APCI-MS base-peak chromatogram of FAMEs obtained from *Punicum granatum* seed oil and the list of identified species.

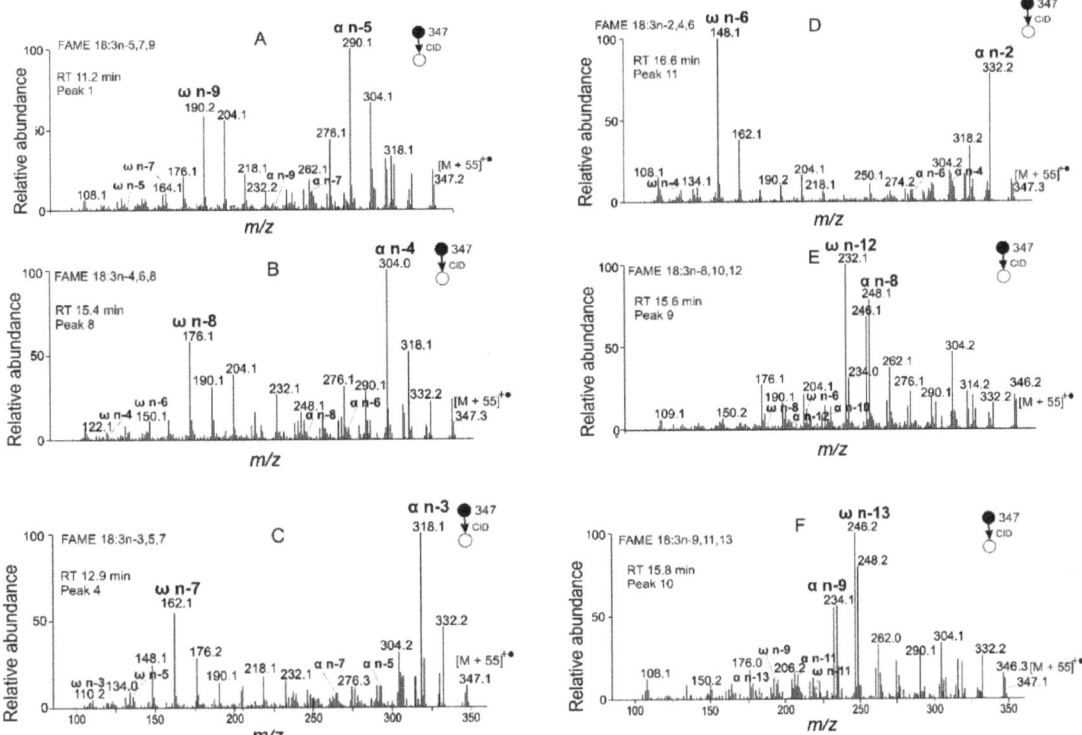

Figure 8. APCI MS/MS spectra of the [M + 55]$^{+\bullet}$ adducts of selected conjugated FAMEs from PSO interpreted as FAME 18:3n-5,7,9 (**A**), 18:3n-4,6,8 (**B**), 18:3n-3,5,7 (**C**), 18:3n-2,4,6 (**D**), 18:3n-8,10,12 (**E**), and 18:3n-9,11,13 (**F**).

2.3.3. FAMEs from *Marrubium vulgare* Seeds

White horehound (*Marrubium vulgare*) is a perennial, aromatic herb native to Europe, northern Africa, and southwestern and central Asia. Like other plants of the Lamiaceae family, it contains FAs with cumulated double bonds (allenic FAs). TGs from white horehound seeds were transesterified, and the resulting mixture of FAMEs analyzed by HPLC/MS (Figure 9). FAMEs with 18 to 21 carbons and up to three double bonds were detected.

The most abundant peak t_R 16.1 min corresponded to FAME 18:2 with the main fragments m/z 194.0 (α n-12) and m/z 248.1 (ω n-13), Figure 10A. The MBR value of 93 indicated two cumulated double bonds. It was interpreted as FAME 18:2n-12,13, most probably laballenic acid, highly abundant in *M. vulgare* seeds [61]. The fragmentation spectrum of FAME 18:2n-12,13 with the allenic system differed conspicuously from other arrangements of double bonds. The α fragment was accompanied by an α + 1 fragment with almost the same intensity, providing a double peak m/z 194/195 (Figure 10A). Analogous fragmentation behavior was also observed for other FAMEs with cumulated double bonds and helped us interpret allenic motifs in FAMEs. For instance, the compound eluting in 18.9 min was interpreted as FAME 19:2n-12,13. Its MS/MS spectrum provided m/z 208.1 (α n-12), m/z 209.0 (α n-12 + 1), and m/z 248.2 (ω n-13), corresponding to an MBR of 93 (Figure 10B). Analogously, peak t_R 22.6 min showing m/z 222.1 (α n-12), m/z 223.1 (α n-12 + 1), and m/z 248.2 (ω n-13) was consistent with 20:2n-12,13 (spectrum not shown).

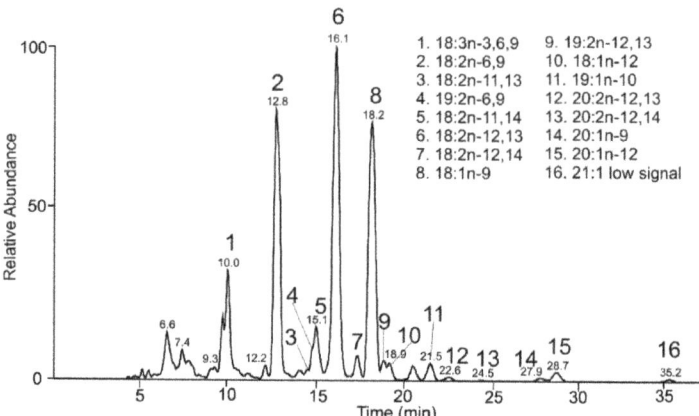

Figure 9. HPLC/APCI-MS base-peak chromatogram of FAMEs obtained from *Marrubium vulgare* seeds and the list of identified species.

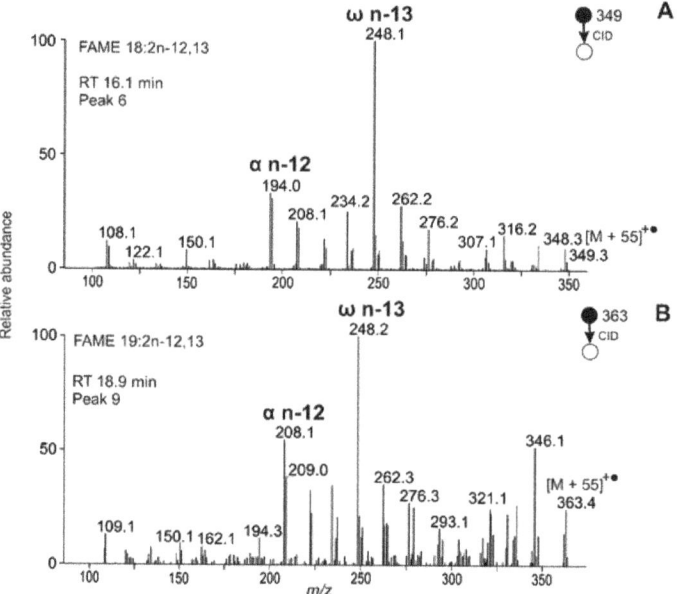

Figure 10. APCI MS/MS spectra of the [M + 55]$^{+\bullet}$ adducts of allenic FAMEs from *Marrubium vulgare* seeds interpreted as FAME 18:2n-12,13 (**A**) and 19:2n-12,13 (**B**).

In addition to allenic species, *M. vulgare* seeds contained FAMEs with conjugated double bonds. For example, the chromatographic peak t_R 14.6 min represented FAME 18:2n-11,13. Its structure was deduced using m/z 182.1 (α n-13), m/z 208.1 (α n-11), m/z 222.1 (ω n-11), and m/z 248.1 (ω n-13), an MBR value of 107 (Figure 11A). Similarly, peak t_R 17.4 min corresponded to FAME 18:2n-12,14 (Figure 11B). Overall, sixteen unsaturated FAMEs were detected in *M. vulgare* seeds, including monounsaturated, diunsaturated with allenic and conjugated double bonds, and triunsaturated species with methylene-interrupted double bonds (Table 4).

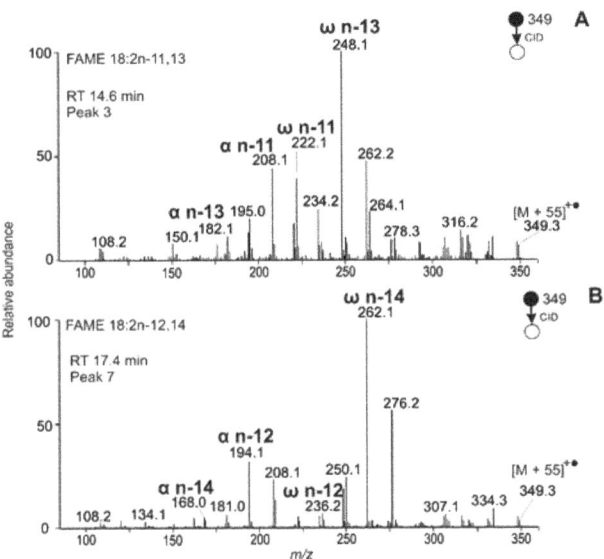

Figure 11. APCI MS/MS spectra of the [M + 55]$^{+\bullet}$ adducts of conjugated FAMEs from *Marrubium vulgare* seeds interpreted as FAME 18:2n-11,13 (**A**) and 18:2n-12,14 (**B**).

Table 4. FAMEs identified in TG fraction of *Marrubium vulgare* seed lipids.

FAME	t_R (min)	Rel. Peak Area (%)	References
18:3n-3,6,9	10.1	6.6	-
18:2n-6,9	12.8	25.4	-
18:2n-11,13	14.6	0.3	-
19:2n-6,9	14.9	1.2	-
18:2n-11,14	15.1	2.7	-
18:2n-12,13	16.1	31.4	[61]
18:2n-12,14	17.4	1.6	-
18:1n-9	18.2	26.5	-
19:2n-12,13	18.9	0.7	-
18:1n-12	19.2	0.6	-
19:1n-10	21.5	1.5	-
20:2n-12,13 *	22.6	0.4	[61]
20:2n-12,14 *	25.1	0.1	-
20:1n-9	27.9	0.2	-
20:1n-12	28.8	0.9	-
21:1 *	35.2	<0.1	-
20:1n-11	-	-	[61]

* Tentative identifications.

2.3.4. FAMEs from *Santalum album* Seeds

Indian sandalwood (*Santalum album*) is a tropical tree native to southern India and Southeast Asia. The oil from its seeds and seeds of other Santalaceae species is a rich source of acetylenic FAs. [96]. FAMEs obtained by the transesterification of the TGs from *Santalum album* seeds provided chromatogram shown in Figure 12.

The most abundant peak t_R 10.3 min corresponded to FAME with 18 carbons and either three double bonds or a double and a triple bond. The MS/MS spectrum (Figure 13A) revealed the latter possibility, i.e., an acetylenic acid methyl ester. Diagnostic fragment m/z 190.1 and its satellite ion m/z 205.1 indicated a triple bond in the n-9 position (ω n-9^{TB}). The corresponding α fragment (α n-9^{TB}) at m/z 236.1 was not accompanied by a significant satellite ion at m/z 251.1, likely because of the triple bond conjugation with the n-7 double bond. The α fragment *m/z* 262.0 and its satellite *m/z* 276.1 indicated a double bond in the position n-7. Low-intensity fragment ω n-7 was detected at *m/z* 166.1. The MBR value of 105 corresponds to a conjugated system of one double and one triple bond. The compound was identified as FAME 18:2n-7,9^{TB}, most probably santalbic acid methyl ester.

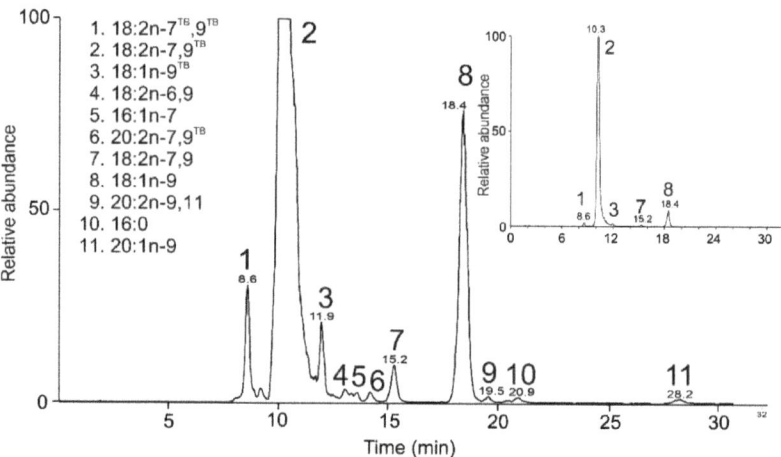

Figure 12. HPLC/APCI-MS base-peak chromatogram of FAMEs obtained from *Santalum album* seeds and the list of identified species. The inset shows a chromatogram of the same sample injected in 10× less amount.

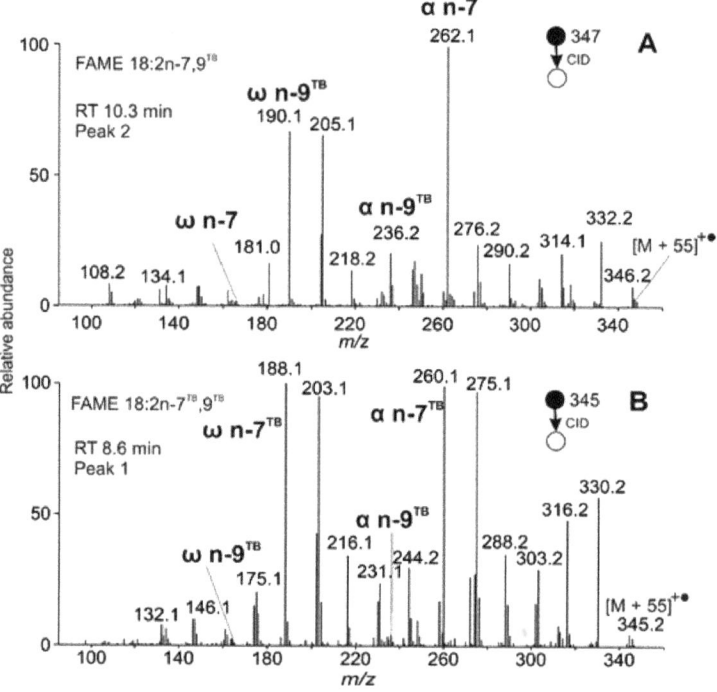

Figure 13. APCI MS/MS spectra of the [M + 55]$^{+\bullet}$ adducts of acetylenic FAME from *Santalum album* seeds interpreted as FAME 18:2n-7,9^{TB} (**A**) and 18:2n-7 ^{TB},9^{TB} (**B**).

The MS/MS spectrum of a peak in 8.6 min revealed another acetylenic FAME with two triple bonds (Figure 13B). The ω fragment m/z 188.1 and its satellite peak m/z 203.1 indicated the triple bond at the position n-9^{TB}, and the α fragment m/z 260.1 and its satellite m/z 275.0 the triple bond in n-7^{TB}. The complementary α (n-9^{TB}) and ω (n-7^{TB}) fragments m/z 236.1 and m/z 164.1, respectively, were of low abundance. The MBR value calculated from the most abundant fragments (m/z 188.1 and m/z 260.1) equaled 103 and was consistent with two conjugated triple bonds. The compound was identified as FAME 18:2n-7^{TB},9^{TB}.

The seeds oil was also found to contain acetylenic acids (FAMEs 18:3n-9^{TB}, 20:2n-7,9^{TB}) and conjugated acids (FAMEs 18:2n-7,9, 20:2n-9,11) not reported previously for *S. album*. In summary, FAMEs with triple bonds conjugated with either double or triple bond were found, together with saturated, monounsaturated, diunsaturated conjugated and methylene interrupted, triunsaturated, and tetraunsaturated species (Table 5).

Table 5. FAMEs identified in TG fraction of *S. album* seed lipids.

FAME	t_R (min)	Rel. Peak Area (%)	Literature Data (%) *
18:2 n-7^{TB},9^{TB}	8.6	1.6	-
18:2n-7,9^{TB}	10.3	89.0	33.5
18:1n-9^{TB}	11.9	0.7	-
18:2n-6,9	13.0	0.2	1.5
16:1n-7	13.6	0.2	0.8
20:2n-7,9^{TB}	14.2	0.2	-
18:2n-7,9	15.2	0.7	-
18:1n-9	18.4	7.0	52.1
20:2n-9,11	19.5	0.1	-
16:0	20.9	0.1	3.7
20:1n-9	28.2	0.2	-
16:1n-9	-	-	0.1
17:1	-	-	0.3
17:2	-	-	0.5
18:0	-	-	1.7
18:1n-7	-	-	1.4
18:3n-3,6,9	-	-	3.1
18:4n-3,6,9,12	-	-	1.3

* Composition of sandalwood oil ethyl esters reported in ref. [97].

3. Experimental

3.1. Chemicals and Materials

MS-grade acetonitrile and methanol (Sigma-Aldrich, St. Louis, MO, USA) were used as received. Chloroform, hexane, and diethyl ether were distilled from analytical-grade solvents (Penta, Czech Republic). Other chemicals, NaCl ($\geq$99%, Sigma-Aldrich, St. Louis, MO, USA), di-tert-butyl-4-methylphenol (Fluka, Buchs, Switzerland), Rhodamine 6G (Sigma-Aldrich, St. Louis, MO, USA), and Diazald (99%, Sigma-Aldrich, St. Louis, MO, USA) were used. The standards of crepenynic acid (99%) and punicic acid methyl ester (purity 98%) were from Larodan (Malmö, Sweden), and 9-octadecynoic acid methyl ester, 9(E),11(E)-octadecadienoic acid methyl ester, and 9(Z),11(Z)-octadecadienoic acid methyl ester (all 98%) were purchased from Cayman Europe (Tallinn, Estonia). The standards were dissolved in chloroform at 1 mg/mL concentrations and stored at −25 °C. *Bombus pratorum* males were collected in the Czech Republic during the spring season and immobilization at −18 °C. Cold-pressed pomegranate seed oil (organic, unrefined) was from Biopurus Ltd. (Ashford, England). Seeds of *Marrubium vulgare* and *Santalum album* were purchased from a local garden center.

3.2. Extraction and Transesterification of Lipids

The samples were treated with organic solvents to obtain total lipid extracts. Briefly, peripheral fat bodies of three *B. pratorum* males were dissected and extracted with CHCl$_3$/CH$_3$OH (1:1, v/v) containing di-tert-butyl-4-methylphenol at a concentration of 25 mg/mL (500 μL each) and sonicated for 15 min. The extract was collected using a Pasteur pipette. *M. vulgare* seeds (approx. 240 pieces; 0.25 g) or *S. album* seeds (5 pieces; 0.94 g) were crushed and extracted in methanol/chloroform (2:1 v/v, 10 mL) for 30 min. After filtration, 5 mL of 0.9% NaCl was added, shaken for few seconds, and the aqueous (upper) phase was removed. The cleaning step was repeated three more times with 2 mL of 0.9% NaCl solution.

Total lipid extracts or seed oil were separated by semipreparative TLC to isolate TGs.

Pre-cleaned, in-house made silica-gel glass TLC plates (60 mm × 76 mm) and hexane/diethyl ether (80:20, by vol.) mobile phase were used. TLC zones were made visible by spraying Rhodamine 6G solution (0.05% in ethanol). A zone corresponding to TGs (*B. pratorum* R_f = 0.36–0.55, pomegranate R_f = 0.20–0.55, *M. vulgare* R_f = 0.33–0.55, *S. album* R_f = 0.30–0.55) was scraped off the plate and extracted with 10 mL freshly distilled diethyl ether. The solvent was evaporated to dryness under a nitrogen stream.

While TGs from *B. pratorum*, pomegranate seed oil, and *M. vulgare* seeds were transesterified in acidic conditions [98], base-catalyzed transesterification [99] was required for *S. album* lipids containing triple bonds. FA standards were methylated by diazomethane (synthesized in-house from Diazald). Diazomethane in diethyl ether was added dropwise to the FA solution in chloroform (10 mg/mL) until the color of the reaction mixture turned light-yellow. Unreacted diazomethane was deactivated by formic acid.

3.3. RP-HPLC/APCI-MS and APCI-MS

The liquid chromatograph consisted of a Rheos Allegro UHPLC pump, Accela autosampler with an integrated column oven, and an LCQ Fleet ion-trap mass spectrometer; the system was controlled by Xcalibur software (all Thermo Fisher Scientific, San Jose, CA, USA). Develosil RP-Aqueous C30 (250 × 4.6 mm, particle size: 5 µm; Nomura Chemical, Seto, Japan) stainless-steel column and isocratic elution with acetonitrile at 0.7 mL/min flow rate [20] were used. The chromatography proceeded at laboratory temperature except for *B. pratorum* sample separated at 40 °C. The injected volume of samples (standards and biological samples, 1 mg/mL and 10–20 mg/mL, respectively) was 10–20 µL. The APCI vaporizer and heated capillary temperatures were set to 380 °C and 180 °C, respectively; the corona discharge current was 2 µA. Nitrogen served both as the sheath and auxiliary gas at a flow rate of 50 and 20 arbitrary units, respectively. The MS spectra of positively charged ions were recorded in the m/z 180–470 range. The CID MS^2 spectra of [M + 55]$^{+\bullet}$ were collected using a data-dependent analysis with an isolation width of 1.7 Da and normalized collision energy of 28%. The m/z range of MS^2 spectra was set automatically, depending on the precursor ion mass. The masses of the acetonitrile adducts for fragmentation were calculated as higher partners of the base peaks (m/z [M + H]$^+$ + 54 Da). The retention times and relative peak areas were obtained from ion chromatograms extracted for [M + H]$^+$. The high-resolution MS data were recorded using an LTQ Orbitrap XL hybrid mass spectrometer (Thermo Fisher Scientific, San Jose, CA, USA) equipped with an APCI ion source operated at the same conditions as for low resolution. The Orbitrap spectra were acquired at a resolution of 100,000 FWHM.

The standard (1 mg/mL) solutions were also analyzed by direct infusion to the mobile phase flow using the same APCI-MS conditions, as described above.

3.4. Fragment Ion Abbreviations and Nomenclature

The diagnostic ions in the MS/MS spectra of [M + 55]$^{+\bullet}$ were denoted "α" if they carried the ester moiety or "ω" if they contained the terminal-carbon end without the ester group. The double bond position was indicated as α n-x and ω n-x, where x is the distance from the terminal end of the hydrocarbon chain. A triple bond was marked by "TB" in superscript.

4. Conclusions

This work demonstrates the applicability of acetonitrile gas-phase chemistry in APCI for characterizing the structure of polyunsaturated FAMEs. The reaction of $C_3H_5N^{+\bullet}$ with double and triple bonds occurs in the ion source, and the reaction products are fragmented to generate diagnostic ions. The method is highly versatile and suitable to many (if not all) arrangements of double and triple bonds in mono- and polyunsaturated chains. It was successfully applied to FAMEs with isolated, cumulated, and conjugated double bonds, triple bonds, and their combinations. The localization of the isolated double and triple bond positions is straightforward because of intense α and ω fragments. Distinguishing a double bond from a triple bond is easy based on the satellite fragments. While the satellite ions appear at +14 Da in the lipids with a double bond, they are found as intense +15 Da fragments in the case of a triple bond. When two or more unsaturated bonds exist in a chain, the spectra predominantly show α and ω fragments related to cleavages of C–C bonds before and after the unsaturated region. This can be utilized for deducing a possible arrangement of unsaturated bonds. A parameter named multiple bond region (MBR) can be calculated using the most abundant fragments and compared to tabulated theoretical values. The type and position of the unsaturated bonds within the unsaturated region can then be inspected in detail after focusing on less intense diagnostic fragments and their satellites. In the case of allenic FAMEs, the α fragment was accompanied by an intense α + 1 fragment, which gave a hint for the cumulated double bonds. When a triple bond was present in a polyunsaturated chain, it manifested itself by the +15 Da satellite peak accompanying the corresponding diagnostic fragment.

The localization of unsaturated bonds by HPLC/APCI-MS/MS with an acetonitrile mobile phase is a simple and convenient method. Since the derivatization occurs in the ion source during ionization, there is no need to perform the chemical modification of the analytes as a separate step before the analysis. Nominal mass resolution spectra were successfully used for the structure elucidation. However, high-resolution MS/MS data could help distinguish α and ω fragments, thus making the interpretation even easier. In this work, unsaturated FAMEs were characterized in *Bombus pratorum*, *Punicum granatum*, *Marrubium vulgare*, and *Santalum album*. The method's power is illustrated by the fact that, in addition to the known lipids, several new FAMEs were discovered. Although the method can also be applied to complex lipids [19,21,22], spectra interpretation is easier for lipids having only one fatty acyl chain.

Supplementary Materials: The following are available online, Figure S1: APCI MS/MS CID spectrum of the $[M + 55]^{+\bullet}$ adduct of ricinenic acid methyl ester (FAME 18:2n-7c,9c); MBR = 107; Table S1: High-resolution data for fragments from APCI MS/MS spectra of FAME standards. Figure S2: APCI MS/MS spectra of the $[M + 55]^{+\bullet}$ adducts of selected conjugated FAMEs from PSO interpreted as FAME 18:3n-5,7,9 (A), 18:3n-4,6,8 (B), 18:3n-2,4,6 (C), 18:3n-5,7,9 (D).

Author Contributions: Conceptualization, J.C.; methodology, V.V. and J.C.; investigation, P.H. and V.V.; resources, J.K.; writing—original draft preparation, P.H.; writing—review and editing, K.S.-P., P.H. and J.C.; supervision, J.C.; funding acquisition, P.H. and J.C. All authors have read and agreed to the published version of the manuscript.

Funding: This research was funded by the Charles University Grant Agency, project number. 10119, the Charles University in Prague, project SVV 260560, and the European Regional Development Fund, OP RDE, No. CZ.02.1.01/0.0/0.0/16_019/0000729.

Institutional Review Board Statement: Not applicable.

Data Availability Statement: The data presented in this study are available on request from the corresponding author.

Acknowledgments: The authors wish to thank Stancho Stanchev for the preparation of diazomethane.

Conflicts of Interest: The authors declare no conflict of interest.

Sample Availability: Samples of the compounds are not available from the authors.

References

1. Dembitsky, V.M. Anticancer activity of natural and synthetic acetylenic lipids. *Lipids* **2006**, *41*, 883–924. [CrossRef] [PubMed]
2. Li, X.-C.; Jacob, M.R.; Khan, S.I.; Ashfaq, M.K.; Babu, K.S.; Agarwal, A.K.; El Sohly, H.N.; Manly, S.P.; Clark, A.M. Potent In Vitro Antifungal Activities of Naturally Occurring Acetylenic Acids. *Antimicrob. Agents Chemother.* **2008**, *52*, 2442–2448. [CrossRef] [PubMed]
3. Vetter, W.; Walther, W.; Vecchi, M. Pyrrolidide als Derivate für die Strukturaufklärung aliphatischer und alicyclischer Carbonsäuren mittels Massenspektrometrie. *Helv. Chim. Acta* **1971**, *54*, 1599–1605. [CrossRef]
4. Yu, Q.T.; Liu, B.N.; Zhang, J.Y.; Huang, Z.H. Location of methyl branchings in fatty acids: Fatty acids in uropygial secretion of shanghai duck by GC-MS of 4,4-dimethyloxazoline derivatives. *Lipids* **1989**, *24*, 160. [CrossRef]
5. Francis, G.W.; Veland, K. Alkylthiolation for the determination of double-bond positions in linear alkenes. *J. Chromatogr. A* **1987**, *219*, 379–384. [CrossRef]
6. Ma, X.; Chong, L.; Tian, R.; Shi, R.; Hu, T.; Ouyang, Z.; Xia, Y. Identification and quantitation of lipid C=C location isomers: A shotgun lipidomics approach enabled by photochemical reaction. *Proc. Natl. Acad. Sci. USA* **2016**, *113*, 2573–2578. [CrossRef]
7. Murphy, R.C.; Okuno, T.; Johnson, C.A.; Barkley, R.M. Determination of Double Bond Positions in Polyunsaturated Fatty Acids Using the Photochemical Paternò-Büchi Reaction with Acetone and Tandem Mass Spectrometry. *Anal. Chem.* **2017**, *89*, 8545–8553. [CrossRef] [PubMed]
8. Xie, X.; Xia, Y. Analysis of Conjugated Fatty Acid Isomers by the Paternò-Büchi Reaction and Trapped Ion Mobility Mass Spectrometry. *Anal. Chem.* **2019**, *91*, 7173–7180. [CrossRef] [PubMed]
9. Zhao, Y.; Zhao, H.; Zhao, X.; Jia, J.; Ma, Q.; Zhang, S.; Zhang, X.; Chiba, H.; Hui, S.-P.; Ma, X. Identification and Quantitation of C=C Location Isomers of Unsaturated Fatty Acids by Epoxidation Reaction and Tandem Mass Spectrometry. *Anal. Chem.* **2017**, *89*, 10270–10278. [CrossRef] [PubMed]
10. Song, C.; Gao, D.; Li, S.; Liu, L.; Chen, X.; Jiang, Y. Determination and quantification of fatty acid C=C isomers by epoxidation reaction and liquid chromatography-mass spectrometry. *Anal. Chim. Acta* **2019**, *1086*, 82–89. [CrossRef] [PubMed]
11. Wan, L.; Gong, G.; Liang, H.; Huang, G. In situ analysis of unsaturated fatty acids in human serum by negative-ion paper spray mass spectrometry. *Anal. Chim. Acta* **2019**, *1075*, 120–127. [CrossRef] [PubMed]
12. Takashima, S.; Toyoshi, K.; Yamamoto, T.; Shimozawa, N. Positional determination of the carbon–carbon double bonds in unsaturated fatty acids mediated by solvent plasmatization using LC–MS. *Sci. Rep.* **2020**, *10*, 12988. [CrossRef] [PubMed]
13. Yang, W.-C.; Adamec, A.J.; Regnier, F.E. Enhancement of the LC/MS Analysis of Fatty Acids through Derivatization and Stable Isotope Coding. *Anal. Chem.* **2007**, *79*, 5150–5157. [CrossRef] [PubMed]
14. Yang, K.; Dilthey, B.G.; Gross, R.W. Identification and Quantitation of Fatty Acid Double Bond Positional Isomers: A Shotgun Lipidomics Approach Using Charge-Switch Derivatization. *Anal. Chem.* **2013**, *85*, 9742–9750. [CrossRef]
15. Thomas, M.C.; Mitchell, T.W.; Harman, D.G.; Deeley, J.M.; Murphy, R.C.; Blanksby, S.J. Elucidation of Double Bond Position in Unsaturated Lipids by Ozone Electrospray Ionization Mass Spectrometry. *Anal. Chem.* **2007**, *79*, 5013–5022. [CrossRef] [PubMed]
16. Mitchell, T.W.; Pham, H.; Thomas, M.C.; Blanksby, S.J. Identification of double bond position in lipids: From GC to OzID. *J. Chromatogr. B* **2009**, *877*, 2722–2735. [CrossRef]
17. Poad, B.L.J.; Marshall, D.L.; Harazim, E.; Gupta, R.; Narreddula, V.R.; Young, R.S.E.; Duchoslav, E.; Campbell, J.L.; Broadbent, J.A.; Cvačka, J.; et al. Combining Charge-Switch Derivatization with Ozone-Induced Dissociation for Fatty Acid Analysis. *J. Am. Soc. Mass Spectrom.* **2019**, *30*, 2135–2143. [CrossRef]
18. Xu, Y.; Brenna, J.T. Atmospheric Pressure Covalent Adduct Chemical Ionization Tandem Mass Spectrometry for Double Bond Localization in Monoene- and Diene-Containing Triacylglycerols. *Anal. Chem.* **2007**, *79*, 2525–2536. [CrossRef]
19. Vrkoslav, V.; Háková, M.; Pecková, K.; Urbanová, K.; Cvačka, J. Localization of Double Bonds in Wax Esters by High-Performance Liquid Chromatography/Atmospheric Pressure Chemical Ionization Mass Spectrometry Utilizing the Fragmentation of Acetonitrile-Related Adducts. *Anal. Chem.* **2011**, *83*, 2978–2986. [CrossRef]
20. Vrkoslav, V.; Cvačka, J. Identification of the double-bond position in fatty acid methyl esters by liquid chromatography/atmospheric pressure chemical ionisation mass spectrometry. *J. Chromatogr. A* **2012**, *1259*, 244–250. [CrossRef]
21. Háková, E.; Vrkoslav, V.; Mikova, R.; Schwarzová, K.; Bosakova, Z.; Cvačka, J. Localization of double bonds in triacylglycerols using high-performance liquid chromatography/atmospheric pressure chemical ionization ion-trap mass spectrometry. *Anal. Bioanal. Chem.* **2015**, *407*, 5175–5188. [CrossRef] [PubMed]
22. Šubčíková, L.; Hoskovec, M.; Vrkoslav, V.; Čmelíková, T.; Háková, E.; Míková, R.; Coufal, P.; Doležal, A.; Plavka, R.; Cvačka, J. Analysis of 1,2-diol diesters in vernix caseosa by high-performance liquid chromatography—Atmospheric pressure chemical ionization mass spectrometry. *J. Chromatogr. A* **2015**, *1378*, 8–18. [CrossRef] [PubMed]
23. Kalužíková, A.; Vrkoslav, V.; Harazim, E.; Hoskovec, M.; Plavka, R.; Buděšínský, M.; Bosáková, Z.; Cvačka, J. Cholesteryl esters of ω-(O-acyl)-hydroxy fatty acids in vernix caseosa. *J. Lipid Res.* **2017**, *58*, 1579–1590. [CrossRef] [PubMed]
24. Van Pelt, C.K.; Carpenter, B.K.; Brenna, J.T. Studies of structure and mechanism in acetonitrile chemical ionization tandem mass spectrometry of polyunsaturated fatty acid methyl esters. *J. Am. Soc. Mass Spectrom.* **1999**, *10*, 1253–1262. [CrossRef]
25. Michaud, A.L.; Diau, G.-Y.; Abril, R.; Brenna, J. Double bond localization in minor homoallylic fatty acid methyl esters using acetonitrile chemical ionization tandem mass spectrometry. *Anal. Biochem.* **2002**, *307*, 348–360. [CrossRef]

26. Michaud, A.L.; Yurawecz, M.P.; Delmonte, P.; Corl, B.A.; Bauman, D.E.; Brenna, J.T. Identification and Characterization of Conjugated Fatty Acid Methyl Esters of Mixed Double Bond Geometry by Acetonitrile Chemical Ionization Tandem Mass Spectrometry. *Anal. Chem.* **2003**, *75*, 4925–4930. [CrossRef] [PubMed]
27. Lawrence, P.; Brenna, J.T. Acetonitrile Covalent Adduct Chemical Ionization Mass Spectrometry for Double Bond Localization in Non-Methylene-Interrupted Polyene Fatty Acid Methyl Esters. *Anal. Chem.* **2006**, *78*, 1312–1317. [CrossRef] [PubMed]
28. Barthélemy, M.; Elie, N.; Pellissier, L.; Wolfender, J.-L.; Stien, D.; Touboul, D.; Eparvier, V. Structural Identification of Antibacterial Lipids from Amazonian Palm Tree Endophytes through the Molecular Network Approach. *Int. J. Mol. Sci.* **2019**, *20*, 2006. [CrossRef]
29. Spitzer, V.; Marx, F.; Maia, J.G.; Pfeilsticker, K. *Curupira tefeensis* II: Occurrence of Acetylenic Fatty Acids. *Fette Seifen Anstrichm.* **1991**, *93*, 169–174. [CrossRef]
30. Spitzer, V.; Bordignon, S.A.D.L.; Schenkel, E.P.; Marx, F. Identification of nine acetylenic fatty acids, 9-hydroxystearic acid and 9,10-epoxystearic acid in the seed oil of *Jodina rhombifolia* Hook et Arn. (Santalaceae). *J. Am. Oil Chem. Soc.* **1994**, *71*, 1343–1348. [CrossRef]
31. Spitzer, V. The mass spectra of the 4,4-dimethyloxazoline derivatives of some conjugated hydroxy ene-yne C17 and C18 fatty acids. *J. Am. Oil Chem. Soc.* **1996**, *73*, 489–492. [CrossRef]
32. Spitzer, V.; Tomberg, W.; Hartmann, R.; Aichholz, R. Analysis of the seed oil of *Heisteria silvanii* (Olacaceae)—A rich source of a novel C18 acetylenic fatty acid. *Lipids* **1997**, *32*, 1189–1200. [CrossRef] [PubMed]
33. Gurr, M.I.; Harwood, J.L.; Frayn, K.N.; Murphy, D.J.; Michell, R.H. *Lipids: Biochemistry, Biotechnology and Health*, 6th ed.; Wiley-Blackwell: Hoboken, NJ, USA, 2016.
34. Litchfield, C.; Greenberg, A.J.; Noto, G.; Morales, R.W. Unusually high levels of C24–C30 fatty acids in sponges of the class Demospongiae. *Lipids* **1976**, *11*, 567–570. [CrossRef] [PubMed]
35. Morales, R.W.; Litchfield, C. Incorporation of 1-14C-Acetate into C26 fatty acids of the marine sponge *Microciona prolifera*. *Lipids* **1977**, *12*, 570–576. [CrossRef] [PubMed]
36. Christie, W.W.; Brechany, E.Y.; Stefanov, K.; Popov, S. The fatty acids of the sponge *Dysidea fragilis* from the black sea. *Lipids* **1992**, *27*, 640–644. [CrossRef]
37. Nechev, J.; Christie, W.W.; Robaina, R.; De Diego, F.; Popov, S.; Stefanov, K. Chemical composition of the sponge *Hymeniacidon sanguinea* from the Canary Islands. *Comp. Biochem. Physiol. Part A Mol. Integr. Physiol.* **2004**, *137*, 365–374. [CrossRef] [PubMed]
38. Kawashima, H. Unusual minor nonmethylene-interrupted di-, tri-, and tetraenoic fatty acids in limpet gonads. *Lipids* **2005**, *40*, 627–630. [CrossRef]
39. Zhukova, N.V. Lipid Classes and Fatty Acid Composition of the Tropical Nudibranch Mollusks *Chromodoris* sp. and *Phyllidia coelestis*. *Lipids* **2007**, *42*, 1169–1175. [CrossRef]
40. Carballeira, N. New advances in fatty acids as antimalarial, antimycobacterial and antifungal agents. *Prog. Lipid Res.* **2008**, *47*, 50–61. [CrossRef] [PubMed]
41. Cvačka, J.; Kofroňová, E.; Vašíčková, S.; Stránský, K.; Jiroš, P.; Hovorka, O.; Kindl, J.; Valterová, I. Unusual Fatty Acids in the Fat Body of the Early Nesting Bumblebee, *Bombus pratorum*. *Lipids* **2008**, *43*, 441–450. [CrossRef] [PubMed]
42. Sehat, N.; Kramer, J.K.G.; Mossoba, M.M.; Yurawecz, M.P.; Roach, J.A.G.; Eulitz, K.; Morehouse, K.M.; Ku, Y. Identification of conjugated linoleic acid isomers in cheese by gas chromatography, silver ion high performance liquid chromatography and mass spectral reconstructed ion profiles. Comparison of chromatographic elution sequences. *Lipids* **1998**, *33*, 963–971. [CrossRef] [PubMed]
43. Yurawecz, M.P.; Roach, J.A.G.; Sehat, N.; Mossoba, M.M.; Kramer, J.K.G.; Fritsche, J.; Steinhart, H.; Ku, Y. A new conjugated linoleic acid isomer, 7 trans, 9 cis-octadecadienoic acid, in cow milk, cheese, beef and human milk and adipose tissue. *Lipids* **1998**, *33*, 803–809. [CrossRef] [PubMed]
44. Ip, C.; Chin, S.F.; Scimeca, J.A.; Pariza, M.W. Mammary cancer prevention by conjugated dienoic derivative of linoleic acid. *Cancer Res.* **1991**, *51*, 6118–6124.
45. Pariza, M.W.; Park, Y.; Cook, M.E. The biologically active isomers of conjugated linoleic acid. *Prog. Lipid Res.* **2001**, *40*, 283–298. [CrossRef]
46. O'Connor, R.; Heinzelman, D.; Freeman, A.; Pack, F. Spectrophotometric Determination of Alpha-Eleostearic Acid in Freshly Extracted Tung Oil Determination of Extinction Coefficients in Oil Solvents. *Ind. Eng. Chem. Anal. Ed.* **1945**, *17*, 467–470. [CrossRef]
47. Özgül-Yücel, S. Determination of conjugated linolenic acid content of selected oil seeds grown in Turkey. *J. Am. Oil Chem. Soc.* **2005**, *82*, 893–897. [CrossRef]
48. Hopkins, C.Y.; Chisholm, M.J. A survey of the conjugated fatty acids of seed oils. *J. Am. Oil Chem. Soc.* **1968**, *45*, 176–182. [CrossRef]
49. Chisholm, M.J.; Hopkins, C.Y. Conjugated fatty acids of tragopogon and calendula seed oils. *Can. J. Chem.* **1960**, *38*, 2500–2507. [CrossRef]
50. Toyama, Y.; Tsuchiya, T. A new stereoisomer of eleostearic acid in pomegranate seed oil. *J. Soc. Chem. Ind. Jpn. B* **1935**, *38*, 182–185.
51. Cao, Y.; Gao, H.-L.; Chen, J.-N.; Chen, Z.-Y.; Yang, L. Identification and Characterization of Conjugated Linolenic Acid Isomers by Ag^+-HPLC and NMR. *J. Agric. Food Chem.* **2006**, *54*, 9004–9009. [CrossRef]
52. Saha, S.S.; Patra, M.; Ghosh, M. In vitro antioxidant study of vegetable oils containing conjugated linolenic acid isomers. *LWT* **2012**, *46*, 10–15. [CrossRef]
53. Aruna, P.; Venkataramanamma, D.; Singh, A.K.; Singh, R. Health Benefits of Punicic Acid: A Review. *Compr. Rev. Food Sci. Food Saf.* **2015**, *15*, 16–27. [CrossRef] [PubMed]

54. De Melo, I.L.P.; de Carvalho, E.B.T.; de Oliveira e Silva, A.M.; Yoshime, L.T.; Sattler, J.A.G.; Pavan, R.T.; Mancini-Filho, J. Characterization of constituents, quality and stability of pomegranate seed oil (*Punica granatum* L.). *Food Sci. Technol.* **2016**, *36*, 132–139. [CrossRef]
55. Costa, A.; Silva, L.; Torres, A. Chemical composition of commercial cold-pressed pomegranate (*Punica granatum*) seed oil from Turkey and Israel, and the use of bioactive compounds for samples' origin preliminary discrimination. *J. Food Compos. Anal.* **2019**, *75*, 8–16. [CrossRef]
56. Benjamin, S.; Spener, F. Conjugated linoleic acids as functional food: An insight into their health benefits. *Nutr. Metab.* **2009**, *6*, 36. [CrossRef]
57. Dubey, K.K.D.; Sharma, G.; Kumar, A. Conjugated Linolenic Acids: Implication in Cancer. *J. Agric. Food Chem.* **2019**, *67*, 6091–6101. [CrossRef] [PubMed]
58. Badami, R.; Patil, K. Structure and occurrence of unusual fatty acids in minor seed oils. *Prog. Lipid Res.* **1980**, *19*, 119–153. [CrossRef]
59. Dembitsky, V.M.; Maoka, T. Allenic and cumulenic lipids. *Prog. Lipid Res.* **2007**, *46*, 328–375. [CrossRef]
60. Bagby, M.O.; Smith, C.R.; Wolff, I.A. Laballenic Acid. A New Allenic Acid from *Leonotis nepetaefolia* Seed Oil1. *J. Org. Chem.* **1965**, *30*, 4227–4229. [CrossRef]
61. Aitzetmüller, D.U.P.D.K.; Tsevegsüren, N.; Vosmann, K. A New Allenic Fatty Acid in *Phlomis* (Lamiaceae) Seed Oil. *Fette Seifen Anstrichm.* **1997**, *99*, 74–78. [CrossRef]
62. Mikolajczak, K.L.; Rogers, M.F.; Smith, C.R.; Wolff, I.A. An octadecatrienoic acid from *Lamium purpureum* L. seed oil containing 5,6-allenic and trans-16-olefinic unsaturation. *Biochem. J.* **1967**, *105*, 1245–1249. [CrossRef]
63. Smith, C. Occurrence of unusual fatty acids in plants. *Prog. Chem. Fats Other Lipids* **1971**, *11*, 137–177. [CrossRef]
64. Bohlmann, F.; Burkhardt, T.; Zdero, C. *Naturally Occurring Acetylenes*; Academic Press: London, UK, 1973; pp. 1–222.
65. Huang, Y.; Zhang, S.-B.; Chen, H.-P.; Zhao, Z.-Z.; Li, Z.-H.; Feng, T.; Liu, J.-K. New acetylenic acids and derivatives from the Basidiomycete *Craterellus lutescens* (Cantharellaceae). *Fitoterapia* **2016**, *115*, 177–181. [CrossRef] [PubMed]
66. Fatope, M.O.; Adoum, O.A.; Takeda, Y. C18 Acetylenic Fatty Acids of *Ximenia americana* with Potential Pesticidal Activity. *J. Agric. Food Chem.* **2000**, *48*, 1872–1874. [CrossRef] [PubMed]
67. Li, X.-C.; Jacob, M.R.; ElSohly, H.N.; Nagle, D.G.; Smillie, T.J.; Walker, L.A.; Clark, A.M. Acetylenic Acids Inhibiting Azole-Resistant *Candida albicans* from *Pentagonia gigantifolia*. *J. Nat. Prod.* **2003**, *66*, 1132–1135. [CrossRef]
68. Carballeira, N.M.; Sanabria, D.; Cruz, C.; Parang, K.; Wan, B.; Franzblau, S. 2,6-hexadecadiynoic acid and 2,6-nonadecadiynoic acid: Novel synthesized acetylenic fatty acids as potent antifungal agents. *Lipids* **2006**, *41*, 507–511. [CrossRef]
69. Xu, T.; Tripathi, S.K.; Feng, Q.; Lorenz, M.; Wright, M.A.; Jacob, M.R.; Mask, M.M.; Baerson, S.R.; Li, X.-C.; Clark, A.M.; et al. A Potent Plant-Derived Antifungal Acetylenic Acid Mediates Its Activity by Interfering with Fatty Acid Homeostasis. *Antimicrob. Agents Chemother.* **2012**, *56*, 2894–2907. [CrossRef] [PubMed]
70. Kilimnik, A.; Kuklev, D.V.; Dembitsky, V.M. Antitumor Acetylenic Lipids. *Mathews J. Pharm. Sci.* **2016**, *1*, 5.
71. Aitzetmüller, K.; Matthäus, B.; Friedrich, H. A new database for seed oil fatty acids—The database SOFA. *Eur. J. Lipid Sci. Technol.* **2003**, *105*, 92–103. [CrossRef]
72. Aitzetmüller, K. Santalbic acid in the plant kingdom. *Plant Syst. Evol.* **2012**, *298*, 1609–1617. [CrossRef]
73. Neff, W.E.; Adlof, R.O.; Konishi, H.; Weisleder, D. High-performance liquid chromatography of the triacylglycerols of *Vernonia galamensis* and *Crepis alpina* seed oils. *J. Am. Oil Chem. Soc.* **1993**, *70*, 449–455. [CrossRef]
74. Neff, W.E.; Adlof, R.O.; El-Agaimy, M. Silver ion high-performance liquid chromatography of the triacylglycerols of *Crepis alpina* seed oil. *J. Am. Oil Chem. Soc.* **1994**, *71*, 853–855. [CrossRef]
75. Sun, J.-Y.; Guo, X.; Smith, M.A. Identification of Crepenynic Acid in the Seed Oil of *Atractylodes lancea* and *A. macrocephala*. *J. Am. Oil Chem. Soc.* **2017**, *94*, 655–660. [CrossRef]
76. Anderson, W.H.; Gellerman, J.L. Acetylenic acids from mosses. *Lipids* **1975**, *10*, 501–502. [CrossRef] [PubMed]
77. Dembitsky, V.M.; Řezanka, T. Distribution of acetylenic acids and polar lipids in some aquatic bryophytes. *Phytochemistry* **1995**, *40*, 93–97. [CrossRef]
78. Kalacheva, G.S.; Sushchik, N.N.; Gladyshev, M.I.; Makhutova, O.N. Seasonal dynamics of fatty acids in the lipids of water moss *Fontinalis antipyretica* from the Yenisei River. *Russ. J. Plant Physiol.* **2009**, *56*, 795–807. [CrossRef]
79. Pejin, B.; Bianco, A.; Newmaster, S.; Sabovljevic, M.; Vujisić, L.; Tešević, V.; Vajs, V.; De Rosa, S. Fatty acids of *Rhodobryum ontariense* (Bryaceae). *Nat. Prod. Res.* **2011**, *26*, 696–702. [CrossRef] [PubMed]
80. Aveldano, M.I.; VanRollins, M.; Horrocks, L.A. Separation and Quantitation of Free Fatty Acids and Fatty Acid Methyl Esters by Reverse Phase High Pressure Liquid Chromatograph. *J. Lipid Res.* **1983**, *24*, 83–93. [CrossRef]
81. Rao, M.S.; Hidajat, K.; Ching, C.B. Reversed-Phase HPLC: The Separation Method for the Characterization and Purification of Long Chain Polyunsaturated Fatty Acids–A Review. *J. Chromatogr. Sci.* **1995**, *33*, 9–21. [CrossRef]
82. Carballeira, N.; Shalabi, F.; Cruz, C.; Rodriguez, J.; Rodríguez, E. Comparative study of the fatty acid composition of sponges of the genus Ircinia. Identification of the new 23-methyl-5,9-tetracosadienoic acid. *Comp. Biochem. Physiol. Part B Comp. Biochem.* **1991**, *100*, 489–492. [CrossRef]
83. Christie, W.W.; Brechany, E.Y.; Marekov, I.N.; Stefanov, K.L.; Andreev, S.N. The fatty acids of the sponge *Hymeniacidon sanguinea* from the Black Sea. *Comp. Biochem. Physiol. Part B Comp. Biochem.* **1994**, *109*, 245–252. [CrossRef]

84. Makarieva, T.N.; Santalova, E.A.; Gorshkova, I.A.; Dmitrenok, A.S.; Guzii, A.G.; Gorbach, V.I.; Svetashev, V.I.; Stonik, V.A. A new cytotoxic fatty acid (5Z,9Z)-22-methyl-5,9-tetracosadienoic acid and the sterols from the far Eastern sponge *Geodinella robusta*. *Lipids* **2002**, *37*, 75–80. [CrossRef]
85. Kullenberg, B.; Bergström, G.; Ställberg-Stenhagen, S. Volatile components of the marking secretion of male bumblebees. *Acta Chim. Scand.* **1970**, *24*, 1481–1483. [CrossRef] [PubMed]
86. Sassano, G.; Sanderson, P.; Franx, J.; Groot, P.; van Straalen, J.; Bassaganya-Riera, J. Analysis of pomegranate seed oil for the presence of jacaric acid. *J. Sci. Food Agric.* **2009**, *89*, 1046–1052. [CrossRef]
87. Elfalleh, W.; Tlili, N.; Nasri, N.; Yahia, Y.; Hannachi, H.; Chaira, N.; Ying, M.; Ferchichi, A. Antioxidant Capacities of Phenolic Compounds and Tocopherols from Tunisian Pomegranate (*Punica granatum*) Fruits. *J. Food Sci.* **2011**, *76*, C707–C713. [CrossRef]
88. Yoshime, L.T.; De Melo, I.L.P.; Sattler, J.A.G.; Torres, R.P.; Mancini-Filho, J. Bioactive compounds and the antioxidant capacities of seed oils from pomegranate (*Punica granatum* L.) and bitter gourd (*Momordica charantia* L.). *Food Sci. Technol.* **2019**, *39*, 571–580. [CrossRef]
89. Hajib, A.; Nounah, I.; Harhar, H.; Gharby, S.; Kartah, B.; Matthäus, B.; Bougrin, K.; Charrouf, Z. Oil content, lipid profiling and oxidative stability of "Sefri" Moroccan pomegranate (*Punica granatum* L.) seed oil. *OCL* **2021**, *28*, 5. [CrossRef]
90. Topkafa, M.; Kara, H.; Sherazi, S.T.H. Evaluation of the Triglyceride Composition of Pomegranate Seed Oil by RP-HPLC Followed by GC-MS. *J. Am. Oil Chem. Soc.* **2015**, *92*, 791–800. [CrossRef]
91. Alcaraz-Mármol, F.; Nuncio-Jáuregui, N.; Calín-Sánchez, Á.; Carbonell-Barrachina, Á.A.; Martínez, J.J.; Hernández, F. Determination of fatty acid composition in arils of 20 pomegranates cultivars grown in Spain. *Sci. Hortic.* **2015**, *197*, 712–718. [CrossRef]
92. Van Nieuwenhove, C.P.; Moyano, A.; Castro-Gómez, P.; Fontecha, J.; Sáez, G.; Zárate, G.; Pizarro, P.L. Comparative study of pomegranate and jacaranda seeds as functional components for the conjugated linolenic acid enrichment of yogurt. *LWT* **2019**, *111*, 401–407. [CrossRef]
93. Fadavi, A.; Barzegar, M.; Azizi, M.H. Determination of fatty acids and total lipid content in oilseed of 25 pomegranates varieties grown in Iran. *J. Food Compos. Anal.* **2006**, *19*, 676–680. [CrossRef]
94. Hernández, F.; Melgarejo, P.; Martínez, R.; Legua, P. Fatty acid composition of seed oils from important Spanish pomegranate cultivars. *Ital. J. Food Sci.* **2011**, *23*, 188–193.
95. Jing, P.; Ye, T.; Shi, H.; Sheng, Y.; Slavin, M.; Gao, B.; Liu, L.; Yu, L. Antioxidant properties and phytochemical composition of China-grown pomegranate seeds. *Food Chem.* **2012**, *132*, 1457–1464. [CrossRef] [PubMed]
96. Hopkins, C.Y.; Chisholm, M.J.; Orgodnik, J.A. Identity and configuration of conjugated fatty acids in certain seed oils. *Lipids* **1969**, *4*, 89–92. [CrossRef]
97. Montañés, F.; Tallon, S.; Catchpole, O. Isolation of Non-methylene Interrupted or Acetylenic Fatty Acids from Seed Oils Using Semi-preparative Supercritical Chromatography. *J. Am. Oil Chem. Soc.* **2017**, *94*, 981–991. [CrossRef]
98. Stránský, K.; Jursík, T. Simple quantitative transesterification of lipids 1. Introduction. *Fette Seifen Anstrichm.* **1996**, *98*, 65–71. [CrossRef]
99. Christie, W.W.; Han, X. *Lipid Analysis—Isolation, Separation, Identification and Structural Analysis of Lipids*; The Oily Press: Bridgwater, UK, 2003; p. 212.

Article

A Newly Developed HPLC-UV/Vis Method Using Chemical Derivatization with 2-Naphthalenethiol for Quantitation of Sulforaphane in Rat Plasma

Kyong-Oh Shin [1,2,3] and Kyungho Park [2,3,*]

1. LaSS Lipid Institute (LLI), LaSS Inc., Chuncheon 24252, Korea; 0194768809@hanmail.net
2. Department of Food Science & Nutrition, and Convergence Program of Material Science for Medicine and Pharmaceutics, Hallym University, Chuncheon 24252, Korea
3. The Korean Institute of Nutrition, Hallym University, Chuncheon 24252, Korea
* Correspondence: Kyungho.Park@hallym.ac.kr; Tel.: +82-33-248-2131

Citation: Shin, K.-O.; Park, K. A Newly Developed HPLC-UV/Vis Method Using Chemical Derivatization with 2-Naphthalenethiol for Quantitation of Sulforaphane in Rat Plasma. *Molecules* **2021**, *26*, 5473. https://doi.org/10.3390/molecules26185473

Academic Editor: Paraskevas D. Tzanavaras

Received: 9 August 2021
Accepted: 6 September 2021
Published: 8 September 2021

Publisher's Note: MDPI stays neutral with regard to jurisdictional claims in published maps and institutional affiliations.

Copyright: © 2021 by the authors. Licensee MDPI, Basel, Switzerland. This article is an open access article distributed under the terms and conditions of the Creative Commons Attribution (CC BY) license (https://creativecommons.org/licenses/by/4.0/).

Abstract: Sulforaphane (SFN), a naturally occurring isothiocyanate, has received significant attention because of its ability to modulate multiple biological functions, including anti-carcinogenic properties. However, currently available analytical methods based on high-performance liquid chromatography (HPLC)-UV/Vis for the quantification of SFN have a number of limitations, e.g., low UV absorbance, sensitivity, or accuracy, due to the lack of a chromophore for spectrometric detection. Therefore, we here employed the analytical derivatization procedure using 2-naphthalenethiol (2-NT) to improve the detectability of SFN, followed by HPLC separation and quantification with UV/Vis detection. The optimal derivatization conditions were carried out with 0.3 M of 2-NT in acetonitrile with phosphate buffer (pH 7.4) by incubation at 37 °C for 60 min. Separation was performed in reverse phase mode using a Kinetex C18 column (150 mm × 4.6 mm, 5 μm) at a flow rate of 1 mL/min, with 0.1% formic acid as a mobile phase A, and acetonitrile/0.1% formic acid solution as a mobile phase B with a gradient elution, with a detection wavelength of 234 nm. The method was validated over a linear range of 10–2000 ng/mL with a correlation of determination (R^2) > 0.999 using weighted linear regression analysis. The intra- and inter-assay accuracy (% of nominal value) and precision (% of relative standard deviation) were within ±10 and <15%, respectively. Moreover, the specificity, recovery, matrix effect, process efficiency, and short-term and long-term stabilities of this method were within acceptable limits. Finally, we applied this method for studying in vivo pharmacokinetics (PK) following oral administration of SFN at doses of 10 or 20 mg/kg. The C_{max} (μg/mL), T_{max} (hour), and AUC_{0-12h} (μg·h/mL) of each oral dose were 0.92, 1.99, and 4.88 and 1.67, 1.00, and 9.85, respectively. Overall, the proposed analytical method proved to be reliable and applicable for quantification of SFN in biological samples.

Keywords: 2-naphthalenethiol; derivatization; sulforaphane; HPLC-UV/Vis; pharmacokinetics

1. Introduction

Sulforaphane (SFN) is a naturally occurring sulfur-containing isothiocyanate enriched in natural products, such as broccoli, brussels sprouts, cauliflower, or cabbage [1,2]. SFN has received significant attention because of its ability to suppress cancer development or metastasis in multiple tissues through different mechanisms as follows: (i) modulation of carcinogen-associated enzymes, leading to suppression of the mutagenic activity within the target genes [3,4]; (ii) inhibition of cell growth and induction of apoptosis, thereby suppressing neoplastic development of the initiated and/or spontaneously transformed cells capable of producing cancer [5]; and (iii) inhibition of consequent angiogenesis for tumor progression and metastasis formation [6]. With this scientific evidence, investigators have been trying to determine the potential of SFN as an agent in cancer prevention and/or therapy [7].

Accordingly, downstream studies for assessing the pharmacokinetics (PK) and pharmacodynamics (PD) profiles of SFN are required for both preclinical and clinical stages of the drug discovery process. Although prior studies have reported methods for quantifying SFN utilizing high-performance liquid chromatography (HPLC)-UV/Vis [8–10] or LC-mass spectrometry (MS) [11,12], the current analytical methods are still not in common use in the quantification of SFN due to a number of reasons: (1) the HPLC-UV/Vis detector method has poor sensitivity because SFN has no UV chromophore; (2) although the LC-MS-based method has better capability in accurate identification and sensitive detection compared with HPLC methods, this methodology requires expensive instruments and highly trained personnel, suggesting that the LC-MS-based analytical method cannot be used for small-scale laboratory research. Therefore, alternative methods that can provide high sensitivity and are cost effective are needed to fulfill the requirements of quantification or qualification in SFN-mediated studies, including PK/PD. Recent studies revealed that chemical derivatization with appropriate reagents before an HPLC-equipped UV/Vis detector is generally used to determine certain analytes lacking chromophores, including SFN. Of different derivatizing reagents, such as Cys-ME [8], mercaptoethanol [11], benzenethiol [13], or 2-naphthalenethiol (2-NT) [14], 2-naphthalenethiol (2-NT) exhibits high molar absorptivity and longer and/or specific wavelengths ($\geq$230 nm; 234, 280, and 320 nm), and sensitive detection with reduced interference from the sample matrix can be achieved in comparison to others [14]. Despite such benefits, no studies have employed 2-NT as a derivatizing reagent for SFN quantification measured by the HPLC-UV/Vis method.

In the present study, therefore, we utilized the chemical derivatization reaction using 2-naphthalenethiol (2-NT) to improve the detectability of SFN, followed by the development of specific HPLC-UV/Vis analytical conditions and justified each validation characteristic within acceptable limits. Moreover, we successfully applied the proposed method to study rat pharmacokinetics (PK) following oral administration of SFN.

2. Results and Discussion

2.1. Method Development

While the reaction of isothiocyanates (ITCs) derived from natural sources, such as SFN, with thiol leads to the formation of a stable dithiocarbamate ester, several prior studies have used certain organic reagents, e.g., mercaptoethanol [11] or N-tBOC-Cys-ME [8] or benzenethiol [13], for the detection and quantification of ITCs. However, as previously developed methods for SFN quantification still have a number of weaknesses, e.g., low sensitivity, accuracy, or cost effectiveness, we here developed a new method for quantitation of SFN using HPLC coupled with a UV/Vis detector after analytical derivatization with 2-naphthalenethiol (2-NT). The derivatization process of SFN or its internal standard (IS) isothiocynate using 2-NT is shown in Figure 1. The derivatization condition was optimized by changing various parameters, i.e., the concentration of the derivatization agent 2-NT (0.05–0.5 M), range of pH (4.0–10.0), incubation temperature (25 to 60 °C), and period of incubation (10–180 min). As such, the optimal derivatization conditions were obtained with 0.3 M of 2-NT in acetonitrile with phosphate buffer (pH 7.4) by incubation for 60 min at 37 °C. A typical chromatogram of SFN and IS after chemical derivatization with 2-NT is shown in Figure 2. We confirmed separated symmetric peaks for 2-NT-SFN and 2-NT-IS at a run time of 7.5 and 8.4 min, respectively. Moreover, the selectivity and specificity of the proposed analytical method were evaluated by the absence of any endogenous interference at the retention times of peaks of interest as evaluated by the chromatograms of the following samples: the blank rat plasma (Figure 2A), blank rat plasma spiked with the internal standard (Figure 2B), blank rat plasma spiked with 0.01 µg/mL of SFN (LLOQ) and the internal standard (Figure 2C), and the plasma collected from a rat 1 h following a single oral administration of 10 mg/kg SFN or IS (Figure 2D). Next, we further confirmed the derivatives of both SFN and IS using ESI-MS/MS in positive mode with 30 eV of collision energy. While fragment ions of 2-NT-SFN were generated at m/z 338, 274, 178, 161, 128, 114, and 72, we identified that ions at m/z 274, 178, 114, 72, and at m/z

161, 128 corresponded to SFN and 2-NT moieties, respectively (Figure 3A). In addition, fragment ions of 2-NT-IS were at *m/z* 338, 161, 128, 74.0, while ions at 161, 128 *m/z* stand for 2-NT (Figure 3B). To quantify the derivatized SFN or IS, the chromatographic condition was examined by altering a number of parameters, such as buffer, organic solvent, and absorption spectra, for the detection. The optimized separation was carried out with 0.1% formic acid (solvent A) and acetonitrile with 0.1% formic acid (solvent B). In addition, the highest signals of the derivatives were specified at the wavelengths of 234 nm under UV/Vis measurements.

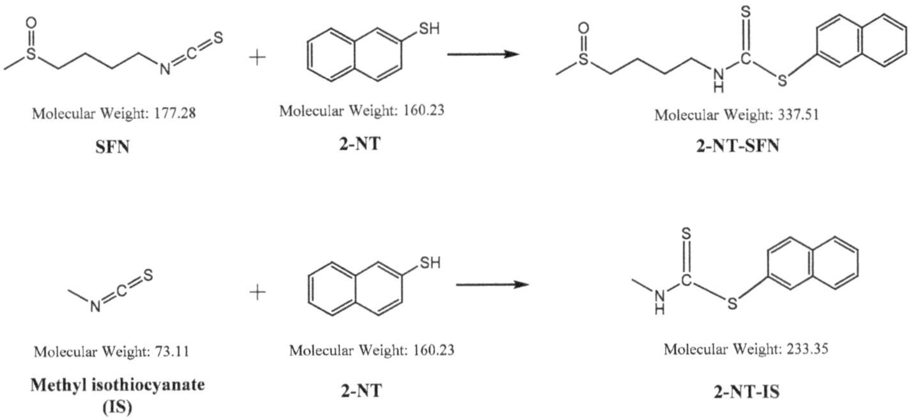

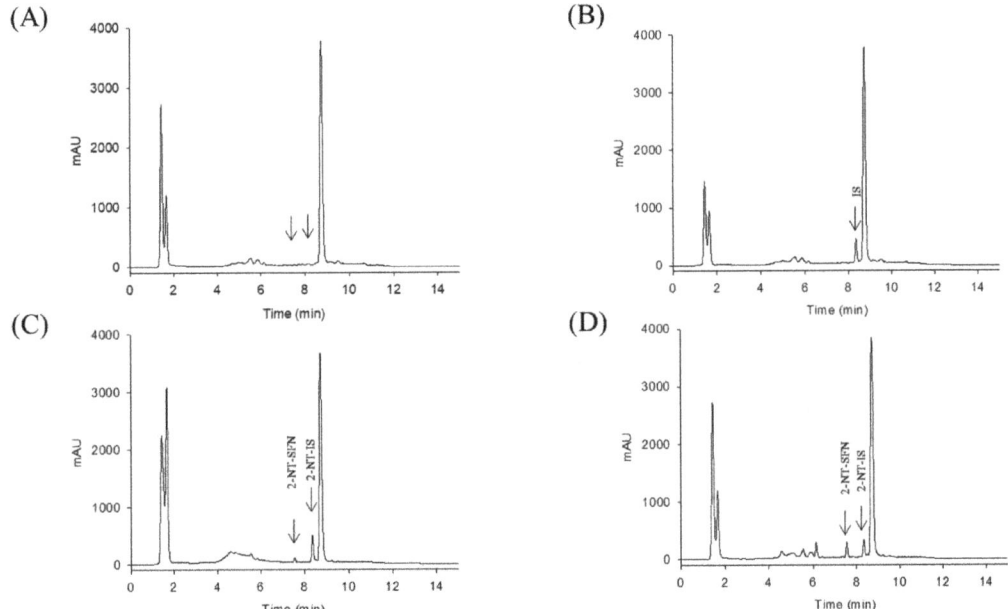

Figure 1. Chemical structures of sulforaphane (SFN) and methyl isothiocynate employed as an internal standard (IS), and their derivatives reacted with 2-naphthalenethiol (2-NT).

Figure 2. Representative HPLC chromatograms of double blank rat plasma (**A**), blank rat plasma spiked with the IS (**B**), blank rat plasma spiked with 0.01 μg/mL of SFN (LLOQ) or the IS (**C**), a plasma sample obtained 1 h after a single oral administration of 10 mg/kg of SFN (**D**). SFN, sulforaphane; 2-NT, 2-naphthalenethiol.

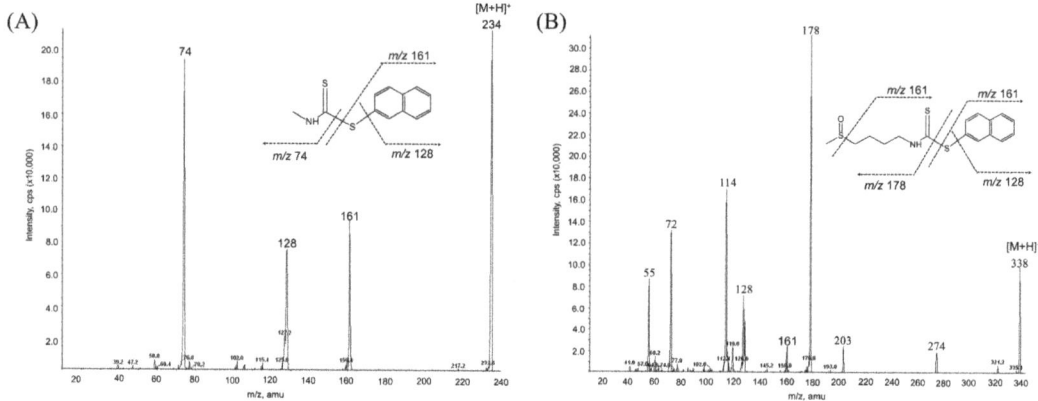

Figure 3. MS/MS spectra of 2-NT-SFN (**A**) and 2-NT-IS (**B**) obtained by ESI in positive mode using a collision energy of 30 eV. IS, internal standard; SFN, sulforaphane; 2-NT, 2-naphthalenethiol.

2.2. Method Validation

2.2.1. Linearity and Calibration Curve

The equations of coefficients of determination (R^2) and linear regression are described in Table 1. R^2 of three replicates was greater than 0.999, and the accuracy of all three calibration points was within ±10% of the nominal concentration, indicating excellent linearity with a detection range of 10–2000 ng/mL. The regression equation was $y = 2.2502x + 0.0499$. The limits of detection (LOD) and limits of quantification (LOQ) were 0.0028 and 0.0091, respectively (Table 1), suggesting that the detection range and sensitivity of calibration curves of our method were sufficient to quantify SFN levels in rat plasma after the single dietary doses of 10 or 20 mg/kg SFN.

Table 1. Summary of the calibration curve, linear range, LOD and LOQ for the quantification for SFN in rat plasma by the HPLC method.

Calibration Curve	R^2	Linear Range (µg/mL)	LOD	LLOQ
y = 2.2502x + 0.0499	0.9995	0.01~2.0	0.0028	0.0091

LOD, limit of detection; LLOQ, lower limit of quantification; R^2, coefficients of determination; SFN, sulforaphane.

2.2.2. Accuracy and Precision

The results of the intra- and inter-assay accuracy and precision are described in Table 2. The accuracy and precision of all the QCs at different levels were within ±10%, and <15%, respectively. In addition, the mean accuracy and precision values obtained were 92.15 and 8.14%, respectively. Our accuracy and precision results satisfied the USFDA guidelines for bioanalytical method validation, suggesting that rat plasma samples with a concentration >10 ng/mL of SFN can be quantified with good enough accuracy and precision.

2.2.3. Recovery and Matrix Effect

Table 3 summarizes the results of the matrix effect and extraction recoveries of SFN and IS from rat plasma samples. The mean extraction recoveries and efficiency at three different levels of SFN and IS were ±15, ±18% or ±20, ±22%, respectively, indicating a good recovery in different rat plasma samples. The mean matrix effects at three different levels of SFN (0.03, 0.8, and 2.0 µg/mL) were −2.87, −2.99, and −2.37%, while the IS at 0.5 µg/mL was −2.58. These results indicated the presence of some matrix effect in terms of the 2-NT-SFN/2-NT-IS response ratio. Nevertheless, because the LLOQ signal intensity was sufficient for the SFN quantification, this method can be applied for downstream analysis, such as the PK study.

Table 2. Intra- and inter-day precision and accuracy for quantification for SFN in rat plasma by the HPLC method.

Spike Amount (µg/mL)	Intra-Day (RSD, %) [1]	Inter-Day (RSD, %)	Intra-Day (Accuracy, %) [2]	Inter-Day (Accuracy, %)
0.01	7.95	8.14	93.41 ± 6.41	94.21 ± 7.14
0.03	7.27	7.69	92.15 ± 6.71	91.97 ± 8.19
0.8	3.04	3.41	100.17 ± 3.27	100.83 ± 3.57
2.0	1.57	2.22	98.94 ± 2.11	99.09 ± 2.58

[1] RSD (%) = standard deviation of the concentration/mean concentration × 100. [2] Accuracy (%) = calculated concentration/theoretical concentration × 100.

Table 3. Recovery, matrix effect, and process efficiency of SFN and the IS from spiked rat plasma ($n = 4$).

Component	Spike Amount (µg/mL)	Recovery [1]	Matrix Effect [2]	Process Efficiency [3]
SFN	0.03	85.31 ± 4.94	−2.87 ± 0.14	82.35 ± 5.89
	0.8	86.71 ± 1.24	−2.99 ± 0.16	84.10 ± 2.71
	2.0	87.41 ± 0.47	−2.37 ± 0.13	85.91 ± 0.83
IS	0.5	80.27 ± 2.18	−2.58 ± 0.21	78.14 ± 2.61

[1] Recovery = (Response before extraction spiked sample/Response post-extracted spiked sample) × 100. [2] Matrix effect = (Response post-extracted spiked sample/Response non-extracted neat sample − 1) × 100. [3] Process efficiency = (Response before extraction spiked sample/Response non-extracted neat sample) × 100. IS, internal standard; SFN, sulforaphane.

2.2.4. Stability

The results of the short-term and long-term stabilities of SFN in rat plasma are summarized in Table 4. SFN stock solutions were stable for at least 3 weeks when prepared in an acetonitrile solution at −20 °C (data not shown). Moreover, nominal% of QCs at three concentrations (0.03, 0.8, or 2 µg/mL) of SFN after 30 days at −20 and −80 °C were within ±19% or ±11%, respectively. These results indicate that SFN is stable for at least 30 days in rat plasma under the described storage conditions (Table 4).

Table 4. Stability test for SFN in rat plasma.

Condition Tested	0.03 µg/mL			0.8 µg/mL			2 µg/mL		
	Mean	RSD [1] (%)	RE [2] (%)	Mean	RSD (%)	RE (%)	Mean	RSD (%)	RE (%)
Short-term stability									
Freeze-thaw (−80 °C, 3 cycle)	0.028	3.74	−7.67	0.797	2.91	−0.38	2.013	1.84	0.65
Refrigerator (4 °C, 1 day)	0.025	3.12	−17.67	0.589	3.17	−26.38	1.731	1.56	−13.45
Freezer (−20 °C, 1 day)	0.027	3.39	−10.00	0.703	2.67	−12.13	1.959	1.58	−2.05
Freezer (−80 °C, 1 day)	0.033	2.95	9.67	0.751	3.10	−6.13	2.003	0.87	0.15
post-preparative stability (4 °C, 1 day)	0.030	2.34	−1.00	0.792	2.36	−1.00	2.084	1.41	4.20
post-preparative stability (4 °C, 1 week)	0.031	3.49	4.33	0.862	2.81	7.75	2.002	2.10	0.10
Long-term stability									
Freezer (−80 °C, 30 days)	0.032	3.81	7.00	0.780	2.90	−2.50	1.777	1.59	−11.15
Freezer (−20 °C, 30 days)	0.024	2.67	−19.67	0.768	3.13	−4.00	1.726	2.88	−13.70

[1] RSD (%) = standard deviation of the concentration/mean concentration × 100. [2] RE (%) = calculated concentration/theoretical concentration × 100.

2.3. In Vivo Pharmacokinetics Study

We finally performed in vivo pharmacokinetics (PK) study of oral SFN using a newly developed method. The mean plasma concentration–time profiles and PK parameters calculated by the non-compartment model of SFN in rat plasma are described in Figure 4 and Table 5, respectively. The concentration of SFN was readily measurable in plasma samples collected up to 12 h post-dose. Consistent with prior findings [9,15], the maximum plasma concentrations of oral SFN (10 mg/kg: 0.92 μg/mL; 20 mg/kg: 1.67 μg/mL) were reached in approximately 1 h at both doses, suggesting that SFN is rapidly absorbed from the gastro-intestinal tract. A rapid absorption rate could be caused by the chemical properties of SFN, i.e., a low molecular weight and higher lipophilicity. Afterwards, SFN at both doses was eliminated with a half-life of approximately 5–6 h, which is consistent with previous findings that SFN is rapidly eliminated after oral absorption [9,15,16]. Finally, our studies revealed that the values of AUC_{0-12h}, $AUC_{0-\infty h}$, AUMC, and C_{max} in orally administered rats increased in a dose-dependent manner, suggesting that a newly established method would be a time-efficient and accurate method for measurement of SFN in rat plasma applicable for in vivo PK study.

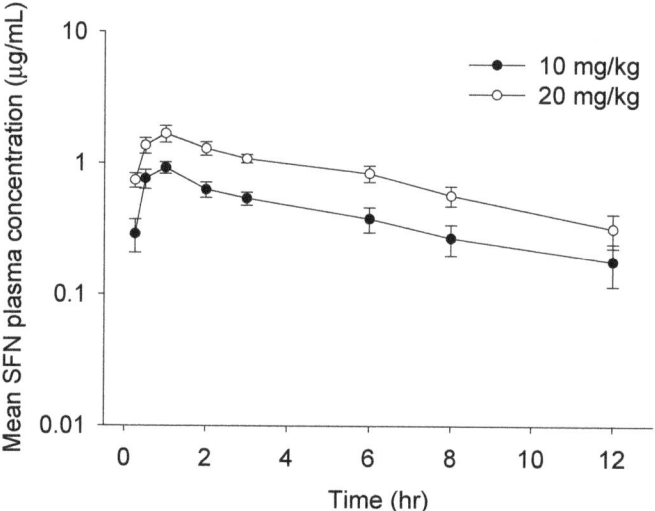

Figure 4. Mean plasma concentration–time plot of sulforaphane (SFN) after a single oral administration of SFN at 10 or 20 mg/kg to rats. All values are mean ± SD (n = 4 rats).

Table 5. Pharmacokinetic parameters of SFN after oral administration at doses of 10 or 20 mg/kg to rats (mean ± S.D., n = 4 rats).

Parameters	10 mg/kg	20 mg/kg
AUC_{0-12h} (μg·h/mL)	4.88 ± 0.89	9.85 ± 1.37
$AUC_{0-\infty h}$ (μg·h/mL)	6.25 ± 1.59	12.42 ± 2.36
AUMC (0–12 h)	21.85 ± 5.11	44.57 ± 7.42
MRT (0–12 h) (h)	4.45 ± 0.24	4.51 ± 0.13
$t_{1/2}$ (h)	5.05 ± 0.91	5.47 ± 0.56
T_{max} (h)	1.00 ± 0.00	1.00 ± 0.00
CLz/F (L/h/kg)	1.67 ± 0.44	1.65 ± 0.32
Vz/F (L/kg)	11.83 ± 0.93	12.84 ± 1.14
C_{max} (μg/mL)	0.92 ± 0.09	1.67 ± 0.24

AUC, area under the curve; AUMC, area under the first moment curve; CLz/F, apparent oral clearance; Cmax, peak plasma concentration; MRT, mean residence time; $t_{1/2}$, terminal; Tmax, time to reach Cmax; Vz/F, apparent volume of distribution.

3. Materials and Methods

3.1. Chemicals and Standards

Sulforaphane (SFN) (>90% purity), methyl isothyocyanate (>97% purity) as an internal standard (IS), formic acid (analytical reagent grade), and 2-naphthalenethiol (99% purity) were purchased from Sigma-Aldrich (St. Louis, MO, USA). Organic solvents, such as methanol, ethyl alcohol, acetonitrile, and chloroform, for both SFN extraction and HPLC analysis were supplied by Merck (Darmstadt, Germany). Purified deionized water was obtained from an in-house purification system (18 MΩ, Millipore, Bedford, MA, USA). Unless otherwise stated, all other chemicals were obtained from Sigma (St. Louis, MO, USA).

3.2. Equipment

The Agilent 1260 series HPLC system (Agilent Technologies, Santa Clara, CA, USA) was equipped with an auto-sampler, degasser, quaternary pump, and UV/Vis detector. Analyte separation was performed in reverse phase mode with a Kinetex C18 column (150 mm × 4.6 mm, 5.0 µm). Data manipulation was performed using the Chemstation B.04.03 software.

3.3. Preparation of the Stock Solution and Working Solutions

SFN was dissolved in acetonitrile to prepare stock solutions at a concentration of 1000 µg/mL. Working solutions at concentrations of 0.01, 0.025, 0.05, 0.1, 0.25, 0.5, 1, and 2 µg/mL were obtained by serial dilution from stock solution by adding an appropriate volume of ethyl alcohol (EtOH). Methyl isothiocyanate, which was employed for the internal standard (IS), was diluted with acetonitrile to obtain the working IS solution at a concentration of 5 µg/mL. All stock solutions were stored at −20 °C and protected from light.

3.4. Calibration Standards and Quality Control (QC) Samples

Calibration standard working solutions were diluted 10-fold with rat blank plasma (1:9, v/v) to obtain 6 calibration standards at 0.01, 0.025, 0.05, 0.1, 0.25, 0.5, 1, and 2 µg/mL. Quality control (QC) working solutions were also diluted 10-fold with rat blank plasma to obtain QCs at 4 different concentration levels, i.e., low, mid, and high levels of 0.01, 0.03, 0.8, and 2.0 µg/mL, respectively.

3.5. SFN Extraction from Rat Plasma and Chemical Derivatization with 2-NT

In total, 100 µL of blank rat plasma, rat plasma spiked with SFN, or plasma collected from rats orally administered SFN at doses of 10 or 20 mg/kg for the pharmacokinetic (PK) study were mixed with 10 µL of the IS working solution, followed by incubation at room temperature for 3 min. The reacted solution was then mixed with the solution of saturated sodium chloride and chloroform (1:4, v/v), and centrifuged at 10,000× g for 5 min. The organic phase was separated from the aqueous layer, and dried using a speed vacuum system (Vision, Seoul, Korea). The residue dissolved in ethanol was further derivatized with 0.3 M 2-naphthalenethiol (NT) by incubation at 37 °C for 60 min, and finally the reacted residue was fltered using a 0.45 µm PTFE filter (Millipore, Billerica, MA, USA) prior to HPLC analysis.

3.6. Derivatives Confirmation by ESI Mass Spectrometry

An API 3200 triple quadruple Mass Spectrometer (Applied Biosystems, Foster City, CA, USA) equipped with an electrospray ionization (ESI) interface in positive ionization mode was employed for the confirmation of SFN derivatives. The mass range was set between 50 and 500 m/z and derivatives were infused directly into the mass spectrometry. The optimized instrument conditions were as follows: source temperature, 400 °C; curtain gas pressure, 20 psi; nebulizing gas (GS1) pressure, 50 psi; and heating gas (GS2) pressure, 40 psi. The first quadrupole (Q1) was set to unit resolution and Q3 to low resolution.

Analyst software (ver. 1.4.2; Applied Biosystems, Foster City, CA, USA) was used for instrument control and data collection.

3.7. Chromatographic Conditions to Quantify SFN in Rat Plasma

The LC chromatographic separation was conducted with a Kinetex C18 column (150 mm × 4.6 mm, 5.0 µm). The mobile phase was delivered at a flow rate of 1.0 mL/min through gradient elution and consisted of pure water with 0.1% formic acid (aqueous mobile phase A) and acetonitrile with 0.1% formic acid (organic mobile phase B). The total analytical run time for each injection was 15 min, including 5 min of re-equilibration.

The initial gradient elution started with 10% mobile phase B, which was maintained for 1 min, followed by gradual elevation of mobile phase B to 70% over 5 min. These experimental conditions were held for 3 min, returned to initial conditions, and the column was re-equilibrated for 5 min. Optimization of chromatographic conditions involved the subsequent evaluation of the following parameters: buffer, mobile phase, organic solvent, gradient elution, flow rate, auto-sampler temperature, column temperature, and injection volume. Finally, the conditions showing the best retention and separation were then selected. Both SFN and IS derivatives' absorbance was detected at 234 nm. Research manuscripts reporting large datasets that are deposited in a publicly available database should specify where the data have been deposited and provide the relevant accession numbers. If the accession numbers have not yet been obtained at the time of submission, please state that they will be provided during review. They must be provided prior to publication.

Intervention studies involving animals or humans, and other studies that require ethical approval, must list the authority that provided approval and the corresponding ethical approval code.

3.8. Method Validation

Full validation of the current method in rat plasma was performed in accordance with the guidelines of the US Food and Drug Administration (FDA) on chromatographic bioanalytical method validation. The validation included the following parameters: linearity, stability, selectivity, sensitivity, and intra-/inter-day accuracy and precision. The linearity of this method was evaluated using calibration standards prepared at six concentrations over a range of 0.01–2.0 µg/mL. Repeatability or intra-assay accuracy were determined by analyzing five individually prepared replicates at each concentration within the same run and five injections of one replicate within another run to evaluate injection repeatability. Inter-assay accuracy and precision were obtained by analyzing five individually prepared replicates at each concentration within five different days. The stability of SFN in rat plasma was determined using five individually prepared replicates of QCs at three concentration levels. The following stability conditions were evaluated: short-term stability (24 h at room temperature, 4 °C and at −80 °C), post-preparation with or without exposure to derivative reagent (24 h or 7 days at 4 °C), freeze thaw (three cycles, −20 °C/room temperature, and 24 h between cycles), and long-term stability (30 days at −20 and −80 °C).

3.9. In Vivo Pharmacokinetic Study

All animal procedures were approved by the Institutional Animal Care and Use Committee (IACUC) of Hallym University (Permit number: Hallym2020-48) and performed in accordance with their guidelines as well as ARRIVE guidelines (Animal Research: Reporting of In Vivo Experiments) (https://www.nc3rs.org.uk/arriveguidelines; accessed on 15 February 2021). Eight-week-old Sprague–Dawley (SD) female rats, weighing about 230–250 g, were purchased from DBL Ltd. (Eumseong, Korea) and used for the pharmacokinetic study. All experimental rats were housed in individual cages at the Hallym University Laboratory Animal Resources Center under specific pathogen-free (SPF) conditions with a controlled consistent temperature (23 ± 2 °C) and lighting environment (12 h/12 h light/dark cycle). At the end of the study, the experimental mice were sacrificed by CO_2 inhalation. A gradual fill rate of 20% chamber volume per minute of displacement was used for CO_2 euthanasia. All efforts were made to minimize the number and suffering of any animals used in these experiments. Animals were randomly divided into two groups after an acclimation period of 1 week. SFN dissolved in water was administered as a single oral dose of either 10 mg/kg or 20 mg/kg. Blood samples (300 µL) were collected in lithium-heparinized tubes from the tail vein before dosing and subsequently at 0.25, 0.5, 1, 2, 3, 6, 8, and 12 h after administration. Blood samples were then centrifuged at 10,000× g for 3 min at 4 °C to separate plasma. The SFN pharmacokinetic parameters processed by non-compartmental analysis of plasma concentration versus time data using the computer program Winnonlin Ver. 5.1 (Pharsight Corporation, Mountain View, CA, USA) were as follows: the area under the plasma concentration–time curve to the last measurable plasma concentration (AUC_{0-t}); the area under the plasma concentration–time curve to time infinity ($AUC_{0-\infty}$); the maximum plasma concentration (C_{max}); the time to reach the maximum concentrations (T_{max}); the elimination half-life ($t_{1/2}$); the mean residence time (MRT); and total plasma clearance (CL). Both C_{max} and T_{max} were obtained directly from the generated curve.

4. Conclusions

A new optimized HPLC-UV/Vis method for the quantification of SFN in rat plasma was developed and validated in accordance with USFDA guidelines. To the best of our knowledge, this is the first study to utilize 2-NT derivatization for a HPLC-UV/Vis-based quantification of SFN. While prior studies established the HPLC-UV/Vis methods without chemical derivatization [9,17–19], which exhibited that UV absorbance at a short wavelength (202–210 nm) causes a relatively high background and low sensitivity, the use of 2-NT derivatization significantly increases the UV absorption at a wavelength of 234 nm, helping to improve the detectability of SFN. Moreover, the advantage of this method lies in its low limit of detection (LOD) of 0.0078 µg/mL compared to a previous HPLC assay with an LOD of 0.01 µg/mL [9]. The assay was fully validated, with good selectivity and linearity over a large range of 0.01–2.0 µg/mL. SFN was stable in the solvent employed in this study and plasma for at least 6 months and 1 month, respectively. Moreover, we successfully applied this method to the in vivo pharmacokinetic study of SFN with single oral dietary doses. A limitation of this study is the absence of clinical application on plasma samples of patients receiving SFN administration. Thus, future studies are needed for assessment of the assay using patients' plasma samples. Overall, the method we developed in the present study could be useful to perform not only clinical PK/PD studies but also to investigate SFN side effects, which will be instructive in the creation of a dosage regimen and optimization of SFN safety and efficiency.

Author Contributions: Conceptualization, K.-O.S. and K.P.; Data curation, K.-O.S. and K.P.; Formal analysis, K.-O.S.; Funding acquisition, K.P.; Investigation, K.-O.S. and K.P.; Methodology, K.-O.S.; Project administration, K.-O.S. and K.P.; Resources, K.-O.S. and K.P.; Supervision, K.P.; Writing—original draft, K.-O.S. and K.P.; Writing—review and editing, K.-O.S. and K.P. Both authors have read and agreed to the published version of the manuscript.

Funding: This study was supported by Main Research Program (E0210600-01) of the Korea Food Research Institute (KFRI) funded by the Ministry of Science and ICT.

Institutional Review Board Statement: The study was conducted according to the guidelines of the Declaration of Helsinki, and approved by the Institutional Review Board (or Ethics Committee) of Hallym University (protocol code: Hallym2020-48, and date of approval: 16 February 2021).

Informed Consent Statement: Not applicable.

Data Availability Statement: Not applicable.

Conflicts of Interest: K.-O.S. is current employee of LaSS Inc. The other author has no conflict of interest to declare.

Sample Availability: Samples of the compounds are not available from the authors.

References

1. Nakagawa, K.; Umeda, T.; Higuchi, O.; Tsuzuki, T.; Suzuki, T.; Miyazawa, T. Evaporative light-scattering analysis of sulforaphane in broccoli samples: Quality of broccoli products regarding sulforaphane contents. *J. Agric. Food Chem.* **2006**, *54*, 2479–2483. [CrossRef] [PubMed]
2. Matusheski, N.V.; Wallig, M.A.; Juvik, J.A.; Klein, B.P.; Kushad, M.M.; Jeffery, E.H. Preparative HPLC method for the purification of sulforaphane and sulforaphane nitrile from Brassica oleracea. *J. Agric. Food Chem.* **2001**, *49*, 1867–1872. [CrossRef] [PubMed]
3. Tortorella, S.M.; Royce, S.G.; Licciardi, P.V.; Karagiannis, T.C. Dietary Sulforaphane in Cancer Chemoprevention: The Role of Epigenetic Regulation and HDAC Inhibition. *Antioxid. Redox Signal.* **2015**, *22*, 1382–1424. [CrossRef] [PubMed]
4. Cheung, K.L.; Kong, A.N. Molecular targets of dietary phenethyl isothiocyanate and sulforaphane for cancer chemoprevention. *AAPS J.* **2010**, *12*, 87–97. [CrossRef] [PubMed]
5. Li, Y.; Zhang, T.; Korkaya, H.; Liu, S.; Lee, H.F.; Newman, B.; Yu, Y.; Clouthier, S.G.; Schwartz, S.J.; Wicha, M.S.; et al. Sulforaphane, a dietary component of broccoli/broccoli sprouts, inhibits breast cancer stem cells. *Clin. Cancer Res.* **2010**, *16*, 2580–2590. [CrossRef] [PubMed]
6. Liu, P.; Atkinson, S.J.; Akbareian, S.E.; Zhou, Z.; Munsterberg, A.; Robinson, S.D.; Bao, Y. Sulforaphane exerts anti-angiogenesis effects against hepatocellular carcinoma through inhibition of STAT3/HIF-1alpha/VEGF signalling. *Sci. Rep.* **2017**, *7*, 12651. [CrossRef] [PubMed]
7. Russo, M.; Spagnuolo, C.; Russo, G.L.; Skalicka-Wozniak, K.; Daglia, M.; Sobarzo-Sanchez, E.; Nabavi, S.F.; Nabavi, S.M. Nrf2 targeting by sulforaphane: A potential therapy for cancer treatment. *Crit. Rev. Food Sci. Nutr.* **2018**, *58*, 1391–1405. [CrossRef] [PubMed]
8. Budnowski, J.; Hanschen, F.S.; Lehmann, C.; Haack, M.; Brigelius-Flohe, R.; Kroh, L.W.; Blaut, M.; Rohn, S.; Hanske, L. A derivatization method for the simultaneous detection of glucosinolates and isothiocyanates in biological samples. *Anal. Biochem.* **2013**, *441*, 199–207. [CrossRef] [PubMed]
9. Ong, C.; Elbarbry, F. A new validated HPLC method for the determination of sulforaphane: Application to study pharmacokinetics of sulforaphane in rats. *Biomed. Chromatogr.* **2016**, *30*, 1016–1021. [CrossRef] [PubMed]
10. Campas-Baypoli, O.N.; Sanchez-Machado, D.I.; Bueno-Solano, C.; Ramirez-Wong, B.; Lopez-Cervantes, J. HPLC method validation for measurement of sulforaphane level in broccoli by-products. *Biomed. Chromatogr.* **2010**, *24*, 387–392. [CrossRef] [PubMed]
11. Wilson, E.A.; Ennahar, S.; Zhao, M.; Bergaentzle, M.; Marchioni, E.; Bindler, F. Simultaneous Determination of Various Isothiocyanates by RP-LC Following Precolumn Derivatization with Mercaptoethanol. *Chromatographia* **2011**, *73*, 137–142. [CrossRef] [PubMed]
12. Dominguez-Perles, R.; Medina, S.; Moreno, D.A.; Garcia-Viguera, C.; Ferreres, F.; Gil-Izquierdo, A. A new ultra-rapid UHPLC/MS/MS method for assessing glucoraphanin and sulforaphane bioavailability in human urine. *Food Chem.* **2014**, *143*, 132–138. [CrossRef] [PubMed]
13. Wilson, E.A.; Ennahar, S.; Marchioni, E.; Bergaentzle, M.; Bindler, F. Improvement in determination of isothiocyanates using high-temperature reversed-phase HPLC. *J. Sep. Sci.* **2012**, *35*, 2026–2031. [CrossRef] [PubMed]
14. Faraji, M.; Hamdamali, M.; Aryanasab, F.; Shabanian, M. 2-Naphthalenthiol derivatization followed by dispersive liquid-liquid microextraction as an efficient and sensitive method for determination of acrylamide in bread and biscuit samples using high-performance liquid chromatography. *J. Chromatogr. A* **2018**, *1558*, 14–20. [CrossRef] [PubMed]
15. Hanlon, N.; Coldham, N.; Gielbert, A.; Kuhnert, N.; Sauer, M.J.; King, L.J.; Ioannides, C. Absolute bioavailability and dose-dependent pharmacokinetic behaviour of dietary doses of the chemopreventive isothiocyanate sulforaphane in rat. *Br. J. Nutr.* **2008**, *99*, 559–564. [CrossRef] [PubMed]
16. Ye, L.; Dinkova-Kostova, A.T.; Wade, K.L.; Zhang, Y.; Shapiro, T.A.; Talalay, P. Quantitative determination of dithiocarbamates in human plasma, serum, erythrocytes and urine: Pharmacokinetics of broccoli sprout isothiocyanates in humans. *Clin. Chim. Acta* **2002**, *316*, 43–53. [CrossRef]
17. Kamal, M.M.; Nazzal, S. Development and validation of a HPLC-UV method for the simultaneous detection and quantification of paclitaxel and sulforaphane in lipid based self-microemulsifying formulation. *J. Chromatogr. Sci.* **2020**, *57*, 931–938. [CrossRef] [PubMed]
18. Sangthong, S.; Weerapreeyakul, N. Simultaneous quantification of sulforaphene and sulforaphane by reverse phase HPLC and their content in Raphanus sativus L. var. caudatus Alef extracts. *Food Chem.* **2016**, *201*, 139–144. [CrossRef] [PubMed]
19. Han, D.; Row, K.H. Separation and purification of sulforaphane from broccoli by solid phase extraction. *Int. J. Mol. Sci.* **2011**, *12*, 1854–1861. [CrossRef] [PubMed]

Review

Analysis of Organophosphorus-Based Nerve Agent Degradation Products by Gas Chromatography-Mass Spectrometry (GC-MS): Current Derivatization Reactions in the Analytical Chemist's Toolbox

Carlos A. Valdez [1,2,3,*] and Roald N. Leif [1,2,3]

1. Physical and Life Sciences Directorate, Lawrence Livermore National Laboratory, Livermore, CA 94550, USA; leif1@llnl.gov
2. Nuclear and Chemical Sciences Division, Lawrence Livermore National Laboratory, Livermore, CA 94550, USA
3. Forensic Science Center, Lawrence Livermore National Laboratory, Livermore, CA 94550, USA
* Correspondence: valdez11@llnl.gov

Citation: Valdez, C.A.; Leif, R.N. Analysis of Organophosphorus-Based Nerve Agent Degradation Products by Gas Chromatography-Mass Spectrometry (GC-MS): Current Derivatization Reactions in the Analytical Chemist's Toolbox. *Molecules* 2021, *26*, 4631. https://doi.org/10.3390/molecules26154631

Academic Editor: Paraskevas D. Tzanavaras

Received: 19 June 2021
Accepted: 29 July 2021
Published: 30 July 2021

Publisher's Note: MDPI stays neutral with regard to jurisdictional claims in published maps and institutional affiliations.

Copyright: © 2021 by the authors. Licensee MDPI, Basel, Switzerland. This article is an open access article distributed under the terms and conditions of the Creative Commons Attribution (CC BY) license (https://creativecommons.org/licenses/by/4.0/).

Abstract: The field of gas chromatography-mass spectrometry (GC-MS) in the analysis of chemical warfare agents (CWAs), specifically those involving the organophosphorus-based nerve agents (OPNAs), is a continually evolving and dynamic area of research. The ever-present interest in this field within analytical chemistry is driven by the constant threat posed by these lethal CWAs, highlighted by their use during the Tokyo subway attack in 1995, their deliberate use on civilians in Syria in 2013, and their use in the poisoning of Sergei and Yulia Skripal in Great Britain in 2018 and Alexei Navalny in 2020. These events coupled with their potential for mass destruction only serve to stress the importance of developing methods for their rapid and unambiguous detection. Although the direct detection of OPNAs is possible by GC-MS, in most instances, the analytical chemist must rely on the detection of the products arising from their degradation. To this end, derivatization reactions mainly in the form of silylations and alkylations employing a vast array of reagents have played a pivotal role in the efficient detection of these products that can be used retrospectively to identify the original OPNA.

Keywords: nerve agents; GC-MS; derivatization; silylation; methylation; chemical warfare agents; sarin; Novichoks

1. Introduction

Gas Chromatography-Mass Spectrometry (GC-MS) has been a central analytical technique in the field of Chemical Warfare Agent (CWA) detection, analysis, and identification. The role that this form of mass spectrometric method has provided to analysts immersed in this specific area is invaluable and continues to grow in importance as the CWA molecular space rapidly expands. CWAs represent some of the most toxic chemical entities ever produced in a laboratory. Their sole purpose is to eliminate a threat or incapacitate it [1]. Among the most notorious members in this class of toxic compounds are the organophosphorus-based nerve agents [2]. The lethal effects of nerve agents stem from their efficient inhibition of the enzyme acetylcholinesterase (AChE), which is an enzyme involved in maintaining homeostasis in the nervous system and controls muscle function [3,4]. Inhibition of AChE leads to the accumulation of the neurotransmitter acetylcholine in synaptic junctions that results in complete muscle relaxation, an event that, when affecting those involved in respiration, leads to the demise of the affected individual by asphyxiation.

Traditional OPNAs fall into two main categories since their introduction in the 20th century, being used against military as well as civilian targets and these are the G-series and the V-series (Figure 1) [5–7]. The G-series were developed in Germany and hence the G-prefix in their notation, and they share a common chemical core, a pentavalent phosphorus center containing a methyl group, a fluorine atom and an alkyl ester group that ultimately distinguishes each member in this class from one another. Thus, Sarin (or GB) possesses the isopropoxy moiety, while Soman (or GD) possesses the chiral, 2,2-dimethyl-2-butoxy moiety and lastly Cyclosarin (or GF) which contains the cyclohexyloxy moiety. The OPNA Tabun (or GA) is considered a G-series agent even though it does not share many common structural features with the other members in the group. GA possesses a cyanide moiety instead of the fluoride which serves as the leaving group in addition to a dimethylamino moiety instead of a methyl group. The second category, the V-series also consists of a pentavalent phosphorus atom featuring an alkoxy and a methyl substituent in addition to a N,N-dialkylaminoethylthiolate group that serves as a leaving group in its reaction with AChE. One of the most famous members in this class is VX (O-ethyl-S-[2-(Diisopropylamino)ethyl] methylphosphonothioate) that consists aside of a methyl and ethoxy substituents, the hallmark N,N-diisopropylaminoethylthiol substituent. Additionally, the other two members in this class, VR (Russian VX) and CVX (Chinese VX) are structural isomers of VX and of each other. In the case of VR, the alkoxy group is the isobutoxy group while the thioalkyl side chain is the N,N-diethylaminoethylthiol substituent. For CVX, the alkoxy group is the n-butoxy group while the thioalkyl side chain is the N,N-diethylaminoethylthiol substituent. A third category that broke into the world's spotlight are the A-series or Novichok agents [8] with their use in the poisoning of Sergei Skripal and his daughter in Salisbury, UK in 2018 [9] and the poisoning of Alexei Navalny, who became ill during a flight from Tomsk to Moscow, Russia in 2020 [10]. The structures of these "newcomers" as their name means are very different from the G- and the V-series and also differ significantly among themselves to the point that they do not share the same structural isomerism the V-series do. Although the structural make-up among these three kinds of OPNAs is different, one common factor that unites them all is the presence of a highly reactive, pentavalent phosphorus center. For the Novichok agents, the role of the leaving group is played by the fluorine atom just as in the case of Sarin or Soman. One last reason for the basis of OPNA's worldwide concern is the fact that established synthetic routes exist for their large-scale production. Fortunately, a great deal of information is known about these routes and the ever-evolving field of chemical forensics is rapidly developing methods involving chemical attribution signature analysis to help forensic analytical chemist determine how a specific OPNA synthesized [11–13].

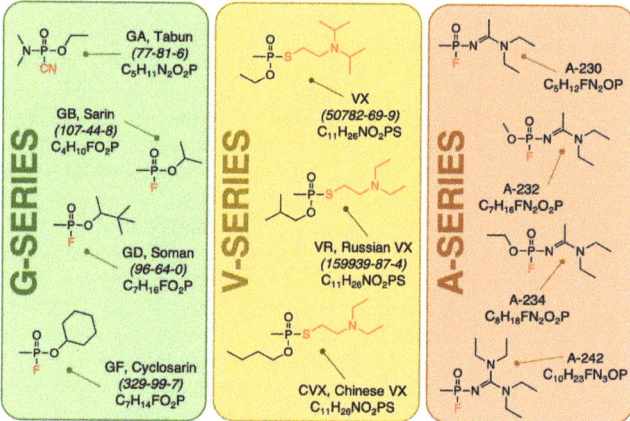

Figure 1. Chemical structures of OPNAs. Degradation products arising from all these three kinds of nerve agents are similar and overlapping in structure.

2. Degradation Pathways for Nerve Agents

As the threat that nerve agents pose is well recognized, several research efforts from various groups in government and private sectors, have focused on the development of more powerful antidotes [14–19], more efficient protective gear for the warfighter [20,21], and more effective methods for their destruction [22]. Due to their different structural features among the G-, V- and A-series, nerve agents do not hydrolyze equally and in some cases one specific set of conditions may lead to the formation of equally toxic by-products. One of the most common, practical, and effective methods for the destruction of nerve agents is the use of bleach (hypochlorite). This oxidative method accomplishes the destruction of most agents in a matter of minutes at ambient temperature [22]. Figure 2 below shows some of the most common pathways for the degradation of the three classes of nerve agents discussed in this review and these involve the normal hydrolysis of these chemicals (i.e., no oxidation). Thus, for the G-series, initial reaction of water with these series leads to the formation of the phosphonic acid monoester via P-F bond scission. A second round of hydrolysis, occurring at a much a faster rate than the first one, involves the hydrolysis of the P-O bond to yield ultimately methylphosphonic acid (MPA). A similar stepwise process can be invoked for the V-series agents, with the first hydrolytic step involving the P-S bond to again yield a phosphonic acid monoester that the undergoes another hydrolytic step to yield MPA. Now, although no published studies involving the hydrolytic pathways for the Novichok agents exists, one can predict that like the G-series, and following the nature of the generated conjugate bases, these agents will undergo scission at the P-F bond initially to generate an amidine/guanidine methylphosphonic acid intermediate that can undergo another round of hydrolysis to ultimately yield MPA.

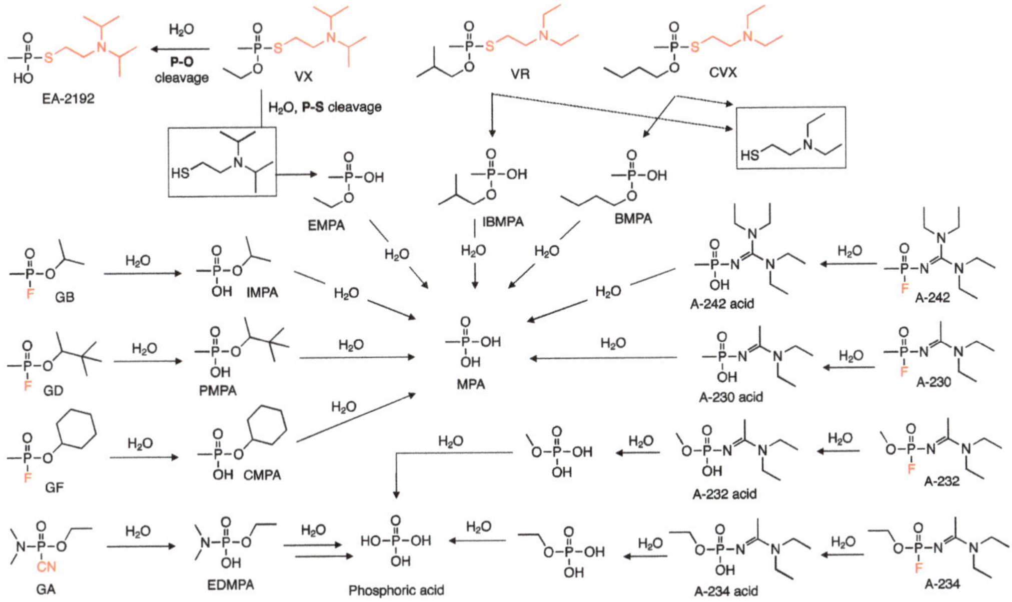

Figure 2. Hydrolytic degradation pathways for OPNAs and intermediary species in the processes. The more reactive leaving group in each agent is denoted in red.

OPNAs possess of range of values when it comes to their persistence in the environment, whether it involves their residence in various soil matrices or their stance in aqueous solutions including those of biological origin. Throughout the years and after the realization of their potential use in mass destruction, several methods have made their appearance into the light showcasing their abilities to efficiently destroy these toxic substances. Two of the original, main approaches that have found numerous uses are their base-mediated and oxidative degradation. As the Novichok agents are relatively new in the open literature, their environmental or biological degradation pathways are unknown, however the degradation pathway sequence outlined in Figure 2 for these OPNAs is very likely to govern the way they will degrade. As a caveat to these degradation methods, one must consider the fact that a method efficient for a class of NA may not necessarily mean that it will be equally efficient on another kind. During the base-mediated degradation protocols, the use of highly basic aqueous solutions (pH > 11) is employed to degrade the agent, a method that relies on the nucleophilicity of the $^-$OH ion and its attack on the phosphorus center of these chemicals [23]. As expected, G-series agents degrade very rapidly under these conditions to yield the alkylmethylphosphonic acids that thrive under the basic conditions and slowly and eventually degrade to methylphosphonic acid (MPA). With regard to the V-series, the same rapid outcome is observed in highly basic solutions but the regioselectivity for the hydrolysis is not uniformly and solely on the P-S linkage but can also affect the P-O linkage (Figure 2). For example, VX undergoes basic hydrolysis to yield the non-toxic product ethylmethylphosphonic acid (EMPA) and EA-2192 which is equally toxic as VX [24–26]. The second method involves the use of bleach as an oxidative source for their degradation and it is the go-to method for their destruction in environmental and laboratory settings. A drawback of the two methods described above is their corrosive nature that needs to be considered when performing the decontamination of very expensive, critical, and specialized army equipment within the military setting. To this end, alternative methods that can accomplish the destruction of OPNAs while minimizing their hydrolytic and oxidative reaction towards the environment have been foci of intense research efforts. Some of these methods include the use of metal-based catalysts that can operate in catalytic fashion at low basic pH ranges [27–34], hydrogen peroxide-based solutions that are not as corrosive as bleach [35–37], and metal oxide-based approaches (e.g., FeO, Fe_2O_3, Al_2O_3, MgO) [38–41] that also degrade the OPNA via oxidative degradation pathways. Another area of research that has yielded unique methods for the degradation of OPNAs is in the materials sciences, efforts that have yielded technologies such as metal organic frameworks (MOFs) [20,42,43] and the aforementioned second skin technologies [20,21] that can find significant application in the field by providing a strong protective layer to the warfighter.

3. GC-MS as an Important Technique in the Analysis of Nerve Agent Degradation Products

OPNAs in their intact form can be analyzed and detected by GC-MS, with their mass spectra easily corroborated by conducting mass spectral searches against databases such as NIST20, from the National Institute of Standards and Technology (NIST) or OCAD, the OPCW Central Analytical Database from the Organisation for the Prohibition of Chemical Weapons (OPCW) [44–46]. However, due to the high reactivity of these chemicals in biological as well as environmental matrices, there is a great chance that analytical chemists will encounter the degradation products rather than the agent itself. Now, detection of these degradation products can be highly useful from a forensic standpoint in retrospectively determining the nature of the original OPNA. Unfortunately, analysis of OPNA degradation products come with its own difficulties as some of these, particularly the low molecular weight phosphonic acids, are difficult to detect by GC-MS in their native state. For this reason, derivatization plays an important role in the analysis of OPNA degradation products as it provides modified analytes with better chromatographic characteristics for GC-MS analysis [47–49]. In the sections below, we will discuss the main derivatization strategies that have been developed and employed over the years in the analysis of these degradation products.

Interest in the field of derivatizations during the analysis of not only OPNA degradation products but also those arising from chemical warfare agent (CWA) in general is always strengthened yearly during proficiency tests (PTs) administered by the Organisation for the Prohibition of Chemical Weapons (OPCW). OPCW is an organization based in the Hague, Netherlands and their primary mission is to monitor, control and eliminate chemical weapons. To this end, OCPW has assigned various laboratories around the globe to serve as hubs where the analysis for these chemicals can take place in the case an event involving these toxic chemicals arises. As our laboratory in the Forensic Science Center (FSC) at the Lawrence Livermore National Laboratory (LLNL) is part of this worldwide consortium of laboratories, we participate in yearly PTs to ensure certification by OCPW. From an OPCW proficiency test (PT) standpoint, a participating laboratory needs to identify a reportable analyte by a minimum of two analytical techniques with underivatized and derivatized versions of a compound being acceptable reporting criteria [50–54]. Multiple derivatization methods are useful because having more than one derivatized version of a given analyte is very important in providing strong, unequivocal evidence of its presence in a given matrix.

4. Silylation Methods

Silylation is one of the most widely employed derivatization techniques in the field of GC-MS analysis [55–57]. The installation of the silyl group or tag onto polar analytes convert it into derivatives that are suitable for the technique and in some cases these derivatives possess enhanced detection relative to the underivatized analyte. Two of the most common silylating reagents are N,O-bis(trimethylsilyl)trifluoroacetamide (BSTFA) and N-tert-butyldimethylsilyl-N-methyltrifluoroacetamide (MTBSTFA) but other approaches to install the silyl moiety have found wide applications specially within the scope of OPCW [58–62] (Figure 3). Reaction with BSTFA results in the installation of the trimethylsilyl (TMS) group while reaction with MTBSTFA results in the installation of the tert-butyldimethylsilyl group (TBDMS). In terms of stability, the TBDMS group offers derivatives with higher stability than the TMS group towards acid and base hydrolysis [63,64].

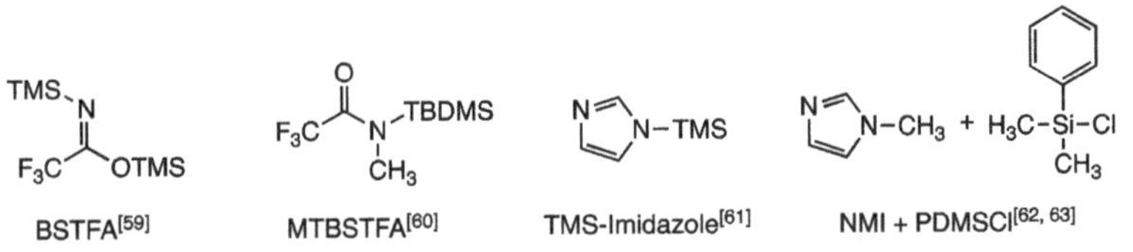

Figure 3. Commonly used silylation reagents and reagent combinations. For MTBSTFA, the reagent is provided with trimethylsilyl chloride (TMCS, 1%) as a catalytic additive. Abbreviations used: TMS = trimethylsilyl-; TBS = tert-butyldimethylsilyl-; NMI = N-methylimidazole; PDMSCl = phenyldimethylsilyl-.

A report appearing in 1989 by Purdon et al. from the Department of National Defence in Ottawa, Canada describes their studies on various CWA-related methylphosphonic acid silylations using the tert-butyldimethylsilyl group [65]. To this end, the authors compare the silylation efficiency of these acids by three different reagent/reagent combinations namely MTBSTFA, MTBSTFA (with 1% TBDMSCl) and the derivatization kit containing TBDMSCl and imidazole (1:2.5) in DMF. The seven acids included in this study were MPA, ethylphosphonic acid (EPA), n-propylphosphonic acid (nPPA), n-butylphosphonic acid (nBPA), EMPA, IMPA and PMPA. The researchers found that all seven acids were quantitatively converted to their respective silyl esters by MTBSTFA and MTBSTFA (with 1% TBDMSCl) at 60 °C and after only 30 min. Conversely, silylation using the derivatization kit (TBDMSCl and imidazole) did not proceed to completion resulting in lower yields of the product available for GC-MS analysis even under more aggressive conditions. The authors found that phosphonic acid quantitation as these derivatives can be carried out down to 500 pg when using the GC coupled to flame-photometric detection (FPD), while when using Electron Impact Gas Chromatography Mass Spectrometry (EI-GC-MS), quantitation values can be lowered to 300–500 pg in full scan mode and even lower, down to 30–60 pg, when using the SIM mode.

In 2007, silylation was also employed for the derivatization of phosphonic acids involved the use of MTBSTFA (with 1% TBDMSCl) to obtain derivatives of various phosphonic acids that were subsequently analyzed by gas chromatography with inductively coupled plasma mass spectrometry (GC-ICPMS) [66]. In this work, Richardson and Caruso describe the analysis of *tert*-butyldimethylsilyl derivatives of seven phosphonic acids that included EMPA (VX acid), IMPA (GB acid), EDPA (GA acid), IBMPA (VR acid), PMPA (GD acid), CMPA (GF acid) and the ultimate end product MPA. ICPMS detection was chosen for this application for its high sensitivity when operated in the ^{31}P channel that led to not only to optimal separation and speciation of all TBDMS-modified phosphonic acids within the short timeframe for the GC run (t = 10 min). Further confirmation of each species was accomplished using time-of-flight (TOF) mass spectrometry. The authors found that the most optimal conditions for the derivatization involved the heating of the MTBSTFA (with 1% TBDMSCl) reagent at 80 °C for 45 min. Detection limits determined for this method were found to be <5 pg and successfully applied to the analysis of all seven phosphonic acids when spiked in an aqueous (Miami river) and a soil matrix.

A paper describing an approach combining solid phase extraction followed by solid phase derivatization of various phosphonic acid markers for nerve agents was introduced in 2009 by the Ostin group at the Swedish Defence Research Agency (FOI) [67]. The report described the use of in vial solid phase derivatization (SPD) of nine phosphonic acids using BSTFA (with 1% trimethylsilyl chloride) followed by analysis by GC-MS. The method was found to be very sensitive for the TMS-derivatized products displaying recovery values between 83–101% and a limit of detection (LOD), under the single-ion monitoring (SIM) mode, down to 0.14 ppb. The approach was successfully tested for its ability to unambiguously identify MPA, PMPA, 1-methylpentyl methylphosphonic acid, 4-methylpentyl methylphosphonic acid and IMPA present in aqueous samples administered during the 19th OPCW PT. The phosphonic acids' concentrations were estimated to range between 5–8 ppm in both aqueous samples showcasing the method's robustness in correctly identifying these in the sample.

In 2012, a report surfaced from the Swedish Defence Research Agency (FOI) and the Department of Chemistry at Umeå University describing the use of BSTFA (with 1% TMCS) to silylate the toxic degradation products of VX namely S-2-(*N, N*-diisopropylaminoethyl) methylphosphonothiolate (EA-2192) and VR namely S-2-(*N, N*-diethylaminoethyl) methylphosphonothiolate (REA) [68]. VX, and by analogy VR, are known to undergo hydrolysis under basic conditions via two pathways with a major one leading to the formation of non-toxic EMPA and a minor one leading to the formation of the equally toxic EA-2192 [22]. EA-2192 is a particularly difficult analyte to derivatize using derivatization agents [69–71] and in this work the authors developed a protocol that involved the initial aqueous extraction of the analyte using a strong anion exchange disk to remove interfering salts. The silylation was carried out after evaporation to dryness and treatment of the residue with BSTFA (with 1% TMCS) in acetonitrile for GC-MS analysis. The LOD values determined for the protocol were 10 ng·mL^{-1} and 100 ng·mL^{-1} for SIMS and full scan mode, respectively. Furthermore, the protocol was employed in the detection of both degradation products in a water sample (W1) from an OPCW PT (19th) and also in spiked water samples (river water and Baltic Bay water).

A recent paper dealing with the derivatization of phosphonic acids using BSTFA was published by Kim et al. from a collaborative effort by the Center of the Cell-Encapsulation Research and the Agency for Defense Development in South Korea [72]. The silylation of the phosphonic acids was done before their efficient extraction using headspace solid-phase microextraction (HS-SPME). The phosphonic acids studied in this work were EMPA, IMPA, CMPA and PMPA and the authors found that the ideal analyte adsorption temperature for their chosen PDMS/DVB SPME fiber was 75 °C. Interestingly, when optimizing the derivatization step, the authors found that increasing the volume of BSTFA resulted in less derivatized product for detection by GC analysis. The method's limit of detection (LOD) was determined to be between 10–20 pg·mL^{-1} depending on the phosphonic acid under study. Furthermore, the authors go on to show the efficacy of the method in the analysis of PMPA in a polyethyleneglycol-rich matrix sample administered during the 35th OPCW PT.

PMPA, present at a 5 µg·mL^{-1} concentration, was detected and unambiguously identified (with the use of a PMPA-TMS standard) in the OPCW sample (sample 351).

5. Methylation Methods

Another method of derivatization used in the field of GC-MS, and not limited to the analysis of phosphonic acids, is methylation. As many alkyl groups can be added to an analyte to enhance its detectability by GC-MS, there is one that has found the widest applicability and that is the methyl group [73]. Currently, the reagent used to accomplish the installation of the methyl group on to molecules is diazomethane (DM) [74]. Diazomethane possesses two favorable characteristics that make it a great derivatizing agent, its high reactivity and the fact that its reaction with analytes does not produce any interfering side products. However, its use comes with a concern linked to the explosive hazard associated with its preparation. As DM's needs to be freshly prepared before its use in sample derivatizations, its explosive hazard becomes a significant concern, although the current methods for its preparation have become fairly safe to conduct [75–77]. The associated explosive hazards with DM's preparation and the need for its constant preparation, have compelled analytical chemists to seek new derivatization agents that can carry out the same transformation (i.e., methylation). To this end, reagents such as trimethylsilyldiazomethane (TMS-DM) [78] and trimethyloxonium tetrafluoroborate (TMO) [79] have found general applicability in GC-MS analyses.

In the field of OPNA degradation product analysis, derivatization in the form of methylation has been widely a useful technique in the GC-MS analysis of these compounds. In 2002, a report by Driskell et al. from the Centers of Disease Control (CDC) in Atlanta described the use of DM for the methylation of five phosphonic acids originating from the degradation of Sarin, Soman, Tabun, Cyclosarin and VX [80]. The authors make use of isotope-dilution GC-MS-MS to analyze all these acids as their methyl esters and report a LOD value of <4 µg·mL^{-1}, with the exception of Tabun acid for which the LOD was determined to be <20 µg·mL^{-1} in urine.

In 2004, another report from the CDC described the use of the same isotope-dilution GC-MS-MS approach to analyze the nerve agent metabolites from Sarin, Soman, Cyclosarin, VX and Russian VX in urine [81]. The phosphonic acid hydrolysis products arising from these agents were all derivatized using DM. The authors report that the method's LOD for all phosphonic acid was <1 µg·L^{-1}, an improvement from their report two years earlier. An interesting methylation approach was introduced in 2011 by scientists at the Organisation for the Prohibition of Chemical Weapons (OCPW) in the Netherlands [82]. In this report, the authors make use of thermally assisted methylation (TAM), using an injector port temperature of 250 °C, followed by silylation using BSTFA to assist in the detection of a panel of phosphonic acids products related to the nerve agents that included EMPA, PMPA, MPA and EPA in addition to benzilic acid which is a marker for the incapacitating agent 3-quinuclidinyl benzilate (BZ) [83,84]. The work compares the performance of TAM using various methylation agents such as trimethylphenylammonium hydroxide (TMPAH) and trimethylsulfonium hydroxide (TMSH). The authors found that methylation using TAM with these reagents provided chromatograms significantly noisier than when using only DM. In addition, the report describes the approach of using TAM in conjunction with silylation (using BSTFA) in order to obtain as much information as possible on a given sample as the use of TMS derivatives is the principal method used for on-site analysis. The method's LOD for TMPAH was found to be <0.5 ng per injection.

Since 2016, our group at the Forensic Science Center (FSC) at the Lawrence Livermore National Laboratory (LLNL) has reported the use of the salt TMO in the derivatization of phosphonic acids related to OPNAs [85–87] (Figure 4). Our interest in this salt originated from our knowledge that alkyloxonium salts, out of which Meerwein's reagent (triethyloxonium tetrafluoroborate) is the most famous, have found a plethora of applications in the field of total synthesis due to their reactivity and thus their ability to alkylate hindered hydroxyl [88–90] and carboxylic acid groups [91–93]. Some key characteristics of the reagent sparked our interest in testing as potential option to DM during analyte methylation reactions for GC-MS analyses. Some of these unique and important characteristics include its salt form, its non-explosive nature, and the generation of minimal interferences in the GC-MS analysis after its use. Thus, in our 2016 report we described the efficient methylation of EMPA, CMPA, and PMPA when present in low concentrations (10 µg·mL^{-1}) in DCM. Furthermore, the method was applied to their methylation when these were spiked (at a 10 µg·mL^{-1} concentration each) in a fatty acid ester-rich organic matrix that was featured during the 38th OCPW PT. An interesting observation in this report was the fact that the marked insolubility of the salt in DCM did not have a negative effect on the overall derivatization. In addition to these three representative phosphonic acids, the methylation of the sulfonic acids deriving from the oxidative degradation of VX and VR, N,N-diethylamino ethanesulfonic acid (VR-SA) and N,N-diisopropylamino ethanesulfonic acid (VX-SA), was also demonstrated.

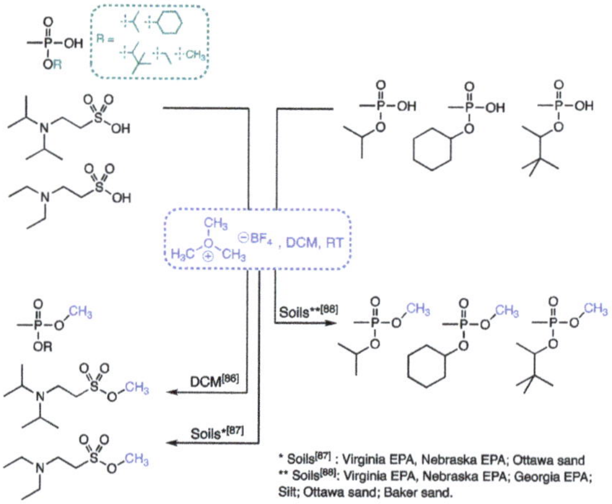

Figure 4. Derivatization approach using trimethyloxonium tetrafluoroborate (TMO) to methylate phosphonic and sulfonic acids related to OPNAs developed in our laboratory.

6. Additional, Alternate Derivatization Methods

Other derivatization approaches for the analysis of OPNA degradation products have been introduced throughout the years and continue to be used in the field in parallel to the more established derivatization protocols described above. In 1991, a report from the US Army Medical Research Institute of Chemical Defense (USAMRICD) described the use of pentafluorobenzyl bromide (PFBBr) to derivatize IMPA, PMPA and CMPA [94]. Analysis and detection of the derivatized phosphonic acids was accomplished using electron ionization as well as chemical ionization GC-MS. The three phosphonic acids were detected in urine and blood samples when spiked at a concentration range of 10–200 ng·mL^{-1}. Interestingly, detection of the PA products was fairly simple in the urine samples, but proper detection of these in the blood sample was only accomplished after the plasma was isolated and analyzed.

A report appearing in 1995 by Fredriksson et al. from the National Defence Research Establishment (NDRE) in Sweden described the use of pentafluorobenzyl bromide to derivatize alkyl methylphosphonic acids related to nerve agents [95]. The panel of acids used in the study included EMPA, IMPA and PMPA. The alkylation of the acids spiked in serum, blood, aqueous matrices, and soil was carried out in acetonitrile and after cleaning of the sample by prewashing through a Bond Elut SAX cartridge. Detection and quantification of the acids was done by using gas chromatography negative ion chemical ionization GC/NIMI-MS and GC/NIMI-MS-MS. The authors report detection levels down to the femtogram levels.

In 1995, the Forensic Science Laboratory from the Osaka Prefectural Police Headquarters in Japan published their studies on the derivatization of various phosphate species that included some related to nerve agents using pentafluorobenzyl bromide (PFBBr) in the presence of various polymeric phase transfer catalysts (PTCs) [96]. The phosphonic acid panel included dimethyl phosphate (DMP), diethyl phosphate (DEP), dimethyl thiophosphate (DMTP), diethyl thiophosphate (DETP), dimethyl dithiophosphate (DMDTP) and diethyl dithiophosphate (DEDTP). A total of five PTCs were evaluated and tri-n-butylmethylphosphonium bromide was found to be most efficient catalyst for the pentafluorobenzylation. The actual derivatization takes place in a three-phase reaction mixture where the phosphonic acids present in an aqueous environment are partitioned between a buffer (phosphate, pH 6.5) and toluene (containing the PFBBr) and treated with the polymer-bound PTC. Therefore, as the phosphonic acid derivatives are generated, they get extracted into the organic phase. The authors complete their work by applying the protocol to the derivatization and detection of these acids in river water and urine matrices.

In 1997, a landmark report describing the direct detection of sarin from blood sample from victims from the Tokyo attack was published by the Netherlands Organisation for Applied Scientific Research (TNO, Prins Maurits Laboratory) [97]. In this work Polhuijs et al. use the fluoride regeneration protocol on the victim's blood samples to produce sarin that was then identified by high resolution SIM GC-MS and GC-MS coupled to an alkali flame ionization detector. The amount of regenerated sarin (from butyrylcholinesterase, BChE) was <4.1 ng·mL^{-1} of plasma, a lower value when compared to the baseline detection of this agent in previous studies involving BChE of ~13 ng·mL^{-1}. A reason for the low concentrations could be the aging of the samples that were not analyzed by TNO until 10 months after the attack.

In 1999, a report by Miki et al. [98] from the Forensic Science Laboratory at the Osaka Prefectural Police Headquarters in Japan appeared using their tri-*n*-butylmethylphosphonium bromide PTC discovered a few years back in their laboratory [96]. In this work, the alkylmethylphosphonic acids were pentafluorobenzylated using their PTC system, a reaction that occurs as discussed above in a triphasic reaction mixture (liquid-liquid-solid). The three phosphonic acids were EMPA, IMPA and PMPA and analysis as their pentafluorobenzylated esters were accomplished by negative ion chemical ionization (NICI/GC-MS). Notably, the limit of detection reported for this protocol was 60 pg·mL^{-1}.

In 2005, a report from the Defence Science and Technology Laboratory (DSTL) in the United Kingdom described the use of a benchtop ion trap GC-MS for the detection of phosphonic acids related to nerve agents as their pentafluorobenzyl esters in spiked urine samples [99]. The panel of phosphonic acids studied included IMPA, IBMPA, PMPA, CMPA and EMPA and after their extraction (SPE) from the urine matrix, they were alkylated using pentafluorobenzyl bromide. The GCMS was operated in the negative ion mode (NICI/GC-MS) thus allowing for a highly sensitive method for the detection of the derivatized acids. The protocol's LOD were determined to be 0.1 ng·mL^{-1} for all acids except for EMPA which was found to be 0.5 ng·mL^{-1}, this was attributed to the poor recovery of the acid during the method's extraction section.

In 2010, a report from FOI explored the use of novel aryldiazomethane-based derivatizing agents to cleverly derivatize phosphonic acids related to OPNAs in a water matrix [100]. The derivatized acids were detected using NICI/GC-MS and further identified by EI-GC-MS. The method's LOD using NICI using SIM was determined to be ~5–10 ng·mL^{-1} in the aqueous sample while for identification using full scan EI was 100 ng·mL^{-1}. The motivation behind the design of these aryldiazomethane agents stems from the idea of producing a reagent that is equally reactive as diazomethane in alkylating the phosphonic acid and in the process installing a fluorinated tag that could be used for enhanced GC-MS detection. In contrast to PFBBr that can be installed in the presence of a base and aided by heating, these aryldiazomethane agents would be so reactive that only mild heating would be needed and analogously to diazomethane would produce very little by-products in the process. Out of the four studied reagents, 1-(diazomethyl)-3,5-bis(trifluoromethyl)benzene was found to be most robust and consistently performed well during their reaction optimization phase and when it was tested for its efficacy in an OPCW PT (19th) aqueous sample (W1) spiked with PMPA, 1-methylpentyl-MPA, 4-methylpentyl-MPA and MPA.

In 2013, Nyholm et al., at FOI disclosed another remarkable report this time involving the use of 3-pyridyldiazomethane to yield picolinyl ester derivatives of a panel of phosphonic acids [101]. Some of the phosphonic acids used in the evaluation of the protocol were direct hydrolysis products of G- and V-series nerve agents and included EMPA (VX), IMPA (GB), BMPA (CVX), IBMPA (VR), PMPA (GD) among others. The reaction was found to be efficient at derivatizing all the phosphonic acids when conducted for 90 min at 60 °C. The detection of the derivatized phosphonic acids was accomplished using EI-GC-MS as well as CI-GC-MS (in both the positive and negative ion modes). After the optimization of the protocol was done, its performance of the protocol was applied to the successful derivatization of MPA (at 5 µg·mL^{-1}) and PMPA (at 6 µg·mL^{-1}) in a water sample administered during the 19th OCPW PT.

In 2013, another report from FOI described the use of 1-(diazomethyl)-3,5-bis (trifluoromethyl)benzene to derivatize a set of six phosphonic acids that included EMPA, IMPA, PMPA, CMPA and MPA [102]. The derivatives were analyzed by GC–MS and NICI/GC-MS-MS. The selectivity and sensitivity of analyses performed by low- and high-resolution single ion monitoring MS-mode were compared with those performed by multiple reaction monitoring MS-MS-mode. The authors found that the MS-MS technique offered the greatest sensitivity and selectivity, with LODs ranging from 0.5–1 ng·mL^{-1} of urine. Lastly, the method's robustness was evaluated using urine samples from the 2nd OPCW biomedical confidence building exercise and all phosphonic acids present in the samples were conclusively identified.

In 2014, a report by Lin et al. from the Academy of Military Sciences in China, demonstrated the established benefit of using pentafluorobenzylation as a means for derivatizing phosphonic acids related to CWAs [103]. In their report, the authors developed and validated a method that detects four phosphonic acids (EMPA, IMPA, IBMPA and PMPA) as their pentafluorobenzyl esters using isotope-dilution GC-MS. The acids were quantified using NICI/GC-MS and found to have a LOD of 0.02 ng·mL^{-1} when dealing with only a 0.2 mL urine sample volume. Table 1 compiles all the derivatization approaches described above and has been assembled as a quick reference and notes on such work for the reader's convenience. Careful attention has been given to provide brief description of each GC-MS method employed as well as the nature of the chemical derivatization used in each work.

Table 1. Summary of derivatization methods for phosphonic acids related to OPNAs discussed in this review.

Derivatizing Agent	Analytes	Matrix	GC-MS Method	Method's Performance	Ref.
MTBSTFA, MTBSTFA (with 1% TBDMSCl) and TBDMSCl:imidazole (1:2.5) in DMF	MPA, ethylphosphonic acid (EPA), n-propylphosphonic acid (nPPA), n-butylphosphonic (nBPA), EMPA, IMPA and PMPA	Aqueous sample	GC-MS/ICPMS	Quantitation down to 500 pg by GC-FPD. Quantitation for EI-GC-MS (300–500 pg (full scan mode) and 30–60 pg (SIM)).	[65]
MTBSTFA (with 1% TBDMSCl)	EMPA, IMPA, EDPA, IBMPA, PMPA, CMPA and MPA	Aqueous (river water) and soil	GC-MS/ICPMS	GC run total run time = 10 min. LOD < 5 pg LOD~0.14 ppb (SIM).	[66]
BSTFA	Nine phosphonic acids that included OPNA-related acids MPA, IMPA and PMPA.	Aqueous (OPCW PT)	GC-MS-SIM	Detection of IMPA, MPA, PMPA in aqueous OPCW PT sample (19th) present in the range 5–8 µg·mL^{-1}.	[67]
BSTFA (with 1% TBDMSCl)	EA-2192, REA	Organic (ACN)	GC-MS-SIM	LOD = 10 ng·mL^{-1} (SIM); 100 ng·mL^{-1} (full scan mode). LOD = 10–20 pg·mL^{-1}.	[68]
BSTFA	EMPA, IMPA, CMPA and PMPA	Organic (DCM)	EI-GC-MS	Detection of PMPA in glycerol-rich matrix OPCW PT sample (35th) present at 5 µg·mL^{-1}.	[69]
DM	Sarin, Soman, Tabun, Cyclosarin and VX	Urine	GC-MS-MS (Isotope dilution).	LOD < 4 µg·mL^{-1}, with the exception of Tabun acid for which the LOD was determined to be < 20 µg·mL^{-1} in urine.	[80]
DM	IMPA, PMPA, CMPA, EMPA, IBMPA	Urine	GC-MS-MS (Isotope dilution).	LOD < 1 µg·L^{-1}.	[81]
DM and TMPAH (TAM) followed by BSTFA	EMPA, PMPA, MPA, EPA and BA	Various organic solvents	EI-GC-MS and GC-FPD	Silylation followed TAM to obtain yet another set of derivatives. LOD (TMAPH) < 0.5 ng.	[82]
TMO·BF$_4$	EMPA, CMPA and PMPA	Organic (DCM).	EI-GC-MS	All three acids detected when spiked in fatty acid-rich matrix from OPCW PT (38th).	[85]
TMO·BF$_4$	EMPA, CMPA, PMPA, VX-SA and VR-SA	Soils (Ottawa sand, Nebraska, and Virginia Type A).	EI-GC-MS	All five acids detected when spiked in all three soils at a 10 µg·g^{-1} concentration.	[86]

Table 1. Cont.

Derivatizing Agent	Analytes	Matrix	GC-MS Method	Method's Performance	Ref.
TMO·BF$_4$	EMPA, CMPA and PMPA	Soils (Ottawa and Baker sands, Nebraska, Virginia Type A, and Georgia soils and silt).	EI-GC-MS	All three acids detected when spiked in all six soils at a 10 µg·g^{-1} concentration. PMPA detected in soil matrix from OPCW PT (44th) present at a 5 µg·g^{-1} concentration.	[87]
PFBBr	EMPA, IMPA, BMPA, PMPA and MPA	Serum, urine, aqueous and soil.	NICI/GC-MS and NICI/GC-MS-MS	Low femtograms levels detected.	[95]
PFBBr w/PTC	DMP, DEP, DMTP, DETP, DMDTP and DEDTP	River water and urine	GC-MS coupled to Flame Ionization and Electron Capture detection	Five PTCs were evaluated 45 °C for thiophosphates 90 °C for non-thiophosphate	[96]
Fluoride Regeneration	Sarin from BChE	Blood from Japan Subway attack victims	SIM-GC-MS and GC-MS	Sarin concentration detected < 4.1 ng·mL^{-1} of plasma	[97]
PFBBr w/PTC	EMPA, IMPA and PMPA	Urine, serum, and saliva	EI-GC-MS and NICI/GC-MS	Detection limits of 50 ng·mL^{-1} (full scan mode), 2.5–10 ng·mL^{-1} (SIM mode), and 60 pg·mL^{-1} for NICI/GC-MS mode.	[98]
PFBBr	IMPA, IBMPA, PMPA, CMPA and EMPA	Urine sample	NICI/GC-MS	SPE on urine sample, followed by GC-MS analysis, LOD = 0.1 ng·mL^{-1} for all acids except for EMPA (0.5 ng·mL^{-1}).	[99]
1-(diazomethyl)-3,5-bis(trifluoromethyl)benzene	PMPA, 1-methylpentyl-MPA, 4-methylpentyl-MPA and MPA	Spiked Aqueous OPCW PT (19th) sample.	NICI/GC-MS; EI-GC-MS	LOD = 5–10 ng·mL^{-1} using NICI-SIM; 100 ng·mL^{-1} using (full scan mode.	[100]
3-Pyridyldiazomethane	EMPA, IMPA, BMPA, IBMPA, PMPA and others	Aqueous OPCW PT (19th) sample.	EI-GC-MS; CI-GC-MS (+/− mode).	MPA (at 5 µg·mL^{-1}) and PMPA (at 6 µg·mL^{-1}) in OPCW sample.	[101]
1-(diazomethyl)-3,5-bis(trifluoromethyl)benzene	EMPA, IMPA, CMPA, PMPA and MPA	Urine (OPCW Biomedical Sample)	NICI/GC-MS-MS	LOD = 0.5–1 ng·mL^{-1}.	[102]
PFBBr	EMPA, IMPA, IBMPA and PMPA	Urine	NICI/GC-MS	LOD = 0.02 ng·mL^{-1} (using 0.2 mL urine)	[103]

7. Outlook and Concluding Remarks

Within the field of GC-MS derivatization reactions continue to appear that provide invaluable assistance to the analytical chemist in the analysis of phosphonic acids and polar compounds related to OPNAs. Established methods such as silylation employing BSTFA and MTBSTFA and methylation using diazomethane still continue to serve as the preferred ways of modifying polar compounds and transforming them into analytes suitable for GC-MS analysis. However, as advances in instrumentation methods for detection improve, so have the chemical methods that can be used to modify analytes in not only a more efficient manner but in a way to enhance their ease of detection particularly when present at low levels (<1 ppm). The appearance of new derivatizing agents as well as the application of reagents commonly used during total synthesis maneuvers, have become a much welcomed event specifically in instances where the presence of a given analyte in a sample must be corroborated independently by several analytical techniques. With regard to this last point, this path of discovery has had a major impact during the OPCW Proficiency Tests where a low-level analyte related to CWAs needs to be not only identified but its existence in a sample vetted by multiple techniques. It is important for not only laboratories participating in these year-round PTs to continue searching for efficient derivatization reactions but also for more efficient methodologies and sample preparation protocols that can aid in reducing the time needed for the unequivocal identification of these important CWA markers.

Author Contributions: C.A.V. and R.N.L. conceptualized the topic for the manuscript and were both involved in the literature compilation for its preparation. Both authors prepared, edited and reviewed the manuscript. All authors have read and agreed to the published version of the manuscript.

Funding: This research received no external funding. This work was performed under the auspices of the U. S. Department of Energy by Lawrence Livermore National Laboratory under Contract DE-AC52-07NA27344. The was funded fully by a Mid-Career Research Grant awarded by the Lawrence Livermore National Laboratory (PLS-21-FS-036) to C.A.V.

Institutional Review Board Statement: Not applicable.

Informed Consent Statement: Not applicable.

Conflicts of Interest: The authors declare no conflict of interest. This document (LLNL-JRNL-823194) was prepared as an account of work sponsored by an agency of the United States government. Neither the United States government nor Lawrence Livermore National Security, LLC, nor any of their employees makes any warranty, expressed or implied, or assumes any legal liability or responsibility for the accuracy, completeness, or usefulness of any information, apparatus, product, or process disclosed, or represents that its use would not infringe privately owned rights. Reference herein to any specific commercial product, process, or service by trade name, trademark, manufacturer, or otherwise does not necessarily constitute or imply its endorsement, recommendation, or favoring by the United States government or Lawrence Livermore National Security, LLC. The views and opinions of authors expressed herein do not necessarily state or reflect those of the United States government or Lawrence Livermore National Security, LLC, and shall not be used for advertising or product endorsement purposes.

Sample Availability: Not applicable.

Abbreviations

AChE	acetylcholinesterase
BSTFA	N,O-bis(trimethylsilyl)trifluoroacetamide
BChE	butyrylcholinesterase
BA	benzylic acid
CDC	Centers for Disease Control and Prevention
CMPA	cyclohexyl methylphosphonic acid (GF acid)
CVX	Chinese VX; O-Butyl-S-[2-(diethylamino)ethyl] methylphosphonothioate
CWA	chemical warfare agent
DETP	diethyl phosphate

DM	diazomethane
DMDTP	diethyl dithiophosphate
DSTL	Defence Science and Technology Laboratory (UK)
EA-2192	S-2-(N, N-diisopropylaminoethyl)methylphosphonothiolate
EMPA	ethyl methylphosphonic acid (VX acid)
EPA	ethylphosphonic acid
FOI	Swedish Defence Research Agency
FPD	flame photometric detection
FSC	Forensic Science Center
GA	Tabun
GB	Sarin
GD	Soman
GF	Cyclosarin
HS-SPME	headspace solid-phase microextraction
ICPMS	inductively coupled plasma mass spectrometry
IMPA	isopropyl methylphosphonic acid (GB acid)
IBMPA	isobutyl methylphosphonic acid (VR acid)
LNLL	Lawrence Livermore National Laboratory
LOD	limit of detection
MPA	methylphosphonic acid
MTBSTFA	N-tert-butyldimethylsilyl-N-methyltrifluoroacetamide
NDRE	National Defence Research Establishment in Sweden
NICI/GC-MS	negative ion chemical ionization gas chromatography
nBPA	n-butylphosphonic acid
nPPA	n-propylphosphonic acid
OPCW	Organisation for the Prohibition of Chemical Weapons
OPNA	organophosphorus-based nerve agents
PFBBr	pentafluorobenzyl bromide
PMPA	pinacolyl methylphosphonic acid (GD acid)
PT	proficiency test (OPCW)
PTC	phase transfer catalysis
REA	S-2-(N, N-diethylaminoethyl)methylphosphonothiolate
SIM	single ion monitoring
TBDMSCl	tert-butyldimethylsilyl chloride
TMCS	trimethylchlorosilane
TMO	trimethyloxonium tetrafluoroborate
TNO	Netherlands Organisation for Applied Scientific Research
TMS-DM	trimethylsilyldiazomethane
USAMRICD	US Army Medical Research Institute of Chemical Defense
VR	Russian VX; O-isobutyl-S-[2-(Diethylamino)ethyl] methylphosphonothioate
VR-SA	N, N-diethylamino ethanesulfonic acid
VX	O-ethyl-S-[2-(Diisopropylamino)ethyl] methylphosphonothioate
VX-SA	N, N-diisopropylamino ethanesulfonic acid

References

1. Ganesan, K.; Raza, S.K.; Vijayaraghavan, R. Chemical warfare agents. *J. Pharm. Bioallied Sci.* **2010**, *2*, 166–178. [CrossRef] [PubMed]
2. Szinicz, L. History of chemical and biological warfare agents. *Toxicology* **2005**, *214*, 167–181. [CrossRef]
3. Friboulet, A.; Rieger, F.; Goudou, D.; Amitai, G.; Taylor, P. Interaction of an organophosphate with a peripheral site on acetylcholinesterase. *Biochemistry* **1990**, *29*, 914–920. [CrossRef] [PubMed]
4. Shih, T.M.; Kan, R.K.; McDonough, J.H. In vivo cholinesterase inhibitory specificity of organophosphorus nerve agents. *Chem. Biol. Interact.* **2005**, *157*, 293–303. [CrossRef]
5. Haines, D.D.; Fox, S.C. Acute and long-term impact of chemical weapons: Lessons from the Iran-Iraq war. *Forensic Sci. Rev.* **2014**, *26*, 97–114. [PubMed]
6. Dolgin, E. Syrian gas attack reinforces need for better anti-sarin drugs. *Nat. Med.* **2013**, *19*, 1194–1195. [CrossRef]
7. Okumura, T.; Hisaoka, T.; Yamada, A.; Naito, T.; Isonuma, H.; Okumura, S.; Miura, K.; Sakurada, M.; Maekawa, H.; Ishimatsu, S.; et al. The Tokyo subway sarin attack-lessons learned. *Toxicol. Appl. Pharmacol.* **2005**, *207*, 471–476. [CrossRef]
8. Kloske, M.; Witkiewicz, Z. Novichoks—The A group of organophosphorus chemical warfare agents. *Chemosphere* **2019**, *221*, 672–682. [CrossRef]

9. Stone, R. U.K. attack puts nerve agent in the spotlight. *Science* **2018**, *359*, 1314–1315. [CrossRef]
10. Howes, L. Novichok compound poisoned Navalny. *Chem. Eng. News* **2020**, *98*, 5. [CrossRef]
11. Höjer-Holmgren, K.; Valdez, C.A.; Magnusson, R.; Vu, A.K.; Lindberg, S.; Williams, A.M.; Alcaraz, A.; Åstot, C.; Hok, S.; Norlin, R. Part 1. Tracing Russian VX to it synthetic routes by multivariate statistics of chemical attribution signatures. *Talanta* **2018**, *186*, 586–596. [CrossRef]
12. Jansson, D.; Wiklund Lindstrom, S.; Norlin, R.; Hok, S.; Valdez, C.A.; Williams, A.M.; Alcaraz, A.; Nilsson, C.; Åstot, C. Part 2: Forensic attribution profiling of Russian VX in food using liquid chromatography-mass spectrometry. *Talanta* **2018**, *186*, 597–606. [CrossRef] [PubMed]
13. Williams, A.M.; Vu, A.K.; Jansson, D.; Mayer, B.P.; Hok, S.; Norlin, R.; Valdez, C.A.; Nilsson, C.; Åstot, C.; Alcaraz, A. Part 3: Solid phase extraction of Russian VX and its chemical attribution signatures in food matrices and their detection by GC-MS and LC-MS. *Talanta* **2018**, *186*, 607–614. [CrossRef] [PubMed]
14. Kalisiak, J.; Ralph, E.C.; Zhang, J.; Cashman, J.R. Amidine-oximes: Reactivators for organophosphate exposure. *J. Med. Chem.* **2011**, *54*, 3319–3330. [CrossRef]
15. Sit, R.K.; Radic, Z.; Gerardi, V.; Zhang, L.; Garcia, E.; Katalinic, M.; Amitai, G.; Kovarik, Z.; Fokin, V.V.; Sharpless, K.B.; et al. New Structural Scaffolds for Centrally Acting Oxime Reactivators of Phosphylated Cholinesterase. *J. Biol. Chem.* **2011**, *286*, 19422–19430. [CrossRef] [PubMed]
16. Kalisiak, J.; Ralph, E.C.; Cashman, J.R. Nonquaternary Reactivators for Organophosphate-Inhibited Cholinesterases. *J. Med. Chem.* **2012**, *55*, 465–474. [CrossRef] [PubMed]
17. Malfatti, M.A.; Enright, H.A.; Be, N.A.; Kuhn, E.A.; Hok, S.; McNerney, M.W.; Lao, V.; Nguyen, T.H.; Lightstone, F.C.; Carpenter, T.S.; et al. The biodistribution and pharmacokinetics of the oxime acetylcholinesterase reactivator RS194B in guinea pigs. *Chem. Biol. Interact.* **2017**, *277*, 159–167. [CrossRef]
18. Bennion, B.J.; Be, N.A.; McNerney, M.W.; Lao, V.; Carlson, E.M.; Valdez, C.A.; Malfatti, M.A.; Enright, H.A.; Nguyen, T.H.; Lightstone, F.C.; et al. Predicting a Drug's Membrane Permeability: A Computational Model Validated With in Vitro Permeability Assay Data. *J. Phys. Chem. B* **2017**, *121*, 5228–5237. [CrossRef]
19. Stone, R. How to defeat a nerve agent. *Science* **2018**, *359*, 23. [CrossRef]
20. Chen, Z.; Ma, K.; Mahle, J.J.; Wang, H.; Syed, Z.H.; Atilgan, A.; Chen, Y.; Xin, J.H.; Islamoglu, T.; Petereson, G.W.; et al. Integration of Metal–Organic Frameworks on Protective Layers for Destruction of Nerve Agents under Relevant Conditions. *J. Am. Chem. Soc.* **2019**, *141*, 20016–20021. [CrossRef]
21. Li, Y.; Chen, C.; Meshot, E.R.; Buchsbaum, S.F.; Herbert, M.; Zhu, R.; Kulikov, O.; McDonald, B.; Bui, N.T.N.; Jue, M.L.; et al. Autonomously Responsive Membranes for Chemical Warfare Protection. *Adv. Mat.* **2020**, *30*, 2000258. [CrossRef]
22. Yang, Y.C.; Baker, J.A.; Ward, J.R. Decontamination of Chemical Warfare Agents. *Chem. Rev.* **1992**, *92*, 1729–1743. [CrossRef]
23. Kim, K.; Tsay, O.G.; Atwood, D.A.; Churchill, D.G. Destruction and Detection of Chemical Warfare Agents. *Chem. Rev.* **2011**, *111*, 5345–5403. [CrossRef]
24. Yang, Y.C. Chemical Detoxification of Nerve Agent VX. *Acc. Chem. Res.* **1999**, *32*, 109–115. [CrossRef]
25. Yang, Y.C.; Szafraniec, L.L.; Beaudry, W.T.; Bunton, C.A. Perhydrolysis of nerve agent VX. *J. Org. Chem.* **1993**, *58*, 6964–6965. [CrossRef]
26. Yang, Y.C.; Berg, F.J.; Szafraniec, L.L.; Beaudry, W.T.; Bunton, C.A.; Kumar, A. Peroxyhydrolysis of nerve agent VX and model compounds and related nucleophilic reactions. *J. Chem. Soc. Perkin Trans. 2* **1997**, 607–613. [CrossRef]
27. Norman, P.R.; Tate, A.; Rich, P. Enhanced hydrolysis of a phosphonate ester by mono-aquo metal cation complexes. *Inorg. Chim. Acta* **1988**, *145*, 211–217. [CrossRef]
28. Hay, R.W.; Govan, N. The [Cu(tmen) (OH) (OH$_2$)]$^+$ promoted hydrolysis of 2,4-dinitrophenyl diethyl phosphate and O-isopropyl methylphosphonofluoridate (Sarin) (tmen = N,N,N',N'-tetramethyl-1,2-diaminoethane). *Polyhedron* **1998**, *17*, 2079–2085. [CrossRef]
29. Hay, R.W.; Govan, N. Kinetic and mechanistic studies of the reaction of a range of bases and metal-hydroxo complexes with the phosphonate ester 2,4-dinitrophenyl ethyl methylphosphonate in aqueous solution. *Transit. Met. Chem.* **1998**, *23*, 133–138. [CrossRef]
30. Lewis, R.E.; Neverov, A.A.; Brown, R.S. Mechanistic studies of La^{3+} and Zn^{2+}-catalyzed methanolysis of O-ethyl O-aryl methylphosphonate esters. An effective solvolytic method for the catalytic destruction of phosphonate CW simulants. *Org. Biomol. Chem.* **2005**, *3*, 4082–4088. [CrossRef]
31. Melnychuk, S.A.; Neverov, A.A.; Brown, R.S. Catalytic Decomposition of Simulants for Chemical Warfare V Agents: Highly Efficient Catalysis of the Methanolysis of Phosphonothioate Esters. *Angew. Chem. Int. Ed.* **2006**, *45*, 1767–1770. [CrossRef]
32. Kuo, L.Y.; Adint, T.T.; Akagi, A.E.; Zakharov, L. Degradation of a VX Analogue: First Organometallic Reagent to Promote Phosphonothioate Hydrolysis Through Selective P−S Bond Scission. *Organometallics* **2008**, *27*, 2560–2564. [CrossRef]
33. Mayer, B.P.; Valdez, C.A.; Hok, S.; Chinn, S.C.; Hart, B.R. ^{31}P-edited diffusion-ordered ^{1}H-NMR spectroscopy for the spectral isolation and identification of organophosphorus compounds related to chemical weapons agents and their degradation products. *Anal. Chem.* **2012**, *84*, 10478–10484. [CrossRef]
34. Kennedy, D.J.; Mayer, B.P.; Baker, S.E.; Valdez, C.A. Kinetics and speciation of paraoxon hydrolysis by zinc(II)–azamacrocyclic catalysts. *Inorg. Chim. Acta* **2015**, *436*, 123–131. [CrossRef]
35. Epstein, J.; Demek, M.M.; Rosenblatt, D.H. Notes—Reaction of Paraoxon with Hydrogen Peroxide in Dilute Aqueous Solution. *J. Org. Chem.* **1956**, *21*, 796–797. [CrossRef]

36. Wagner, G.W.; Yang, Y.C. Rapid Nucleophilic/Oxidative Decontamination of Chemical Warfare Agents. *Ind. Eng. Chem. Res.* **2002**, *41*, 1925–1928. [CrossRef]
37. Hopkins, F.B.; Gravett, M.R.; Self, A.J.; Wang, M.; Hoe-Chee, C.; Sim, N.L.H.; Jones, J.T.A.; Timperley, C.M.; Riches, J.R. Chemical analysis of bleach and hydroxide-based solutions after decontamination of the chemical warfare agent O-ethyl S-2-diisopropylaminoethyl methylphosphonothiolate (VX). *Anal. Bioanal. Chem.* **2014**, *406*, 5111–5119. [CrossRef]
38. Templeton, M.K.; Weinberg, W.H. Adsorption and Decomposition of Dimethyl Methylphosphonate on Metal Oxides. *J. Am. Chem. Soc.* **1985**, *107*, 97–108. [CrossRef]
39. Henderson, M.A.; Jin, T.; White, J.M. A TPD/AES study of the interaction of dimethyl methylphosphonate with iron oxide (.alpha.-Fe$_2$O$_3$) and silicon dioxide. *J. Phys. Chem.* **1986**, *90*, 4607–4611. [CrossRef]
40. Li, Y.X.; Klabunde, K.J. Nano-scale metal oxide particles as chemical reagents. Destructive adsorption of a chemical agent simulant, dimethyl methylphosphonate, on heat-treated magnesium oxide. *Langmuir* **1991**, *7*, 1388–1393. [CrossRef]
41. Li, Y.X.; Schlup, J.R.; Klabunde, K.J. Fourier transform infrared photoacoustic spectroscopy study of the adsorption of organophosphorus compounds on heat-treated magnesium oxide. *Langmuir* **1991**, *7*, 1394–1399. [CrossRef]
42. Islamoglu, T.; Ortuno, M.A.; Proussaloglou, E.; Howarth, A.J.; Vermeulen, N.A.; Atilgan, A.; Asiri, A.; Cramer, C.J.; Farha, O.K. Presence versus proximity: The role of pendant amines in the catalytic hydrolysis of a nerve agent simulant. *Angew. Chem. Int. Ed.* **2018**, *57*, 1949–1953. [CrossRef] [PubMed]
43. Park, H.J.; Jang, J.K.; Kim, S.-Y.; Ha, J.-W.; Moon, D.; Kang, I.-N.; Bae, Y.-S.; Kim, S.; Hwang, D.-H. Synthesis of a Zr-based metal-organic framework with spirobifluorenetetrabenzoic acid for the effective removal of nerve agent simulants. *Inorg. Chem.* **2017**, *56*, 12098–12101. [CrossRef]
44. Stein, S. Mas Spectral Reference Libraries: An Ever-Expanding Resource for Chemical Identification. *Anal. Chem.* **2012**, *84*, 7274–7282. [CrossRef] [PubMed]
45. Mallard, G.W. AMDIS in the Chemical Weapons Convention. *Anal. Bioanal. Chem.* **2014**, *406*, 5075–5086. [CrossRef]
46. Valdez, C.A.; Leif, R.N.; Hok, S.; Alcaraz, A. Assessing the reliability of the NIST library during routine GC-MS analyses: Structure and spectral data corroboration for 5,5-diphenyl-1,3-dioxolan-4-one during a recent OPCW proficiency test. *J. Mass Spectrom.* **2018**, *53*, 419–422. [CrossRef]
47. Valdez, C.A.; Leif, R.N.; Hok, S.; Hart, B.R. Analysis of chemical warfare agents by gas chromatography-mass spectrometry: Methods for their direct detection and derivatization approaches for the analysis of their degradation products. *Rev. Anal. Chem.* **2018**, *37*, 1–25. [CrossRef]
48. Witkiewicz, Z.; Neffe, S. Chromatographic analysis of chemical warfare agents and their metabolites in biological samples. *TrAC Trend. Anal. Chem.* **2020**, *130*, 115960. [CrossRef]
49. Witkiewicz, Z.; Neffe, S.; Sliwka, E.; Quagliano, J. Analysis of the Precursors, Simulants and Degradation Products of Chemical Warfare Agents. *Crit. Rev. Anal. Chem.* **2018**, *48*, 337–371. [CrossRef]
50. Blum, M.-M.; Murty, M.R.V.S. Analytical chemistry and the Chemical Weapons Convention. *Anal. Bioanal. Chem.* **2014**, *406*, 5067–5069. [CrossRef]
51. Üzümcü, A. The Chemical Weapons Convention—Disarmament, science and technology. *Anal. Bioanal. Chem.* **2014**, *406*, 5071–5073. [CrossRef] [PubMed]
52. Murty, M.R.V.S.; Prasada Raju, N.; Prabhakar, S.; Vairamani, M. Chemical ionization mass spectral analysis of pinacolyl alcohol and development of derivatization method using p-tolyl isocyanate. *Anal. Methods* **2010**, *2*, 1599–1605. [CrossRef]
53. Konopski, L.; Liu, P.; Wuryani, W.; Sliwakowski, M. OPCW Proficiency Test: A Practical Approach Also for Interlaboratory Test on Detection and Identification of Pesticides in Environmental Matrices. *Sci. World J.* **2014**, *542357*. [CrossRef] [PubMed]
54. Valdez, C.A.; Corzett, T.H.; Leif, R.N.; Fisher, C.L.; Hok, S.; Koester, C.J.; Alcaraz, A. Acylation as a successful derivatization strategy for the analysis of pinacolyl alcohol in a glycerol-rich matrix by GC-MS: Application during an OPCW Proficiency Test. *Anal. Bioanal. Chem.* **2021**, *413*, 3145–3151. [CrossRef] [PubMed]
55. Halket, J.M.; Zaikin, V.G. Review: Derivatization in mass spectrometry—1. Silylation. *Eur. J. Mass Spectrom.* **2003**, *9*, 1–21. [CrossRef]
56. Poole, C.F. Alkylsilyl derivatives for gas chromatography. *J. Chromatogr. A* **2013**, *1296*, 2–14. [CrossRef]
57. Junker, J.; Chong, I.; Kamp, F.; Steiner, H.; Giera, M.; Müller, C.; Bracher, F. Comparison of Strategies for the Determination of Sterol Sulfates via GC-MS Leading to a Novel Deconjugation-Derivatization Protocol. *Molecules* **2019**, *24*, 2353. [CrossRef]
58. Fine, D.D.; Breidenbach, G.P.; Price, T.L.; Hutchins, S.R. Quantitation of estrogens in ground water and swine lagoon samples using solid-phase extraction, pentafluorobenzyl/trimethylsilyl derivatizations and gas chromatography-negative ion chemical ionization tandem mass spectrometry. *J. Chromatogr. A* **2003**, *1017*, 167–185. [CrossRef] [PubMed]
59. Schummer, C.; Delhomme, O.; Appenzeller, B.M.R.; Wennig, R.; Millet, M. Comparison of MTBSTFA and BSTFA in derivatization reactions of polar compounds prior to GC/MS analysis. *Talanta* **2009**, *77*, 1473–1482. [CrossRef]
60. Garland, S.; Goheen, S.; Donald, P.; McDonald, L.; Campbell, J. Application of Derivatization Gas Chromatography/Mass Spectrometry for the Identification and Quantitation of Pinitol in Plant Roots. *Anal. Lett.* **2009**, *42*, 2096–2105. [CrossRef]
61. Valdez, C.A.; Leif, R.N.; Hart, B.R. Rapid and mild silylation of β-amino alcohols at room temperature mediated by N-methylimidazole for enhanced detectability by gas chromatography/electron ionization mass spectrometry. *Rapid Commun. Mass Spectrom.* **2014**, *28*, 2217–2221. [CrossRef]

62. Albo, R.L.F.; Valdez, C.A.; Leif, R.N.; Mulcahy, H.A.; Koester, C. Derivatization of pinacolyl alcohol with phenyldimethylchlorosilane for enhanced detection by gas chromatography-mass spectrometry. *Anal. Bioanal. Chem.* **2014**, *406*, 5231–5234. [CrossRef]
63. Greene, T.; Wuts, P.G.M. *Protective Groups in Organic Synthesis*, 2nd ed.; John Wiley & Sons, Inc.: New York, NY, USA, 1991; pp. 68–87.
64. Davies, G.S.; Higginbotham, L.C.L.; Tremeer, E.J.; Brown, C.; Treadgold, R.C. Protection of hydroxy groups by silylation: Use in peptide synthesis and as lipophilicity modifiers for peptides. *J. Chem. Soc. Perkin Trans. 1* **1992**, 3043–3048. [CrossRef]
65. Purdon, J.; Pagotto, J.; Miller, R. Preparation, stability and quantitative-analysis by gas-chromatography and gas-chromatography electron-impact mass-spectrometry of tert-butyldimethylsilyl derivatives of some alkylphosphonic and alkyl methylphosphonic acids. *J. Chromatogr.* **1989**, *475*, 261–272. [CrossRef]
66. Richardson, D.D.; Caruso, J.A. Derivatization of organophosphorus nerve agent degradation products for gas chromatography with ICPMS and TOF-MS detection. *Anal. Bioanal. Chem.* **2007**, *388*, 809–823. [CrossRef]
67. Subramaniam, R.; Åstot, C.; Nilsson, C.; Ostin, A. Combination of solid phase extraction and *in vial* solid phase derivatization using a strong anion exchange disk for the determination of nerve agent markers. *J. Chromatogr. A* **2009**, *1216*, 8452–8459. [CrossRef]
68. Subramaniam, R.; Åstot, C.; Juhlin, L.; Nilsson, C.; Ostin, A. Determination of S-2-(N,N-diisopropylaminoethyl)- and S-2-(N,N-diethylaminoethyl) methylphosphonothiolate, nerve agent markers, in water samples using strong anion-exchange disk extraction, *in vial* trimethylsilylation, and gas chromatography-mass spectrometry analysis. *J. Chromatogr. A* **2012**, *1229*, 86–94. [CrossRef]
69. Pardasani, D.; Purohit, A.; Mazumder, A.; Dubey, D.K. Gas chromatography-mass spectrometric analysis of trimethylsilyl derivatives of toxic hydrolyzed products of nerve agent VX and its analogues for verification of Chemical Weapons Convention. *Anal. Methods* **2010**, *2*, 661–667. [CrossRef]
70. Black, R.M.; Muir, B. Derivatisation reactions in the chromatographic analysis of chemical warfare agents and their degradation products. *J. Chromatogr. A* **2003**, *1000*, 253–281. [CrossRef]
71. Creasy, W.R.; Stuff, J.R.; Williams, B.; Morrissey, K.; Mays, J.; Duevel, R.; Durst, H.D. Identification of chemical-weapons-related compounds in decontamination solutions and other matrices by multiple chromatographic techniques. *J. Chromatogr. A* **1997**, *774*, 253–263. [CrossRef]
72. Kim, H.; Cho, Y.; Lee, B.S.; Choi, I.S. In-situ derivatization and headspace solid-phase microextraction for gas chromatography-mass spectrometry analysis of alkyl methylphosphonic acids following solid-phase extraction using thin film. *J. Chromatogr. A* **2019**, *1599*, 17–24. [CrossRef] [PubMed]
73. Knapp, D.R. *Handbook of Analytical Derivatization Reactions*; John Wiley & Sons, Inc.: New York, NY, USA, 1979.
74. Halket, J.M.; Zaikin, V.G. Derivatization in Mass Spectrometry—3. Alkylation (Arylation). *Eur. J. Mass Spectrom.* **2004**, *10*, 1–19. [CrossRef]
75. Green, S.P.; Wheelhouse, K.M.; Payne, A.D.; Hallett, J.P.; Miller, P.W.; Bull, J.A. Thermal Stability and Explosive Hazard Assessment of Diazo Compounds and Diazo Transfer Reagents. *Org. Process. Res. Dev.* **2020**, *24*, 67–84. [CrossRef]
76. Kadoum, A.M. Extraction and cleanup methods to determine malathion and its hydrolytic products in stored grains by gas-liquid chromatography. *J. Agric. Food Chem.* **1969**, *17*, 1178–1180. [CrossRef]
77. Verweij, A.; Degenhardt, C.E.A.M.; Boter, H.L. The occurrence and determination of PCH_3-containing compounds in surface water. *Chemosphere* **1979**, *8*, 115–124. [CrossRef]
78. Van't Erve, T.J.; Rautianen, R.H.; Robertson, L.W.; Luthe, G. Trimethylsilyldiazomethane: A safe non-explosive, cost effective and less-toxic reagent for phenol derivatization in GC applications. *Environ. Int.* **2010**, *36*, 835–884. [CrossRef]
79. Pagliano, E. Versatile derivatization for GC-MS and LC-MS: Alkylation with trialkyloxonium tetrafluoroborates for inorganic anions, chemical warfare agent degradation products, organic acids, and proteomic analysis. *Anal. Bioanal. Chem.* **2020**, *412*, 1963–1971. [CrossRef]
80. Driskell, W.; Shih, M.; Needham, L.; Barr, D. Quantitation of organophosphorus nerve agent metabolites in human urine using isotope dilution gas chromatography tandem mass spectrometry. *J. Anal. Toxicol.* **2002**, *26*, 6–10. [CrossRef]
81. Barr, J.R.; Driskell, W.J.; Aston, L.S.; Martinez, R.A. Quantitation of Metabolites of the Nerve Agents Sarin, Soman, Cyclohexylsarin, VX and Russian VX in Human Urine Using Isotope-Dilution Gas Chromatography-Tandem Mass Spectrometry. *J. Anal. Toxicol.* **2004**, *28*, 372–378. [CrossRef]
82. Amphaisri, K.; Palit, M.; Mallard, G. Thermally assisted methylation and subsequent silylation of scheduled acids of chemical weapon convention for on-site analysis and its comparison with the other methods of methylation. *J. Chromatogr. A* **2011**, *1218*, 972–980. [CrossRef] [PubMed]
83. Byrd, G.D.; Paule, R.C.; Sander, L.C.; Sniegoski, L.T.; White, V.E.; Bausum, H.T. Determination of 3-Quinuclidinyl Benzilate (QNB) and Its Major Metabolites in Urine by Isotope Dilution Gas Chromatography/Mass Spectrometry. *J. Anal. Toxicol.* **1992**, *16*, 182–187. [CrossRef]
84. Valdez, C.A.; Leif, R.N.; Salazar, E.P.; Vu, A.K. Trocylation of 3-quinuclidinol, a key marker for the chemical warfare agent BZ, for its enhanced detection at low levels in complex soil matrices by Electron Ionization Gas Chromatography-Mass Spectrometry. *Rapid Commun. Mass Spectrom.* **2021**, *35*, e9123. [CrossRef]

85. Valdez, C.A.; Leif, R.N.; Alcaraz, A. Effective methylation and identification of phosphonic acids relevant to chemical warfare agents mediated by trimethyloxonium tetrafluoroborate for their qualitative detection by gas chromatography-mass spectrometry. *Anal. Chim. Acta* **2016**, *933*, 134–143. [CrossRef]
86. Valdez, C.A.; Marchioretto, M.K.; Leif, R.N.; Hok, S. Efficient derivatization of methylphosphonic and aminoethylsulfonic acids related to nerve agents simultaneously via methylation employing trimethyloxonium tetrafluoroborate for their detection and identification in soils by EI-GC-MS and GC-FPD. *Forensic Sci. Int.* **2018**, *288*, 159–168. [CrossRef] [PubMed]
87. Valdez, C.A.; Leif, R.N.; Hok, S.; Vu, A.K.; Salazar, E.P.; Alcaraz, A. Methylation Protocol for the Retrospective Detection of Isopropyl-, Pinacolyl- and Cyclohexylmethylphosphonic Acids, Indicative Markers for the Nerve Agents Sarin, Soman and Cyclosarin, at Low Levels in Soils Using EI-GC-MS. *Sci. Total Environ.* **2019**, *683*, 175–184. [CrossRef] [PubMed]
88. Rein, K.; Goicoechea-Pappas, M.; Anklekar, T.V.; Hart, G.C.; Smith, G.A.; Gawley, R.E. Chiral dipole-stabilized anions: Experiment and theory in benzylic and allylic systems. Stereoselective deprotonations, pyramidal inversions, and stereoselective alkylations of lithiated (tetrahydroisoquinolyl)oxazolines. *J. Am. Chem. Soc.* **1989**, *111*, 2211–2217. [CrossRef]
89. Trzoss, L.; Xu, J.; Lacoske, M.H.; Theodorakis, E.A. Synthesis of the tetracyclic core of Illicium sesquiterpenes using an organocatalyzed asymmetric Robinson annulation. *Beilstein J. Org. Chem.* **2013**, *9*, 1135–1140. [CrossRef]
90. Milite, C.; Feoli, A.; Horton, J.R.; Rescigno, D.; Cipriano, A.; Pisapia, V.; Viviano, M.; Pepe, G.; Amendola, G.; Novellino, E.; et al. Discovery of a Novel Chemotype of Histone Lysine Methyltransferase EHMT1/2 (GLP/G9a) Inhibitors: Rational Design, Synthesis, Biological Evaluation, and Co-crystal Structure. *J. Med. Chem.* **2019**, *62*, 2666–2689. [CrossRef]
91. Raber, D.J.; Gariano Jr., P.; Brod, A.O.; Gariano, A.; Guida, W.C.; Guida, A.R.; Herbst, M.D. Esterification of carboxylic acids with trialkyloxonium salts. *J. Org. Chem.* **1979**, *44*, 1149–1154. [CrossRef]
92. Hansen, D.W., Jr.; Pilipauskas, D. Chemoselective N-ethylation of Boc amino acids without racemization. *J. Org. Chem.* **1985**, *50*, 945–950. [CrossRef]
93. Hung, K.; Condakes, M.L.; Morikawa, T.; Maimome, T.J. Oxidative Entry into the Illicium Sesquiterpenes: Enantiospecific Synthesis of (+)-Pseudoanisatin. *J. Am. Chem. Soc.* **2016**, *138*, 16616–16619. [CrossRef]
94. Shih, M.L.; Smith, J.R.; McMonagle, J.D.; Dolzine, T.W.; Gresham, V.C. Detection of Metabolites of Toxic Alkylmethylphosphonates in Biological Samples. *Biol. Mass Spectrom.* **1991**, *20*, 717–723. [CrossRef] [PubMed]
95. Fredriksson, S.-Å.; Hammarström, L.-G.; Henriksson, L.; Lakso, H.-Å. Trace Determination of Alkyl Methylphosphonic Acids in Environmental and Biological Samples Using Gas Chromatography/Negative-ion Chemical Ionization Mass Spectrometry and Tandem Mass Spectrometry. *J. Mass Spectrom.* **1995**, *30*, 1133–1134. [CrossRef]
96. Miki, A.; Katagi, M.; Tsuchihashi, H.; Yamashita, M. Gas chromatographic determination and gas chromatographic-mass spectrometric determination of dialkyl phopphates via extractive pentafluorobenzylation sing a polymeric phase-transfer catalyst. *J. Chromatogr. A* **1995**, *718*, 383–389. [CrossRef]
97. Polhuijs, M.; Langenberg, J.P.; Benschop, H.P. New method for retrospective detection of exposure to organophosphorus anticholinesterases: Application to alleged sarin victims of Japanese terrorists. *Toxicol. Appl. Pharmacol.* **1997**, *146*, 156–161. [CrossRef]
98. Miki, A.; Katagi, M.; Tsuchihashi, H.; Yamashita, M. Determination of alkylmethylphosphonic acids, the main metabolites of organophosphorus nerve agents, in biofluids by Gas Chromatography-Mass Spectrometry and Liquid-Liquid-Solid-Phase-Transfer-Catalyzed Pentafluorobenzylation. *J. Anal. Toxicol.* **1999**, *23*, 86–93. [CrossRef]
99. Riches, J.; Morton, I.; Read, R.W.; Black, R.M. The trace analysis of alkyl alkylphosphonic acids in urine using gas chromatography-ion trap negative ion tandem mass spectrometry. *J. Chromatogr. B* **2005**, *816*, 251–258. [CrossRef]
100. Subramaniam, R.; Åstot, C.; Juhlin, L.; Nilsson, C.; Ostin, A. Direct Derivatization and Rapid GC-MS Screening of Nerve Agent Markers in Aqueous Samples. *Anal. Chem.* **2010**, *82*, 7452–7459. [CrossRef]
101. Nyholm, J.R.; Gustafsson, T.; Ostin, A. Structural determination of nerve agent markers using gas chromatography mass spectrometry after derivatization with 3-pyridyldiazomethane. *J. Mass Spectrom.* **2013**, *48*, 813–822. [CrossRef]
102. Subramaniam, R.; Ostin, A.; Nilsson, C.; Åstot, C. Direct derivatization and gas chromatography–tandem mass spectrometry identification of nerve agent biomarkers in urine samples. *J. Chromatogr. B* **2013**, *928*, 98–105. [CrossRef]
103. Lin, Y.; Chen, J.; Yan, L.; Guo, L.; Wu, B.; Li, C.; Feng, J.; Liu, Q.; Xie, J. Determination of nerve agent metabolites in human urine by isotope-dilution gas chromatography-tandem mass spectrometry after solid phase supported derivatization. *Anal. Bioanal. Chem.* **2014**, *406*, 52113–55220. [CrossRef] [PubMed]

Article

Development of a Derivatization Reagent with a 2-Nitrophenylsulfonyl Moiety for UHPLC-HRMS/MS and Its Application to Detect Amino Acids Including Taurine

Shusuke Uekusa [1,2], Mayu Onozato [1], Tatsuya Sakamoto [1], Maho Umino [1], Hideaki Ichiba [1], Kenji Okoshi [3] and Takeshi Fukushima [1,*]

[1] Department of Analytical Chemistry, Faculty of Pharmaceutical Sciences, Toho University, Chiba 274-8510, Japan; shuusuke.uekusa@phar.toho-u.ac.jp (S.U.); mayu.onozato@phar.toho-u.ac.jp (M.O.); tatsuya.sakamoto@phar.toho-u.ac.jp (T.S.); 3020001u@st.toho-u.jp (M.U.); ichiba@phar.toho-u.ac.jp (H.I.)

[2] Department of Clinical Pharmacy, Faculty of Pharmaceutical Sciences, Toho University, Chiba 274-8510, Japan

[3] Department of Environmental Science, Faculty of Science, Toho University, Chiba 274-8510, Japan; kenji.okoshi@env.sci.toho-u.ac.jp

* Correspondence: t-fukushima@phar.toho-u.ac.jp; Tel./Fax: +81-47-472-1504

Abstract: Taurine (Tau) has some important ameliorating effects on human health and is present in bivalve. For the selective analysis of Tau with other amino acids, we designed a derivatization reagent, 2,5-dioxopyrrolidin-1-yl(4-(((2-nitrophenyl)sulfonyl)oxy)-6-(3-oxomorpholino)quinoline-2-carbonyl)pyrrolidine-3-carboxylate (Ns-MOK-β-Pro-OSu). After derivatization with Ns-MOK-β-Pro-OSu, amino acids with Tau in Japanese littleneck clams were determined through ultra-high-performance-liquid chromatography with high-resolution tandem mass spectrometry (UHPLC-HRMS/MS) using an octadecyl silica column. We could detect 18 amino acids within 10 min. Tau, valine, glutamine, glutamic acid, and arginine in the clams were determined in the negative ion mode using the characteristic fragment ion, $C_6H_4N_1O_5S$, which corresponded to the 2-nitrobenzenesulfonylate moiety. The fragment ion, $C_6H_4N_1O_5S$, was recognized as a common feature regardless of the amino acid to be derivatized, and it was convenient for detecting amino acid derivatives with high selectivity and sensitivity. Therefore, highly selective quantification using UHPLC-HRMS/MS was possible using Ns-MOK-β-Pro-OSu.

Keywords: taurine; glutamine; derivatization; clams; high-resolution mass spectrometry

1. Introduction

Taurine (Tau) or 2-aminoethanesulfonic acid is an endogenous sulfur-containing compound that is produced from L-cysteine via hypotaurine. Tau acts as a neuroprotective agent in the central nervous system [1] and plays an important role in endogenous antioxidant activity, brain development, and retinal function [2]. In addition, Tau can be used for the biosynthesis of taurocholate, a bile acid important for digestive functions.

Tau deficiency is associated with the pathogenesis of some diseases [2,3]. Recently, we found a significant decrease in serum taurine levels in the prodromal stage of psychosis, an at-risk mental state (ARMS) through a metabolomic study [4]. However, it was difficult to separate amino acids without pre-column derivatization in an octadecyl silica (ODS) column because they were hardly retained.

Considering the physiological usefulness of Tau, the intake of Tau through food can also be important for maintaining healthy conditions. Tau is abundantly present in octopus, squid [5], and bivalve [6]. Among them, the Japanese littleneck clam, *Asari*, *Ruditapes philippinarum* is an important fishery resource edible bivalve usually cooked as an indispensable seafood dish in many countries. Some studies have examined the profiles of the amino acids present in the Japanese littlenecked clam in terms of their nutritional or taste aspects [7,8].

Previous studies analyzing amino acids in clam have used dansyl chloride [6] for derivatization, followed by separation using high-performance liquid chromatography (HPLC). Since recent years, high-resolution mass spectrometry (HRMS) is being widely utilized in environmental, biological, and pharmaceutical research [9,10].

Previously, we developed a diastereomer derivatization reagent, Ns-MOK-(R)- or -(S)-Pro-OSu (Figure 1a), which was used for derivatization with the amino group of an anti-epileptic drug, vigabatrin, which is a gamma-amino acid [11]. For ease of reaction with the amino group of the target drug, the phenolic hydroxyl group in the derivatization reagent was protected with a 2-nitrobenzenesulfonyl (nosyl) group. However, the reaction mixture was heated for more than 60 min to cause de-nosylation, releasing the nosyl group. This long heating time may be attributed to the steric hindrance of the reaction site with the derivatizing agent.

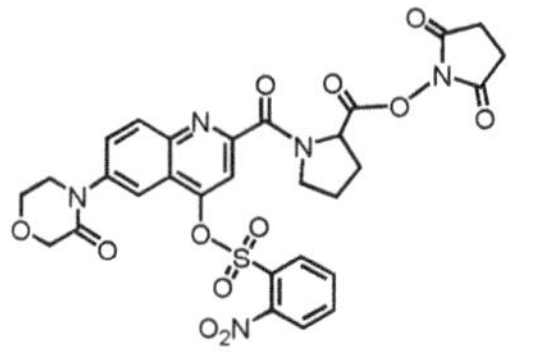

Figure 1. Chemical structure of Ns-MOK-Pro-OSu (a) and Ns-MOK-β-Pro-OSu (b).

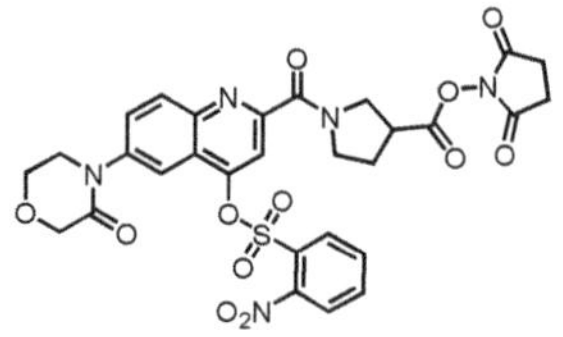

Thus, in the present study, we designed and developed a structural analog, Ns-MOK-β-Pro-OSu (Figure 1b), in which β-proline (β-Pro) was used instead of Pro. The activated ester moiety, 2,5-dioxopyrrolidin-1-yl, attached to β-Pro may react easily with the amino groups of analytes owing to less steric hindrance. The reagent was designed with less steric hindrance for enabling easy reaction with the amino groups.

The time-course profile showing the progress of the reaction with certain amino acids was examined and compared with that of Ns-MOK-β-Pro-OSu. The mass fragmentation patterns of some amino acids and Tau derivatized with Ns-MOK-β-Pro-OSu in collision cells were examined.

Finally, rapid analysis of the amino acid components including Tau in clams within 10 min was conducted using ultra-high-performance liquid chromatography (UHPLC)-HRMS/MS instrument equipped with an ODS column.

2. Results

2.1. Synthesis and Evaluation of 2,5-Dioxopyrrolidin-1-yl(4-(((2-nitrophenyl)sulfonyl)oxy)-6-(3-oxomorpholino)quinoline-2-carbonyl)pyrrolidine-3-carboxylate (Ns-MOK-β-Pro-OSu)

As in our previous reagent, Ns-MOK-Pro-OSu, the succinimide moiety was attached to the carboxyl group to produce an activated ester for labeling the primary amino group of the amino acid containing Tau. In addition, a 2-nitrobenzenesulfonyl (Ns-, nosyl) group was bound to a phenolic hydroxyl group at position 5 in quinoline ring to facilitate easy reactions of the amino group in the amino acid containing Tau. Comparing the retention times of Ns-MOK-Pro-OSu and Ns-MOK-β-Pro-OSu showed that Ns-MOK-(S)-β-Pro-OSu eluted before Ns-MOK-(S)-Pro-OSu from the ODS column (Figure S1), suggesting that rapid analysis using an ODS column is preferred for Ns-MOK-β-Pro-OSu.

2.2. Derivatization of Amino Acid with Ns-MOK-(R)- or -(S)-β-Pro-OSu

Figure 2 shows time-course profiles of the derivatization reaction of the amino acid with Ns-MOK-β-Pro-OSu. According to the LC-TOF-MS data, which showed the [M + H]$^+$ of the reaction product, the main products of the derivatization reaction were Ns-MOK-(S)-Pro- and (S)-β-Pro-amino acids. Regarding the reaction time, the peak area of the derivatives with Ns-MOK-(S)-β-Pro-OSu plateaued after 30 min. Conversely, in Ns-MOK-(S)-Pro-OSu, particularly in Glu, a longer reaction time was required to reach a plateau.

Therefore, we decided to use Ns-MOK-(S)-β-Pro-OSu, which proved beneficial for reducing the reaction time and set the reaction time to 30 min.

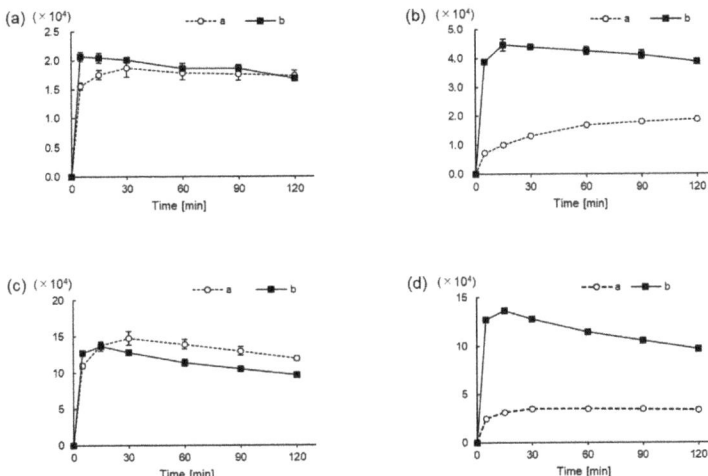

Figure 2. Time-course profiles of derivatization of amino acid: (**a**) Tau, (**b**) Glu, (**c**) Arg, and (**d**) Gln with Ns-MOK-(S)-Pro-OSu (dotted line with circles) and Ns-MOK-(S)-β-Pro-OSu (solid line with squares) by LC-TOF-MS.

2.3. Fragmentation of Amino Acid Derivatives with Ns-MOK-(S)-β-Pro-OSu

Figure 3 and Figure 5a show the MS/MS spectra of the fragment ions, namely Val, Glu, Gln, Arg, and Tau derivatives, obtained in the ESI (+) mode using UHPLC-HRMS. For all the amino acids, fragment ions cleaved at the β-Pro-attached amide bond were commonly observed under relatively small collision energies (CEs) of approximately 30 to 40%. The main fragments observed under each CE are given in Tables S1–S5.

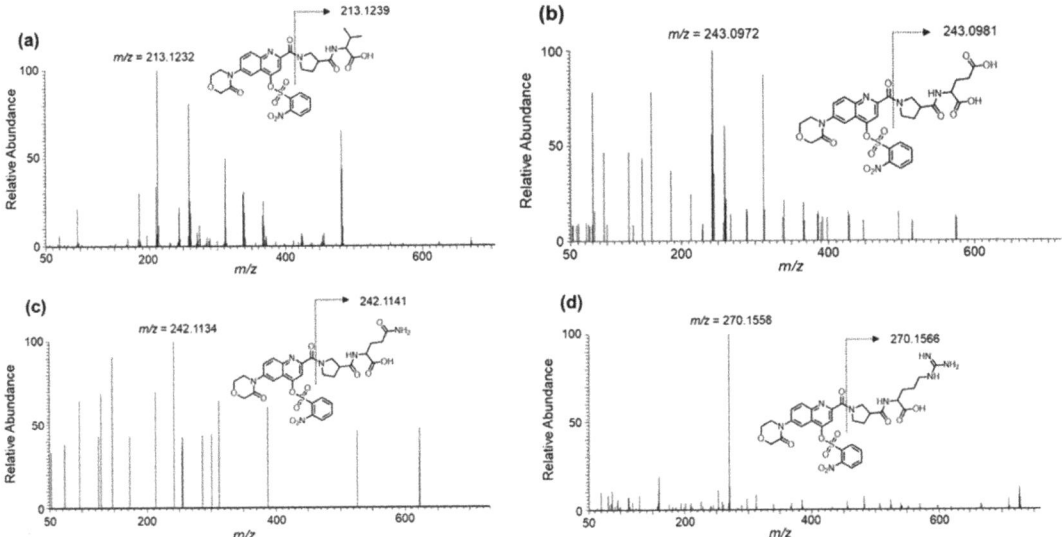

Figure 3. MS/MS spectra of derivatized amino acids: (**a**) Val, collision energy (CE) 30%; (**b**) Glu, CE 30%; (**c**) Gln, CE 30%; and (**d**) Arg, CE 30% with Ns-MOK-(S)-Pro-OSu using ESI (+) mode.

Figures 4 and 5b show the mass spectra of the fragment ions, Val, Glu, Gln, Arg, and Tau derivatives, obtained in the ESI (−) mode. Fragment ions cleaved at the nosylate moiety, $C_6H_4NO_5S$, were commonly observed in all the amino acids. Furthermore, for Tau, in addition to this cleavage, fragment ions cleaved at the β-Pro-attached amide bond were observed. The main fragments in each CE are given in Tables S6–S10.

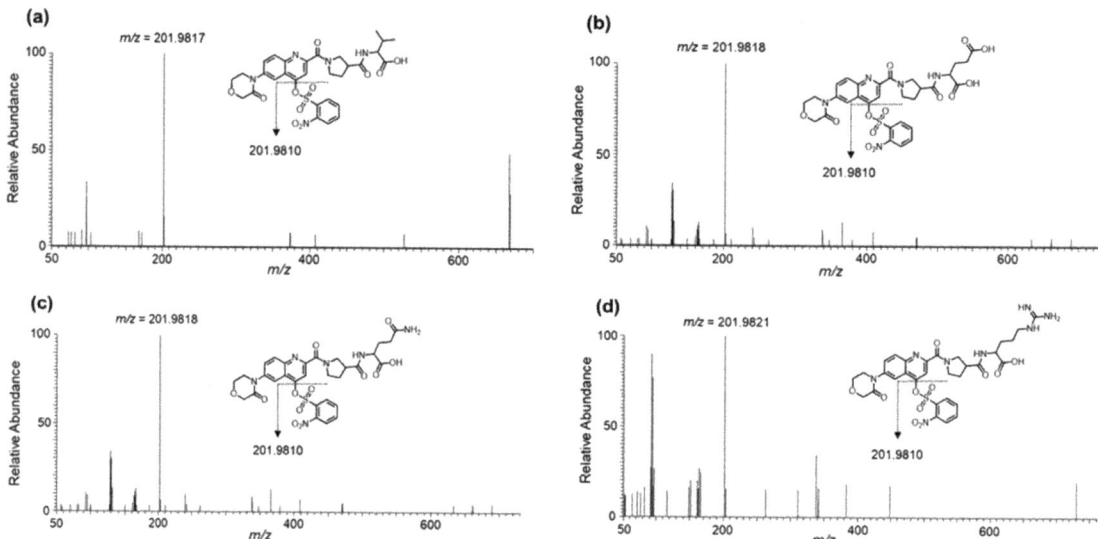

Figure 4. MS/MS spectra of derivatized amino acids: (**a**) Val, CE 20%; (**b**) Glu, CE 30%; (**c**) Gln, CE30%; and (**d**) Arg, CE 40%) with Ns-MOK-(S)-Pro-OSu using ESI (−) mode.

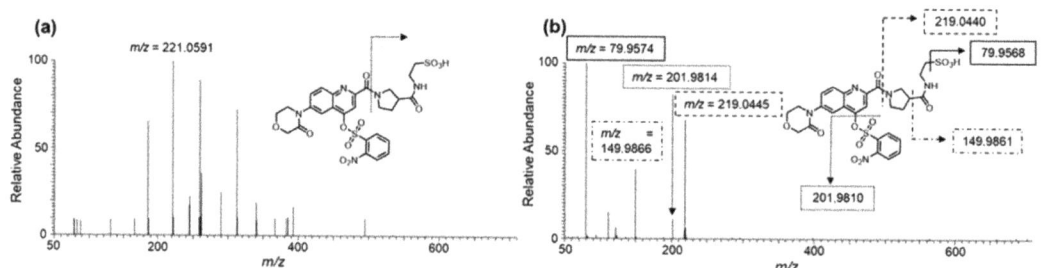

Figure 5. MS/MS spectra of derivatized Tau with Ns-MOK-(S)-Pro-OSu using ESI (+) CE 30% (**a**) and ESI (−) CE 30% modes (**b**).

2.4. Detection of Amino Acids

When the amino acid derivatives were separated by an ODS column and analyzed by UHPLC-HRMS, it was possible to detect 18 isotope-labeled amino acids using SIM mode within 10 min (Figure 6). The retention times of Leu and Ile differed by approximately 0.05 min, but these could not be separated.

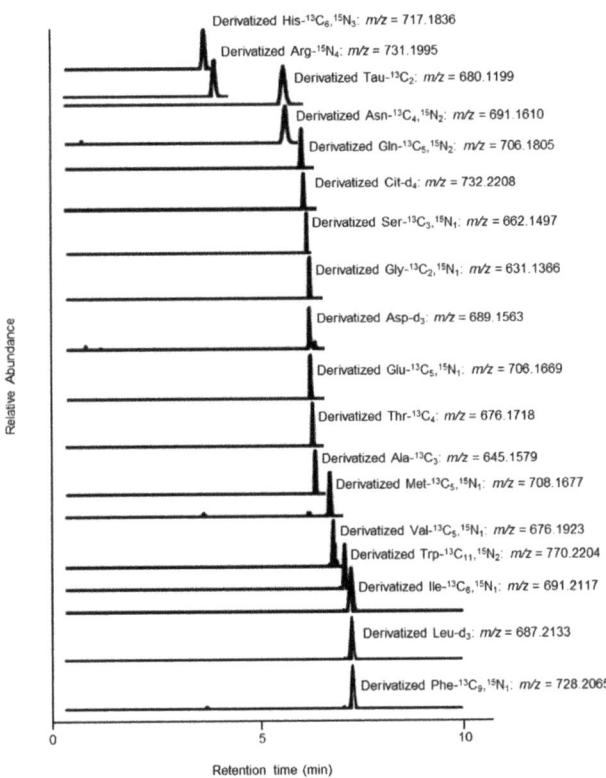

Figure 6. SIM chromatogram showing 18 stable isotope-labeled amino acid derivatives.

2.5. Detection of Amino Acids in Bivalve

In our previous study, high levels of Glu and low levels of Tau were identified as risk factors in ARMS [4]. Therefore, we targeted bivalves containing a large amount of Tau. In addition to Tau and Glu, Gln and Arg, both of which are linked with Glu metabolism, and Val, which is abundant in bivalve, were quantified.

Calibration curves of the Val, Glu, Gln, Arg, and Tau derivatives obtained using the PRM mode are shown in Figure S7. Because the calibration curves showed $R^2 > 0.99$, linearity was obtained in the tested concentration range. All the coefficients of variation in the calibration curve range were within 10% (Tables S11–S15). As shown in Table 1, the amino acids in the bivalve could be determined using the calibration curves. In the present method, the LOQs of Val, Glu, Gln, Arg, and Tau were in the range of 0.042–0.167 pmol. Tables S16–S20 present the quantification data for each amino acid (content per gram of the edible portion), indicating that differences in the amino acid contents existed among the bivalve.

Table 1. Amino acid contents in the bivalve were determined using the proposed LC-HRMS/MS method.

Amino Acid Content per 1 g of the Edible Portion (mg)	Mean	±	sd
Tau	5.33	±	1.91
Val	0.28	±	0.10
Glu	1.16	±	0.49
Gln	0.20	±	0.12
Arg	0.66	±	0.33

Positive correlations were found between the weight of the edible portion and the Tau and Glu contents per gram of the edible portion ($r = 0.708$, $p = 0.00315$ in Figure 7a and $r = 0.682$, $p = 0.00514$ in Figure 7b, respectively).

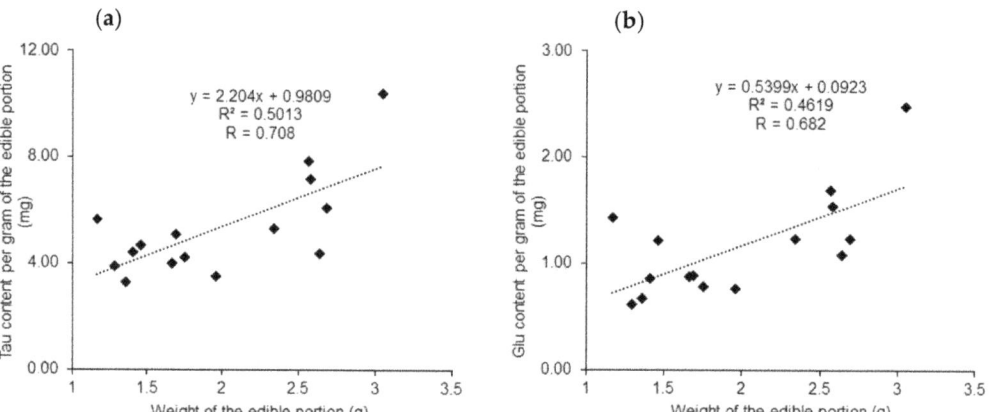

Figure 7. Scatter plot of the weight of the edible portion and Tau (**a**) or Glu (**b**) content per 1 g of the edible portion. Vertical axis: Tau (**a**) or Glu (**b**) content per gram of edible portion (mg); Horizontal axis: weight of the edible portion (g).

3. Discussion

For comparison with Ns-MOK-Pro-OSu, the derivatization reagent designed in our previous study [11], 2,5-dioxopyrrolidin-1-yl(4-(((2-nitrophenyl)sulfonyl)oxy)-6-(3-oxomorpholino)quinoline-2-carbonyl)pyrrolidine-3-carboxylate (Ns-MOK-β-Pro-OSu), a structural analogue, was synthesized.

Consequently, a time-course study revealed that Ns-MOK-β-Pro-OSu reacted with amino acids faster than Ns-MOK-Pro-OSu at room temperature. Furthermore, the peak area values of Glu and Gln were remarkably high when Ns-MOK-β-Pro-OSu was used (Figure 3). The reaction proceeded at room temperature with no release of the nosylate moiety in the produced derivative. Therefore, we could observe proper fragmentation of the nosylate moiety, $C_6H_4N_1O_5S$, in the MS/MS spectrum of the derivative with Ns-MOK-β-Pro-OSu.

In our previous study, Ns-MOK-Pro-OSu was reacted with amino groups under heating conditions to eliminate the nosyl group [11]. In contrast, Ns-MOK-β-Pro-OSu reacted effectively with amino acids at room temperature, and no nosyl elimination product was observed. This difference in the reaction with amino acids may be due to the lower steric hindrance of β-Pro-OSu than that of Pro-OSu. Thus, Ns-MOK-β-Pro-OSu was a more suitable reagent for MS/MS detection than Ns-MOK-Pro-OSu.

Generally, derivatization reagents for LC-MS/MS detection possess suitable structures for efficient fragmentation [12]. As mentioned above, Ns-MOK-β-Pro-OSu, exhibited unique fragmentations in the ESI (+) and ESI (−) modes, as observed in the MS/MS spectra (Figures 4–6).

In addition, as Ns-MOK-(S)-β-Pro-OSu eluted earlier than Ns-MOK-(S)-Pro-OSu, we inferred that the analysis time could be shortened using Ns-MOK-(S)-β-Pro-OSu (Figure S1). Under the mobile phase conditions used in the present study, amino acids containing Tau could be analyzed within 10 min (Figure 7).

In the ESI (+) mode, cleavage occurred regularly at the β-Pro-attached amide bond; however, the signal-to-noise ratio of the ion fragment was not extremely high. In contrast, the nosylate moiety, $C_6H_4N_1O_5S$, was recognized as an intense fragment with a common structure regardless of the amino acid in the ESI (−) mode. The fragment ions of the nosylate moiety, $C_6H_4N_1O_5S$, were used for quantifying the amino acids. The NCE values set for each amino acid were between 20% and 40%, in which $C_6H_4N_1O_5S$ was the most efficiently cleaved moiety. Therefore, this cleavage pattern could be used to detect the derivative with Ns-MOK-β-Pro-OSu with high selectivity and identify whether it was a derivative of Ns-MOK-β-Pro-OSu.

For detecting the Tau derivative, fragment ions cleaved at the SO_3 moiety were observed. When searching for Tau on the *m/z* cloud (https://www.mzcloud.org/, accessed on 1 December 2020), SO_3 fragment ions were observed from approximately 70 to 80% of the NCE. However, in the present study, fragment ions were observed under conditions where the NCE was relatively low. The SO_3 fragment ions detected in MS^2 at an NCE of approximately 30% may be useful to discriminate other compounds with sulfate groups.

Next, the present LC-HR-MS/MS technique using Ns-MOK-β-Pro-OSu was successfully applied to determine amino acids in the samples from the collected bivalve. As observed in Table 1, extremely high levels of Tau were found in the *R. philipinarum* compared to other amino acids. High levels of Tau in *R. philipinarum* have already been reported by Wang et al. [10], who also reported seasonal variations in the amino acid content of clams using LC-HR-MS/MS [10].

As observed in Tables S11–S15 and Figure 8, the difference in the weights of the edible parts could be attributed to the differences in the amino acid contents because significant correlations were observed between the amino acid content and the weight of the edible part. Thus far, several therapeutic applications of Tau supplements have been proposed, including the treatment of diabetes [13], hypertension [14], and heart failure [15]. A recent study reported that Tau increased the effectiveness of antipsychotics in the first episode of schizophrenia [16].

In the present study, the variations in the contents of the five amino acids were large (Table 1). The cause of these variations could be the differences in the sizes of the edible parts (Tables S16–S20). The present data suggest that consuming bivalves with large edible parts is effective for ensuring adequate intake of Tau. Moreover, a bivalve with large edible parts also contains a large amount of Glu. Excessive Glu intake may exacerbate the depressive symptoms of obese schizophrenia [17].

4. Materials and Methods

4.1. Chemicals

LCMS-grade methanol, HPLC-grade formic acid, and APDSTAG® Wako Amino Acids internal standard mixture solution composed of 25 stable-isotope labeled amino acids were purchased from FUJIFILM Wako Pure Chemical Corporation (Osaka, Japan). $CHCl_3$ and LCMS-grade CH_3CN were purchased from Kanto Chemical Co., Inc. (Tokyo, Japan). The other reagents used are mentioned in Supplementary Materials, Section 1.

4.2. Preparation of 2,5-Dioxopyrrolidin-1-yl(4-(((2-nitrophenyl)sulfonyl)oxy)-6-(3-oxomorpholino)quinoline-2-carbonyl)pyrrolidine-3-carboxylate (Ns-MOK-β-Pro-OSu)

Ns-MOK-β-Pro-OSu was synthesized in our laboratory according to the synthetic route shown in Figure 8. The preparation of the intermediate compounds from 4-(3-oxomorpholino)aniline to yield Ns-MOK-β-Pro-OSu is described in Supplementary Materials, Section 5. The structure of the synthesized compound was confirmed through nuclear magnetic resonance (NMR) studies.

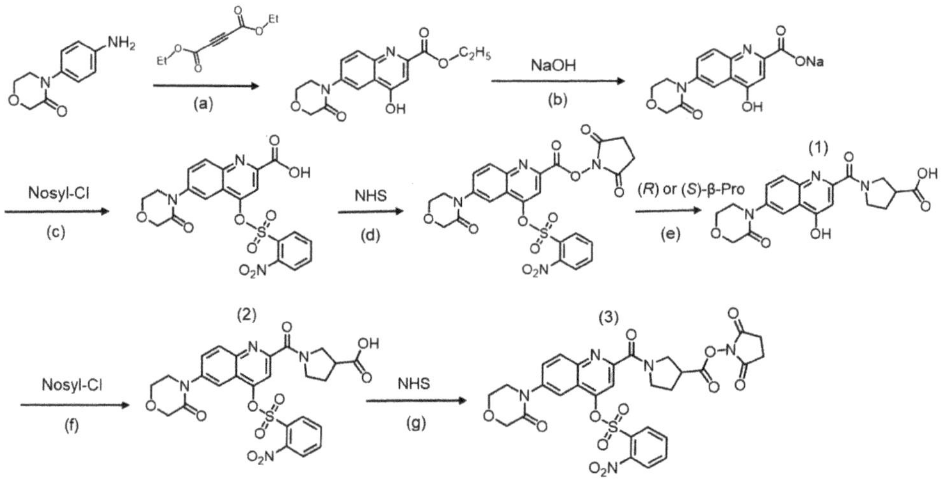

Figure 8. Synthetic route of Ns-MOK-β-Pro-OSu from 4-(3-oxomorpholino)aniline. (**a**) i: MeOH 80 °C 4 h, ii: Ph$_2$O 220 °C 2 h; (**b**) MeOH/H$_2$O r.t. 12 h, (**c**) NaHCO$_3$ r.t. 2 h, (**d**) 1-(3-dimethylaminopropyl)-3-ethylcarbodiimide hydrochloride (EDC) in CH$_3$CN r.t. 2 h, (**e**) DMF, NaHCO$_3$ in H$_2$O r.t. 2 h, (**f**) NaHCO$_3$ r.t. 2 h, and (**g**) EDC in CH$_3$CN r.t. 2 h; (**1**) 1-(4-hydroxy-6-(3-oxomorpholino)quinoline-2-carbonyl)pyrrolidine-3-carboxylic acid; (**2**) 1-(4-(((2-nitrophenyl)sulfonyl)oxy)-6-(3-oxomorpholino)quinoline-2-carbonyl)pyrrolidine-3-carboxylic acid; (**3**) Ns-MOK-β-Pro-OSu.

4.3. Time-Course Study on Derivatization of Amino Acids with Ns-MOK-Pro-OSu and Ns-MOK-β-Pro-OSu

Aliquots containing 5 µL of an amino acid mixture including Tau, (*S*)-glutamate, (*S*)-glutamine, (*S*)-arginine, and (*S*)-valine (each 100 µM) in phosphate-buffered saline (PBS) were added to 5 µL of 20 mM Ns-MOK-(*S*)-Pro-OSu in CH$_3$CN or 20 mM Ns-MOK-(*S*)-β-Pro-OSu in CH$_3$CN, and 5 µL of 10 mM DMAP in CH$_3$CN. The solutions were allowed to react at room temperature for 5, 15, 30, 60, 90, and 120 min. Each reacted solution was diluted with 35 µL of 0.2% HCO$_2$H in H$_2$O/MeOH (1:1, v/v) and subjected to LC-TOF-MS (JMS-T100LP, JEOL Ltd., Tokyo, Japan) (Supplementary Materials, Section 2.2). The observed peak areas were plotted against the derivatization times.

4.4. Pre-Treatment Procedure

First, 10 µL of the amino acid mixture, 5.0 µL of APDSTAG® (solution A:solution B = 65:5, v/v) as an internal standard (IS) solution, and 135 µL of CH$_3$CN/MeOH (1:1, v/v) were mixed and precipitated for 30 min at 4 °C to yield proteins. Next, the mixture was centrifuged at 2500× *g* for 5 min at 4 °C and filtered with Millex®-LG (0.2 µm, Merck Ltd. Tokyo, Japan). The supernatant (100 µL) was evaporated under reduced pressure. After evaporation, the residue was dissolved in 10 µL of PBS, and vortexed with 10 µL of 20 mM Ns-MOK-(*S*)-β-Pro-OSu dissolved in CH$_3$CN and 10 µL of 10 mM DMAP in CH$_3$CN. The solution was reacted at room temperature for 30 min. After the reaction, 70 µL of 0.2% HCO$_2$H in (H$_2$O/MeOH (3:2, v/v)) was added. Then, a 10 µL aliquot was taken from this solution and diluted with 0.05% HCO$_2$H in (H$_2$O/MeOH (9:1, v/v)) (90 µL) and analyzed using ultra-high-performance liquid chromatography-high-resolution mass spectrometry (UHPLC-HRMS/MS, Vanquish UHPLC system and Q-Exactive^{TM}, standalone benchtop Quadrupole-Orbitrap high-resolution mass spectrometer, Thermo Fisher Scientific, Bremen, Germany) (Supplementary Materials, Sections 3.1 and 3.2). The measurement mode of Q-Exactive^{TM} was set to "Full MS-ddMS^2" (Supplementary Materials, Section 3.3).

4.5. Assessment of Fragmentation Patterns of Derivatives

To optimize the fragmentation pattern of the derivatives, the normalized collision energy (NCE) during UHPLC-HRMS/MS was changed by 10% (from 10 to 90%) in the parallel reaction monitoring (PRM) mode used for the measurements (Supplementary Materials, Sections 3.1, 3.2, and 3.4). The precursor ion was a proton adduct of each derivatized amino acid.

4.6. Linearity and Limit of Quantification

Calibration curves were constructed by plotting the peak area corresponding to the IS peak ratio against the amino acid concentrations of 1, 2.5, 5, 10, 25, 50, and 100 µM. The limit of quantification (LOQ) of the amino acid derivatives determined using the proposed UHPLC-HRMS/MS technique was designated as the concentration at which the linearity of the calibration curve was lost. The NCEs were set at 20% for Val derivatives, 30% for Glu derivatives, 30% for Gln derivatives, 40% for Arg derivatives, and 30% for Tau derivatives (Supplementary Materials, Sections 8.1–8.10). The other parameters are described in Supplementary Materials, Sections 3.1, 3.2, and 3.4.

The peak areas of the amino acid derivatives were calculated using the fragment ion, $C_6H_4N_1O_5S$, corresponding to the characteristic fragment ion of Ns-MOK-(S)-β-Pro-OSu in the ESI (−) mode.

4.7. Pre-Treatment of Bivalve

Japanese littleneck clams, *Ruditapes philippinarum* (*R. philippinarum*), were collected from the intertidal and subtidal zones of the artificial tidal flats in Mangoku-ura Lagoon (Miyagi, Japan), and in the estuary of Udagawa in Matsukawa-ura Lagoon (Fukushima, Japan) in November 2020. The time when the clams were collected was not the spawning season.

After placing the clams in locally sampled seawater, the edible soft part was removed from the shell. Then, the water on the surface was wiped off, and the edible part of each clam was weighed.

Water was added to form a solution of the edible part with a concentration of 1 g/10 mL. After homogenization with a kitchen mixer, approximately 10 mL of the solution was taken and centrifuged at $3500 \times g$ for 15 min to obtain approximately 1 mL of supernatant. Furthermore, the supernatant was centrifuged twice at $13,200 \times g$ and 4 °C for 15 min to obtain a clear supernatant.

5. Conclusions

A novel derivatizing reagent, Ns-MOK-β-Pro-OSu, was developed for pre-column derivatization to determine amino acids containing Tau in bivalve using reversed-phase UHPLC-HR-MS/MS. Characteristic fragmentations under both polarities were observed in the MS/MS spectra. Therefore, highly selective quantification was possible using the characteristic fragmentation of Ns-MOK-β-Pro-OSu. The proposed method could detect 18 types of amino acids in 10 min. Considering diet therapy for ARMS based on the present results regarding the amino acid contents in the bivalve, it seems necessary to search for bivalves with a high Tau/Glu ratio. Thus, further studies are needed to investigate the Tau/Glu ratios in other species of bivalve.

Supplementary Materials: 1. Chemicals and Reagents; 2. LC-Time of Flight (TOF)-Mass Spectrometry: 2.1. Comparison of Retention Times of Ns-MOK-β-Pro-OSu and Ns-MOK-Pro-OSu and 2.2. Time-Course Study on Derivatization of Amino Acids with Ns-MOK-Pro-OSu and Ns-MOK-β-Pro-OSu; **3. Measurement Conditions of UHPLC-High-Resolution Mass Spectrometry (UHPLC-HRMS: Vanquish UHPLC System and Q-Exactive™):** 3.1. Instrumentation, 3.2. Electrospray Ionization (ESI), 3.3. Parameters of Full MS-Data-Dependent MS² (ddMS²) Mode, and 3.4. Parameters of Parallel Reaction Monitoring (PRM) Mode; **4. NMR Measurements**; **5. Preparation of 2,5-Dioxopyrrolidin-1-yl 1-(4-(((2-nitrophenyl)sulfonyl)oxy)-6-(3-oxomorpholino)quinoline-2-carbonyl)pyrrolidine-3-carboxylate (Ns-MOK-β-Pro-OSu):** 5.1. 1-(4-Hydroxy-6-(3-oxomorpholino) quinoline-2-carbonyl)pyrrolidine-3-carboxylic acid (1), 5.2. 1-(4-(((2-Nitrophenyl)sulfonyl)oxy)-6-(3-oxomorpholino)quinoline-2-carbonyl)pyrrolidine-3-carboxylic acid (2), and 5.3. 2,5-Dioxopyrrolidin-1-yl 1-(4-(((2-nitrophenyl)sulfonyl)oxy)-6-(3-oxomorpholino)quinoline-2-carbonyl)pyrrolidine-3-carboxylate (Ns-MOK-β-Pro-OSu) (3); **6. Optical Rotation Measurements of Ns-MOK-(*R*)- or (*S*)-Pro-OSu; 7. Supplementary Figures:** Figure S1. Chromatograms of Ns-MOK-(*S*)-Pro-OSu (a) and Ns-MOK-(*S*)-β-Pro-OSu (b) obtained using LC-TOF-MS; Figure S2. Calibration curves of amino acid derivatives: (a) Tau, (b) Val, (c) Gln, (d) Glu, and (e) Arg; **8. Supplementary Tables:** Table S1. Main fragmentation of Val derivatives ($C_{30}H_{31}N_5O_{11}S$, 669.1740) as a function of NCE (ESI (+) mode), Table S2. Main fragmentation of Glu derivatives ($C_{30}H_{29}N_5O_{13}S$, 699.1482) as a function of NCE (ESI (+) mode), Table S3. Main fragmentation of Gln derivatives ($C_{30}H_{30}N_6O_{12}S$, 698.1642) as a function of NCE (ESI (+) mode), Table S4. Main fragmentation of Arg derivatives ($C_{31}H_{34}N_8O_{11}S$, 726.2067) as a function of NCE (ESI (+) mode), Table S5. Main fragmentation of Tau derivatives ($C_{27}H_{27}N_5O_{12}S_2$, 677.1097) as a function of NCE (ESI (+) mode), Table S6. Main fragmentation of Val derivatives ($C_{30}H_{31}N_5O_{11}S$, 669.1740) as a function of NCE (ESI (−) mode), Table S7. Main fragmentation of Glu derivatives ($C_{30}H_{29}N_5O_{13}S$, 699.1482) as a function of NCE (ESI (−) mode), Table S8. Main fragmentation of Gln derivatives ($C_{30}H_{30}N_6O_{12}S$, 698.1642) as a function of NCE (ESI (−) mode), Table S9. Main fragmentation of Arg derivatives ($C_{31}H_{34}N_8O_{11}S$, 726.2067) as a function of NCE (ESI (−) mode), Table S10. Main fragmentation of Tau derivatives ($C_{27}H_{27}N_5O_{12}S_2$, 677.1097) as a function of NCE (ESI (−) mode), Table S11. Precision of Tau/IS ratio, Table S12. Precision of the Val/IS ratio, Table S13. Precision of Gln/IS ratio, Table S14. Precision of Glu/IS ratio, Table S15. Precision of Arg/IS ratio, Table S16. Tau content in bivalve, Table S17. Val content in bivalve, Table S18. Gln content in bivalve, Table S19. Glu content in bivalve, and Table S20. Arg content in bivalve.

Author Contributions: Conceptualization, T.F.; formal analysis, S.U.; resources; investigation, M.O., M.U., T.S., H.I. and K.O.; writing—review & editing, S.U. All authors have read and agreed to the published version of the manuscript.

Funding: This research received no external funding.

Institutional Review Board Statement: Not applicable.

Informed Consent Statement: Not applicable.

Data Availability Statement: Research data are not shared.

Acknowledgments: The authors would like to thank Y. Takegawa and S. Kasai, ThermoFisher Scientific Inc. (Kanagawa, Japan) for their technical support in using Q-Exactive®, and A. Saito, Toho University, for the use of a kitchen mixer.

Conflicts of Interest: The authors declare no conflict of interest.

Sample Availability: Samples of the compounds used are available from the authors.

References

1. Wu, J.Y.; Prentice, H. Role of taurine in the central nervous system. *J. Biomed. Sci.* **2010**, *17*, S1. [CrossRef] [PubMed]
2. Ripps, H.; Shen, W. Review: Taurine: A "very essential" amino acid. *Mol. Vis.* **2012**, *18*, 2673–2686. [PubMed]
3. Menzie, J.; Pan, C.; Prentice, H.; Wu, J.Y. Taurine and central nervous system disorders. *Amino Acids* **2014**, *46*, 31–46. [CrossRef] [PubMed]
4. Uekusa, S.; Onozato, M.; Umino, M.; Sakamoto, T.; Ichiba, H.; Tsujino, N.; Funatogawa, T.; Tagata, H.; Nemoto, T.; Mizuno, M.; et al. Increased inosine levels in drug-free individuals with at-risk mental state: A serum metabolomics study. *Early Interv. Psychiatry* **2021**. [CrossRef] [PubMed]
5. D'Aniello, A.; Nardi, G.; De Santis, A.; Vetere, A.; di Cosmo, A.; Marchelli, R.; Dossena, A.; Fisher, G. Free L-amino acids and D-aspartate content in the nervous system of cephalopoda. A comparative study. *Comp. Biochem. Physiol. B Biochem. Mol. Biol.* **1995**, *112*, 661–666. [CrossRef]
6. Watanabe, K.; Konosu, S. Presence of taurine in the extract of hard clam. *Nippon Suisan Gakkai Shi* **1972**, *38*, 1311. [CrossRef]
7. Mazzucco, E.; Gosetti, F.; Bobba, M.; Marengo, E.; Robotti, E.; Gennaro, M.C. High-performance liquid chromatography-Ultraviolet detection method for the simultaneous determination of typical biogenic amines and precursor amino acids. Applications in Food Chemistry. *J. Agric. Food Chem.* **2010**, *58*, 127–134. [CrossRef] [PubMed]
8. Fuke, S.; Konosu, S. Taste-active components in some foods: A review of Japanese research. *Physiol. Behavior* **1991**, *49*, 863–868. [CrossRef]
9. Hernández, F.; Ibáñez, M.; Bade, R.; Bijlsma, L.; Sancho, J.V. Investigation of pharmaceuticals and illicit drugs in waters by liquid chromatography-high-resolution mass spectrometry. *TrAC Trends Anal. Chem.* **2014**, *63*, 140–157. [CrossRef]

10. Wang, X.Z.; Wu, H.; Li, N.; Cheng, Y.; Wen, H.M.; Liu, R.; Chai, C. Rapid determination of free amino acids, nucleosides, and nucleobases in commercial clam species harvested at different seasons in Jiangsu, China, using UFLC-MS/MS. *Food Anal. Methods* **2016**, *9*, 1520–1531. [CrossRef]
11. Uekusa, S.; Onozato, M.; Sakamoto, T.; Umino, M.; Ichiba, H.; Fukushima, T. Fluorimetric determination of the enantiomers of vigabatrin, an antiepileptic drug, by reversed-phase HPLC with a novel diastereomer derivatisation reagent. *Biomed. Chromatogr.* **2021**, *35*, e5060. [CrossRef] [PubMed]
12. Santa, T. Derivatization in liquid chromatography for mass spectrometric detection. *Drug Discov. Ther.* **2013**, *7*, 9–17. [CrossRef] [PubMed]
13. Franconi, F.; Loizzo, A.; Ghirlanda, G.; Seghieri, G. Taurine supplementation and diabetes mellitus. *Curr. Opin. Clin. Nutr. Metab. Care* **2006**, *9*, 32–36. [CrossRef] [PubMed]
14. Militante, J.D.; Lombardini, J.B. Treatment of hypertension with oral taurine: Experimental and clinical studies. *Amino Acids* **2002**, *23*, 381–393. [CrossRef] [PubMed]
15. Sole, M.J.; Jeejeebhoy, K.N. Conditioned nutritional requirements and the pathogenesis and treatment of myocardial failure. *Curr. Opin. Clin. Nutr. Metab. Care* **2000**, *3*, 417–424. [CrossRef] [PubMed]
16. O'Donnell, C.P.; Allott, K.A.; Murphy, B.P.; Yuen, H.P.; Proffitt, T.M.; Papas, A.; Moral, J.; Pham, T.; O'Regan, M.K.; Phassouliotis, C.; et al. Adjunctive taurine in first-episode psychosis: A phase 2, double-blind, randomized, placebo-controlled study. *J. Clin. Psychiatry* **2016**, *77*, e1610–e1617. [CrossRef] [PubMed]
17. Kumar, P.; Kraal, A.Z.; Prawdzik, A.M.; Ringold, A.E.; Ellingrod, V. Dietary glutamic acid, obesity, and depressive symptoms in patients with schizophrenia. *Front. Psychiatry* **2020**, *11*, 620097. [CrossRef] [PubMed]

Article

Simultaneous Determination of Human Serum Albumin and Low-Molecular-Weight Thiols after Derivatization with Monobromobimane

Katarzyna Kurpet [1,2,*], Rafał Głowacki [2] and Grażyna Chwatko [2,*]

1 Doctoral School of Exact and Natural Sciences, University of Lodz, Banacha 12/16, 90-237 Lodz, Poland
2 Department of Environmental Chemistry, Faculty of Chemistry, University of Lodz, Pomorska 163, 90-236 Lodz, Poland; rafal.glowacki@chemia.uni.lodz.pl
* Correspondence: katarzyna.kurpet@edu.uni.lodz.pl (K.K.); grazyna.chwatko@chemia.uni.lodz.pl (G.C.); Tel.: +48-42-635-58-03 (K.K.); +48-42-635-58-43 (G.C.)

Citation: Kurpet, K.; Głowacki, R.; Chwatko, G. Simultaneous Determination of Human Serum Albumin and Low-Molecular-Weight Thiols after Derivatization with Monobromobimane. *Molecules* 2021, 26, 3321. https://doi.org/10.3390/molecules26113321

Academic Editor: Paraskevas D. Tzanavaras

Received: 23 April 2021
Accepted: 27 May 2021
Published: 1 June 2021

Publisher's Note: MDPI stays neutral with regard to jurisdictional claims in published maps and institutional affiliations.

Copyright: © 2021 by the authors. Licensee MDPI, Basel, Switzerland. This article is an open access article distributed under the terms and conditions of the Creative Commons Attribution (CC BY) license (https://creativecommons.org/licenses/by/4.0/).

Abstract: Biothiols are extremely powerful antioxidants that protect cells against the effects of oxidative stress. They are also considered relevant disease biomarkers, specifically risk factors for cardiovascular disease. In this paper, a new procedure for the simultaneous determination of human serum albumin and low-molecular-weight thiols in plasma is described. The method is based on the pre-column derivatization of analytes with a thiol-specific fluorescence labeling reagent, monobromobimane, followed by separation and quantification through reversed-phase high-performance liquid chromatography with fluorescence detection (excitation, 378 nm; emission, 492 nm). Prior to the derivatization step, the oxidized thiols are converted to their reduced forms by reductive cleavage with sodium borohydride. Linearity in the detector response for total thiols was observed in the following ranges: 1.76–30.0 mg mL^{-1} for human serum albumin, 0.29–5.0 nmol mL^{-1} for α-lipoic acid, 1.16–35 nmol mL^{-1} for glutathione, 9.83–450.0 nmol mL^{-1} for cysteine, 0.55–40.0 nmol mL^{-1} for homocysteine, 0.34–50.0 nmol mL^{-1} for N-acetyl-L-cysteine, and 1.45–45.0 nmol mL^{-1} for cysteinylglycine. Recovery values of 85.16–119.48% were recorded for all the analytes. The developed method is sensitive, repeatable, and linear within the expected ranges of total thiols. The devised procedure can be applied to plasma samples to monitor biochemical processes in various pathophysiological states.

Keywords: low-molecular-weight thiols; human serum albumin; α-lipoic acid; blood plasma; derivatization; monobromobimane; reduction; sodium borohydride; high-performance liquid chromatography; fluorescence detection

1. Introduction

Thiols constitute a class of organic sulfur compounds characterized by the presence of a sulfhydryl functional group (–SH), also known as a thiol group [1–4]. In a biological system, thiols are present as albumin thiols, protein-bound thiols, and low-molecular-weight thiols such as cysteine (Cys), homocysteine (Hcy), glutathione (GSH), and cysteinylglycine (Cys-Gly) [2,4–6]. These free thiols are metabolically related [4–6]. Hcy can be catalyzed to Cys, which in turn is a precursor of GSH—a highly important tripeptide in biological terms. Cys-Gly is an intermediate metabolite in GSH metabolism and is the second most abundant low-molecular-weight thiol in human plasma after Cys [5,7,8]. Reduced, free oxidized, and protein-bound thiols, i.e., Cys, Hcy, Cys-Gly, and GSH, comprise an antioxidant buffer that maintains the reduction–oxidation balance inside the cell and tissues [9,10]. Under physiological conditions, thiols are strong reductants that can undergo reversible or irreversible oxidation processes through one- or two-electron reaction mechanisms, resulting in a wide range of products [11]. Some of them, including disulfides, can be reverted to free thiols in the presence of suitable reducing agents. Such thiol–disulfide homeostasis plays a role in cellular defense against toxic substances, free radicals, and reactive oxygen species, as well as in apoptosis, transcription, enzyme activity regulation, and in the maintenance of the proper structure and function of proteins [5,12,13].

Other biological aminothiols that widely occur in animal tissues and fluids include N-acetyl-L-cysteine (NAC) and α-lipoic acid (α-LA). The former, NAC, is exclusively present in urine and derives from N-acetylation of Cys in the kidney [14,15]. This compound is a commonly used mucolytic drug that alleviates mucus retention by reducing highly cross-linked mucus glycoproteins [3,4,16–19]. In addition, NAC increases the activity of glutathione S-transferase, has anti-inflammatory effects, can break down the pathogenic biofilm, and is widely used in the treatment of acetaminophen overdose. Moreover, the administration of NAC provides Cys as a substrate for the intracellular synthesis of GSH, which is one of the most important naturally occurring antioxidants. The drug may circulate in a free or protein-bound form in plasma, exhibiting the half-life of several dozen minutes after oral administration, which is due to extensive first-pass metabolism in the body [18,20]. Interestingly, NAC causes a substantial decline in plasma low-molecular-weight thiols, including Cys, Cys-Gly, and Hcy, by increasing urinary excretion [21]. NAC administration is also beneficial in systemic sclerosis, cancer chemotherapy, HIV infection, and septic shock [22].

The second aminothiol mentioned above, α-LA, is a naturally occurring cofactor of several multienzyme complexes involved in energy production [23–25]. It can be synthesized from octanoic acid and Cys, but the quantity produced is negligible. In human cells, α-LA is reduced to dithiol, i.e., dihydrolipoic acid, which has two thiol groups per molecule derived from a dithiolane ring. Owing to such a structure, this compound maintains its protective functions in both oxidized and reduced forms [22,24]. *De novo* synthesis is not the only source of α-LA in mammalian cells. The body acquires it with food, mainly of animal origin. This acid exists naturally– via a covalent amide bond—in conjunction with the ε-amino group of lysine. Together, they form lipoyllysine (LLYS), which—when taken with food—can be hydrolyzed in the blood to α-LA by the enzyme called lipoamidase [26]. Even though the content of LLYS in food is sufficient for metabolic processes to take place, the concentration of α-LA that can be obtained during the hydrolysis of this compound in the blood is insufficient for therapeutic purposes [27]. Therefore, α-LA is an extensively used nutraceutical to combat the negative effects of oxidative stress since it meets all the criteria for a perfect antioxidant. It is applicable in various fields of medicine, including in the treatment of diabetic neuropathy, neurodegenerative and cardiovascular diseases, as well as in the fight against obesity, poisoning, cancer, or body aging [25,27]. The half-life of α-LA elimination from the blood, regardless of dose and method of administration, is approximately 30 min [28].

The human plasma has relatively low concentrations of low-molecular-weight thiols but is characterized by the presence of human serum albumin (HSA) as the most abundant protein thiol [11]. The main biological roles of HSA include regulation of plasma pressure and transport of various ligands, such as drugs, hormones, xenobiotics, fatty acids, and metal ions [29–31]. Moreover, HSA is an effective extracellular antioxidant. It scavenges reactive oxygen species, which are responsible for the development of many diseases associated with oxidative stress [30]. The antioxidant activity of HSA results from the presence of a thiol group derived from Cys-34, which is not involved in the formation of intra-chain disulfide bonds and comprises approximately 80% of total free thiols in plasma [1,29–31]. HSA has been used for over 50 years to treat several conditions related to hypovolemia [11]. In addition, this protein is also of interest to pharmacists as a drug carrier.

Altered thiol levels in physiological fluids have been associated with specific pathological conditions and are closely related to several human diseases, especially premature atherosclerosis [32], vascular disease [33], diabetes [34], cancer [35], rheumatoid arthritis [36], leukemia [14], chronic kidney disease [37], acquired immunodeficiency syndrome [32], multiple sclerosis [12], amyotrophic lateral sclerosis [12], and neurodegenerative diseases [12,38] such as Parkinson's and Alzheimer's disease. Due to the importance of the dynamic thiol–disulfide homeostasis and the potential use of plasma levels as valuable information about specific pathological conditions associated with several human diseases, there is a need for measurements of aminothiols in biological samples to understand their physiological role. Nevertheless, the development of highly sensitive and selective analytical methods is difficult since biological thiols do not have specific physicochemical properties which are required for high detection sensitivity. Moreover, thiols are unstable in isolated plasma and have a tendency to oxidize to disulfides. Another challenge in the analysis of thiols is that the high polarity and water solubility of these compounds make them difficult to extract from complex biological matrices such as human blood plasma.

Despite these difficulties, numerous methods have been described for the determination of total thiols in biological fluids. The most commonly used techniques are high-performance liquid chromatography (HPLC) with different detection modes, mainly ultraviolet [4–6,30,39,40] and fluorescence [41–47], capillary electrophoresis [48,49], or spectrophotometry [12,50]. Although several techniques have been used to determine thiols, there are still some problems caused by the need for complicated instrumentation, complexity of the procedure, the time-consuming nature of the methods, and the number of thiols quantified simultaneously [14,30,51,52]. Among the available techniques, the most commonly used method for total thiol determination—due to its high selectivity and sensitivity—is the approach based on pre- or post-column derivatization followed by separation and fluorescence detection. For this reason, various types of labeling reagents such as bimanes, ortho-phthalaldehyde, N-substituted maleimides, halides, and halogenbenzofurazans have been developed for the selective determination of sulfhydryl group-containing compounds. Since concentrations of total thiols are important biomarkers, the development of new fluorescence labeling reagents for more sensitive thiol detection is ongoing.

A new reversed-phase (RP) HPLC-based method for the simultaneous determination of HSA, α-LA, NAC, and metabolically related major plasma low-molecular-weight aminothiols Cys, GSH, Hcy, and Cys-Gly is described herein. The method is based on the derivatization (Figure 1) of analytes with a thiol-specific labeling reagent, monobromobimane (mBBr), and fluorescence detection of stable derivatives at the excitation and emission wavelengths of 378 nm and 492 nm, respectively. The disulfides are converted to their reduced counterparts by reductive cleavage with sodium borohydride (NaBH$_4$) prior to the derivatization step. The usefulness of the developed method has been proven by its application in real plasma samples from 10 apparently healthy individuals.

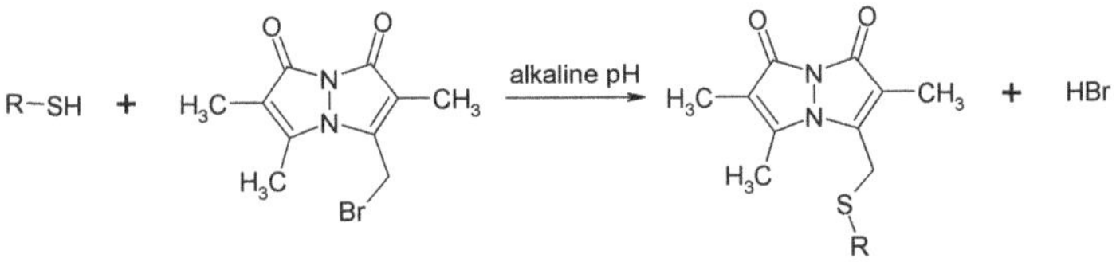

Figure 1. General equation of a chemical derivatization reaction of thiols with mBBr.

The mBBr has been previously used for the determination of low-molecular-weight aminothiols [41,43,53–56]. Now, its application has been extended to the derivatization of proteins such as albumin. The novelty of the presented procedures consists in the development of new chromatographic conditions for simultaneous determination of low-molecular-weight aminothiols and albumin and demonstration of the short-term stability of thiol-bimane adducts at acidic pH.

2. Results and Discussion

2.1. Sample Preparation

Determination of thiols in biological samples is an extremely difficult task due to their high oxidoreductive activity. Moreover, most endogenous thiols lack the structural properties necessary to provide a signal compatible with standard detectors such as spectrophotometry or fluorescence. If these detection methods are used, an additional sample preparation step, i.e., derivatization, must be initiated. However, this involves a longer overall analytical process and is associated with the risk of increasing the overall analytical error. In biological systems, thiols occur mainly in oxidized or protein-bound forms, and therefore the reduction of disulfide bonds to thiol groups must be performed prior to the derivatization reaction. $NaBH_4$ [5,41,54], dithiothreitol (DTT) [42,43,52], tris(2-carboxyethyl)phosphine (TCEP) [4,30,39,57], or 2-mercaptoethanol (2-ME) [6,52] are the most commonly used substances to reduce disulfide bonds. The choice of an appropriate reductant depends on the analytical procedure to be followed, especially on the type of the derivatizing reagent used and the detection method. The use of reagents containing a sulfhydryl group, such as 2-ME or DTT, is almost impossible when using a derivatizing reagent that is highly selective towards the thiol group. The derivatives formed lead to increased consumption of the derivatizing reagent, which can lead to underestimation of the results. In addition, these derivatives interfere with the peaks of the analytes by forming additional very large signals in the chromatogram, making interpretation difficult. Limitations in the use of trialkyl phosphines relate primarily to their toxicity, but also to their irritating and pungent odor. Moreover, the most commonly used phosphine, TCEP, can interfere with certain derivatizing agents such as mBBr [58,59] or 5,5'-dithiobis-2-nitrobenzoic acid [60]. $NaBH_4$ seems to be the optimal reducing reagent. It is non-toxic, readily available, and the excessive amount can be easily and quickly removed by adding hydrochloric acid or acetone. The major advantage of this reducing agent is the short reaction time at high reagent concentrations. Since $NaBH_4$ is unstable in aqueous solutions, it must be prepared immediately before use. The addition of an organic solvent such as dimethyl sulfoxide (DMSO) increases the stability of $NaBH_4$. The main disadvantage of using $NaBH_4$ is its tendency to form aerosols during the reaction due to the intense release of hydrogen. However, this phenomenon can be eliminated by the addition of surfactants such as n-octanol. The thiols obtained in the reduction step can be re-oxidized, so the resulting product should be immediately subjected to a derivatization reaction to block the labile sulfhydryl group. Additionally, the metal cation-catalyzed oxidation of thiols to disulfides can be slowed down by adding a complexing compound such as ethylenediaminetetraacetic acid disodium salt ($EDTA-Na_2$) to the sample. In HPLC coupled with fluorescence detection, such derivatizing reagents as mBBr [41,43,53–56], ammonium 4-fluoro-2,1,3-benzoxadiazole-7-sulfonate [46,47], 4-aminosulfonyl-7-fluoro-2,1,3-benzoxadiazole [44,45], o-phthalaldehyde [61,62], or 4-bromomethyl-6,7-dimethoxycoumarin [63,64], have been widely used for the derivatization of thiols in biological samples. Most of these reagents have an active halogen in their structure, which is exchanged for the sulfur atom of the thiol group of the analyte in a nucleophilic substitution reaction to form a stable thioether.

2.1.1. Disulfide Bonds Cleavage

The determination of total thiol requires the reduction of disulfide bonds since most low-molecular-weight plasma thiols are oxidized in the form of symmetrical or mixed disulfides and bound to proteins, mainly albumin. This fact makes the thiol groups inaccessible to the derivatizing reagent, and therefore plasma samples must be treated with a reducing reagent to release the free thiols. $NaBH_4$ was used for this purpose. We examined the time and molar excess of the borohydride response curves to determine the conditions necessary for complete reduction of the disulfide bonds. Maximum reduction was achieved after five minutes of incubation at room temperature. The highest peak areas for all the analytes were observed when 30 μL of 6 mol L^{-1} $NaBH_4$ were used for the reduction (data not shown). The plasma Hcy, Cys-Gly, and Cys assay methods developed by Fiskerstrand et al. [53] used a lower concentration of $NaBH_4$ (4 mol L^{-1}) for a 3-min reduction, but the addition of dithioerythritol was necessary for maximum yield of all thiol components. Another approach was used by Mansoor et al. [41], where samples were incubated simultaneously with 2 mol L^{-1} $NaBH_4$ and mBBr for 10 min at room temperature in the dark while total plasma thiols were determined. In both methods, the volume of NaBH$_4$ added to the sample was also 30 µL. In our method, 6 µL of 3 mol L^{-1} HCl were added to the thiol-containing sample to remove excess reductant. To avoid foaming of the reaction mixture due to the strong hydrogen evolution, 20 µL of n-octanol were added to the plasma samples according to the previously described methods [41,53]. These reduction conditions were adopted for a routine plasma analysis for total thiols, including HSA.

2.1.2. Derivatization via the –SH Group

Since thiols do not contain a fluorophore in their structure that allows their direct fluorimetric detection, it is necessary to perform a derivatization reaction during the sample preparation step. For this purpose, we used mBBr, which reacts violently with thiols at alkaline pH and room temperature. The resulting highly fluorescent and stable thioethers can be easily detected, even at low analyte concentrations. The reagent itself also fluoresces and its peak is visible in the chromatogram along with the mBBr hydrolysis reaction products (Figure 2). Monobromobimane is readily photodegraded, so it is necessary to protect it from light and perform the reaction in the dark. The yield of the derivatization reaction was optimized with respect to pH (Figure S1), excess reagent (Figure S2), and time (Figure S3). The results indicate that the optimum reaction pH for the derivatization of endogenous plasma thiols with mBBr occurred at pH 9.5, and the reaction was completed after 10 min with 70-fold excess reagent in the dark and at room temperature. The obtained conditions are similar to those described earlier, where the reaction was carried out in N-ethylmorpholine at pH 9 for 10 min [41] or 20 min [54].

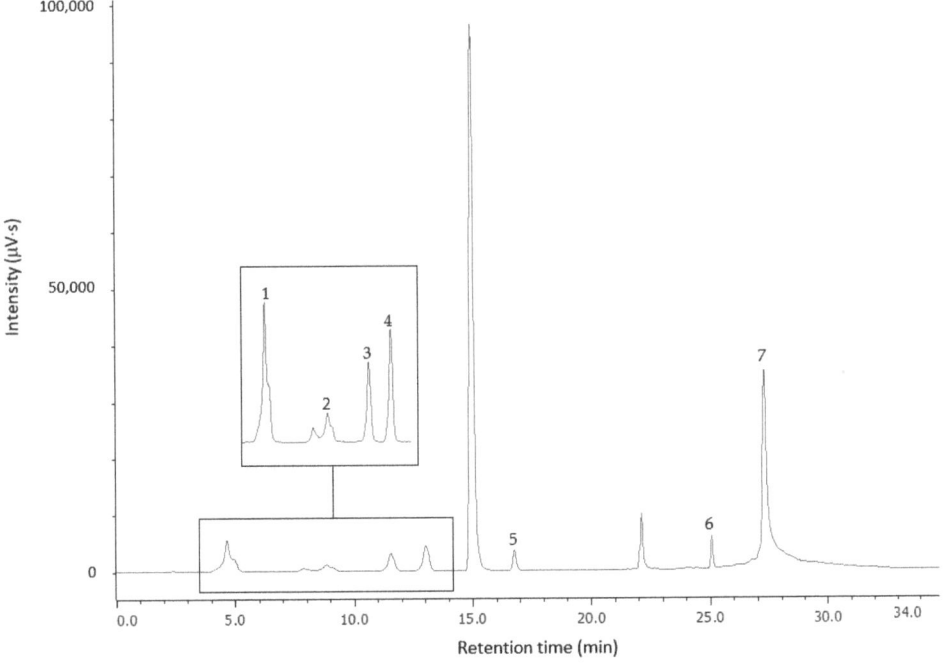

Figure 2. Representative chromatogram for total forms of 10 nmol mL^{-1}: Cys (1), Cys-Gly (2), Hcy (3), GSH (4), NAC (5), α-LA (6), and 3 mg mL^{-1} HSA (7) in standard water solutions after reduction with NaBH$_4$ and derivatization with mBBr. Chromatographic conditions as described in Section 3.2. The unsigned peaks are derived from mBBr and its hydrolysis reaction products.

Finally, the derivatization reaction was quenched by adding 200 µL of 1 mol L^{-1} HCl to reach pH of 3. Other acids used to complete the derivatization of low-molecular-weight thiols include perchloric acid [41] or glacial acetic acid [53], however, their use in this procedure would precipitate HSA. The short-term stability of thiol–bimane adducts is not described in the literature, therefore it was tested during the research. For all the analytes, the thiol–bimane derivatives were stable in an acidic environment at room temperature over the time period studied (Figure 3). This means that the final solution kept at room temperature in the autosampler can be safely analyzed for up to 15 h which allows for long unattended runs.

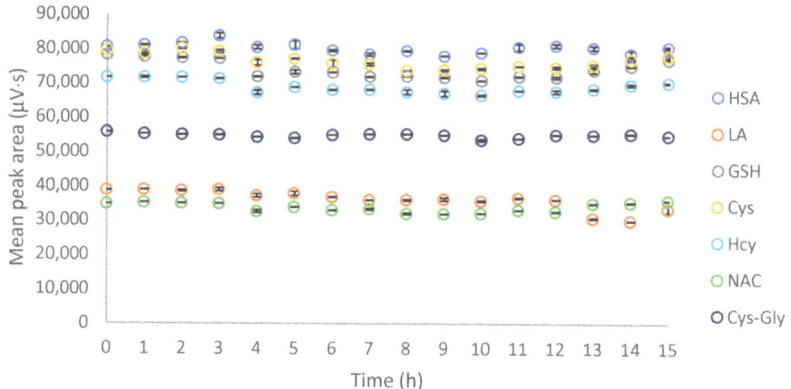

Figure 3. Stability of the analyzed thiol–bimane derivatives in an acidic environment, n = 3. In the figure, the standard deviation bars of peak areas are presented.

2.2. Chromatography

Preliminary experiments aimed at developing a method for the simultaneous determination of low-molecular-weight thiols and HSA in plasma were performed in standard water solutions. In order to achieve good separation of the analyzed thiol–bimane adducts, various parameters of the chromatographic conditions were tested, including the concentration of trifluoroacetic acid (TFA), acetonitrile (ACN) content, pH of the mobile phase, concentration of the organic modifier, and gradient profile. The experiments mentioned briefly above led to the establishment of optimum chromatographic conditions for all the analytes, which are described in Section 3.2. Under these conditions, the thiol–bimane derivatives eluted after 5.14 min (RSD, 0.86%, n = 3) for Cys, 9.63 min (RSD, 0.77%, n = 3) for Cys-Gly, 12.30 min (RSD, 0.61%, n = 3) for Hcy, 13.76 min (RSD, 0.42%, n = 3) for GSH, 17.07 min (RSD, 0.74%, n = 3) for NAC, 25.08 min (RSD, 0.07%, n = 3) for α-LA, and 27.82 min (RSD, 0.05%, n = 3) for HSA. As can be seen from the chromatogram in Figure 2, the four bimane derivatives of low-molecular-weight thiols elute in pairs: GSH in close proximity to Hcy and Cys in close proximity to Cys-Gly, but in this case, the peaks are separated by a reagent peak. These results are in agreement with those obtained for CMQT derivatives [5,30]. The longer hydrophobic chain of NAC and reduced α-LA causes these compounds to elute later than the remaining small thiols. The HSA derivative exhibits the highest hydrophobicity and elutes last. The retention factors for all the analytes varied depending on the concentration and pH of the TFA used in the aqueous phase as a consequence of a change in the hydrophobicity of the eluent and the degree of ionization of the analyte. Despite the high particle size discrepancy of the analyzed biothiols, optimization of the separation conditions resulted in the good separation of the mBBr derivatives of Cys, Cys-Gly, Hcy, GSH, NAC, α-LA, and HSA.

2.3. Method Validation

The method was validated according to the guidelines for the analysis of biological samples [65,66]. The validation parameters of linearity, precision, accuracy, and limits of detection (LOD) and quantification (LOQ) were checked and tested. The standard addition method was used to calibrate the method.

2.3.1. Linearity

The linearity of the method was determined by analyzing plasma samples spiked with the standard solution of thiols prepared as described in Section 3.5.1. Six-point calibration curves were prepared for six low-molecular-weight thiols and HSA in triplicate (Appendix A). Each sample was processed according to the recommended procedure. The validation parameters for all fitted calibration plots were satisfactory. Detailed method calibration data are presented in Table 1.

Table 1. Validation data.

Analyte	Regression Equation	R^2	Linear Range (nmol mL^{-1})	Precision (%)		Recovery (%)		LOD (nmol mL^{-1})	LOQ (nmol mL^{-1})
				Min.	Max.	Min.	Max.		
HSA	y = 150,925x + 10,000,000	0.9953	1.76–30.0 [a]	0.40	5.80	92.94	113.67	0.59 [a]	1.76 [a]
α-LA	y = 3487.6x + 4187.2	0.9977	0.29–5.0	0.26	6.48	83.38	106.88	0.10	0.29
GSH	y = 6153x + 18,335	0.9991	1.16–35.0	0.24	10.92	84.71	99.15	0.39	1.16
Cys	y = 6765.9x + 806,715	0.9994	9.83–450.0	0.89	6.24	87.27	110.74	3.28	9.83
Hcy	y = 6621.9x + 34,488	0.9994	0.55–40.0	0.90	11.93	83.89	115.77	0.18	0.55
NAC	y = 34,376x + 379.58	0.9999	0.34–50.0	0.17	1.26	81.87	106.07	0.11	0.34
Cys-Gly	y = 3370.3x + 54,775	0.9998	1.45–45.0	0.90	5.33	96.34	114.56	0.48	1.45

[a] The unit of measurement for linear range, LOD and LOQ for HSA is mg mL^{-1}.

2.3.2. Limits of Detection and Quantification

LOD is defined as the lowest concentration of the analyte that can be detected with a given measuring device, but without its quantitative determination. LOQ is the lowest concentration of a substance that can be quantified with a given analytical method with the assumed accuracy and precision. The measured validation parameters were evaluated by applying the calibration standard of thiols in proxy matrix (0.9% NaCl in 10 mmol L^{-1} phosphate buffer, pH 7.4) solutions obtained on this occasion. LOD was calculated according to the following formula:

$$LOD = \frac{3.3\,s}{b} \quad (1)$$

where b is the slope of the calibration curve, and s is the standard deviation of the intercept of the regression line.

LOQ values for the analytes tested were calculated based on the LOD using the equation below:

$$LOQ = 3 LOD \quad (2)$$

The limits of detection and quantification for HSA, Cys, Hcy, Cys-Gly, NAC, α-LA, and GSH are presented in Table 1.

2.3.3. Precision and Accuracy

To assess the quality of the devised method for the simultaneous determination of plasma thiols, including HSA, precision and accuracy were determined. The intra- and inter-day precision and accuracy values were measured in triplicate in plasma samples spiked with standard solutions of thiols to obtain four concentrations representing the full range of calibration curves. The measured concentrations were determined by applying the calibration curves obtained on this occasion. Precision was expressed as relative standard deviation (RSD) while accuracy was expressed as percentage recovery of the analyte using the following formula:

$$Accuracy\ (\%) = \frac{measured\ amount - endogenous\ content}{added\ amount} \times 100\% \quad (3)$$

The inter-day precision and recovery values were evaluated on three consecutive days in a week. Detailed intra-day and inter-day precision and accuracy data are presented in Table 2. The estimated validation parameters met the requirements applicable to the analysis of biological samples.

Table 2. Intra-day and inter-day precision and accuracy evaluation for thiols in human plasma samples based on the proposed method, $n = 3$.

Analyte	Concentration (nmol mL^{-1})	Precision (%)		Accuracy (%)	
		Intra-Day	Inter-Day	Intra-Day	Inter-Day
HSA [a]	1.76	0.35	1.49	119.48	99.31
	5.0	0.10	4.83	96.77	109.77
	15.0	0.35	1.99	97.49	94.78
	30.0	0.37	2.47	100.65	101.05
α-LA	0.29	0.55	0.18	97.16	84.30
	0.9	0.34	0.03	104.90	95.83
	1.75	0.60	7.34	97.31	107.24
	4.0	0.20	9.18	100.28	98.91
GSH	1.16	1.50	2.87	89.10	116.57
	4.0	0.71	1.25	100.67	100.87
	15.0	0.05	0.28	101.31	97.14
	30.0	0.13	0.22	99.68	100.68
Cys	9.83	2.23	4.25	85.56	97.18
	30.0	2.66	2.22	93.51	115.62
	100.0	0.57	6.94	104.43	94.38
	350.0	0.10	10.71	99.70	100.35
Hcy	0.55	3.85	14.57	97.45	82.17
	1.7	5.34	2.55	88.30	90.84
	15.0	0.43	0.27	102.69	103.24
	30.0	0.01	3.69	99.37	99.22
NAC	0.34	0.25	0.11	85.24	98.98
	1.0	0.48	2.05	100.83	115.95
	20.0	0.29	0.73	100.42	98.48
	40.0	0.04	0.10	99.90	100.37
Cys-Gly	1.45	1.43	5.09	108.52	93.58
	4.5	0.19	1.58	85.16	82.99
	20.0	0.19	0.36	105.41	108.83
	35.0	0.70	0.19	98.46	97.41

[a] Unit of measurements for HSA is mg mL^{-1}.

2.3.4. Matrix Effect

The matrix effect was evaluated using the slope comparison method. First, the slope coefficients of the calibration curves obtained for thiol standard solutions and the calibration curves obtained for a plasma sample spiked with a known amount of the analyte were compared. The differences between the slope coefficients of the calibration curves for each analyte are presented with reference to RSD, which is, respectively, 34.50% for HSA, 16.86% for α-LA, 11.18% for Cys, 19.73% for Hcy, 15.71% for GSH, 16.24% for NAC, and 141.30% for Cys-Gly. These data indicate that there was a significant matrix effect on the results of endogenous thiol determination in plasma samples. Next, we investigated whether the matrix effect on assay results occurred between plasma samples from five different subjects. The RSD for the coefficients of the calibration curves was as follows: 54.31% for HSA, 31.75% for α-LA, 15.87% for GSH, 17.16% for Cys, 20.50% for Hcy, 26.55% for NAC, and 23.10% for Cys-Gly. Furthermore, in this case, the obtained results prove the existence of the matrix effect in the determination of thiol compounds in plasma samples collected from different individuals. Such an excessive impact of the matrix may result from the fact that plasma samples from randomly selected individuals of different ages were used in the study, without preselection of the subjects for medications, chronic diseases, and other factors that may influence the content of the tested compounds in the bloodstream.

2.4. Application to Real Plasma Samples

The devised and validated method was applied to the simultaneous determination of HSA and low-molecular-weight thiols in plasma samples from 10 apparently healthy subjects (Figure 4). Due to the matrix effect, the method of simple standard addition was used for the determination of analytes [67]. The concentration of the analyte in each sample was calculated using the following formula:

$$C_x = \frac{Y_x \times C_s}{Y_s - Y_x} \tag{4}$$

where C_x—concentration of the analyte in the plasma sample; Y_x—analytical signal for the sample containing only the analyte; C_s—concentration for the sample with the addition of a known amount of the standard; Y_s—analytical signal for the sample with the addition of a known amount of the standard.

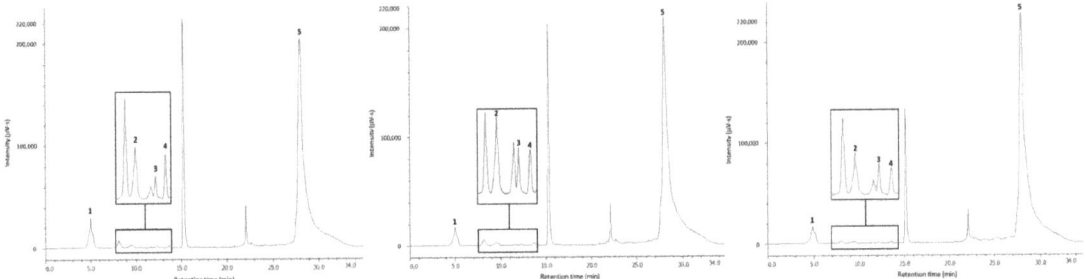

Figure 4. Chromatograms of plasma samples derived from potentially healthy volunteers (from the left: Nos. 1, 3, and 9 Figure 3. after reduction with NaBH₄ and derivatization with mBBr. The peaks labeled 1, 2, 3, 4, and 5 correspond to the thiol–bimane signals Cys, Cys-Gly, Hcy, GSH, and HSA, respectively. Chromatographic conditions as described in Section 3.2. The unsigned peaks are derived from mBBr and its hydrolysis reaction products.

The expected concentration ranges for total thiols in human plasma were as follows: 3.27–13.37 nmol mL^{-1} for Hcy [56], 135.80–266.50 nmol mL^{-1} for Cys [56], 19.24–39.74 nmol mL^{-1} for Cys-Gly [56], 4.90–7.30 nmol mL^{-1} for GSH [11], 35.9–48.3 mg mL^{-1} for HSA [30], and 1.33–5.8 nmol mL^{-1} for α-LA (in the case of administration of 20 or 100 mg LA supplement) [30,45]. We also expected that, without administration, NAC could not be detected in plasma from healthy volunteers at a concentration exceeding the method's detection limit as reported earlier [15]. The mean total thiol content ± standard deviation (SD) in human plasma was as follows: 23.98 ± 1.60 mg mL^{-1} for HSA, 62.99 ± 4.31 nmol mL^{-1} for Cys, 3.93 ± 0.28 nmol mL^{-1} for Hcy, 5.29 ± 0.31 nmol mL^{-1} for GSH, and 50.20 ± 3.28 nmol mL^{-1} for Cys-Gly. The concentrations of α-LA and NAC were not detected in the plasma samples examined. This may be due to the fact that the volunteers did not take specimens containing these compounds. Detailed analytical data for total Cys, Hcy, GSH, Cys-Gly, and HSA are shown in Table 3. These results are mostly in agreement with those previously reported [5,11,30,39,56,57]. The elaborated method can be successfully applied in large populations to monitor changes in thiol concentration in different physiological conditions.

Table 3. Total low-molecular-weight thiols (nmol mL^{-1}) and HSA (mg mL^{-1}) in plasma from 10 subjects, $n = 3$.

No.	HSA		GSH		Cys		Hcy		Cys-Gly	
	Mean (SD)	RSD (%)	Mean (SD)	RSD (%)	Mean (SD)	RSD (%)	Mean (SD)	RSD (%)	Mean (SD)	RSD (%)
1	42.62 (3.01)	7.06	4.14 (0.25)	6.00	65.54 (2.08)	3.22	3.07 (0.28)	9.05	30.71 (0.29)	0.96
2	17.58 (1.51)	8.61	7.93 (0.67)	8.44	54.11 (3.06)	5.66	2.99 (0.11)	3.70	62.55 (3.40)	5.43
3	11.73 (0.72)	6.11	5.80 (0.28)	4.85	74.72 (7.48)	10.01	4.76 (0.22)	4.59	46.49 (4.98)	10.71
4	40.34 (2.64)	6.53	6.00 (0.40)	6.67	89.34 (4.60)	5.14	5.46 (0.54)	9.95	80.28 (3.14)	3.91
5	10.71 (0.49)	4.56	4.77 (0.22)	4.64	67.38 (5.56)	8.25	3.72 (0.25)	6.67	55.48 (4.38)	7.89
6	13.63 (1.25)	9.18	5.22 (0.33)	6.29	70.00 (3.38)	4.83	3.42 (0.14)	4.03	90.16 (7.00)	7.76
7	14.92 (0.84)	5.64	5.25 (0.23)	4.32	36.85 (3.55)	9.64	2.48 (0.07)	2.62	42.90 (1.40)	3.27
8	24.99 (1.02)	4.08	5.69 (0.24)	4.21	53.95 (4.28)	7.94	5.47 (0.48)	8.83	45.75 (3.14)	6.86
9	38.12 (3.18)	8.33	5.33 (0.41)	7.66	41.59 (3.31)	7.95	3.32 (0.31)	9.23	19.63 (1.83)	9.31
10	25.15 (1.31)	5.22	2.73 (0.10)	3.79	77.44 (5.79)	7.48	4.56 (0.45)	9.83	28.08 (3.22)	11.48

3. Materials and Methods

3.1. Chemicals and Reagents

Dimethyl sulfoxide (DMSO), 1-octanol, trifluoroacetic acid (TFA), oxidized glutathione (GSSG), cysteinylglycine (Cys-Gly), homocysteine (Hcy-Hcy), N-acetyl-L-cysteine (NAC), α-lipoic acid (LA), human serum albumin (HSA), and monobromobimane (mBBr) were obtained from Sigma Aldrich (St. Louis, MO, USA). L-cystine (Cys-Cys) was purchased from Reanal (Budapest, Hungary). Ethylenediaminetetraacetic acid disodium salt (EDTA-Na$_2$), sodium hydroxide (NaOH), and HPLC-grade acetonitrile (ACN) were bought from J.T. Beaker (Deventer, The Netherlands). Hydrochloric acid (HCl) was supplied by POCH (Gliwice, Poland). Sodium borohydride (NaBH$_4$) was delivered by Merck (Darmstadt, Germany), and 2-amino-2-(hydroxymethyl)-1,3-propanediol (Tris base) was from BioShop (Canada). Deionized water was prepared in our laboratory using the Mili-QRG system (Millipore, Vienna, Austria). The pH of the buffers was adjusted by potentiometric titration using a system calibrated with standard pH solutions.

3.2. HPLC Instrumentation and Chromatographic Conditions

All the analyses were performed on an integrated LC-4000 Series JASCO RHPLC system (JASCO, Tokyo, Japan) equipped with a quaternary pump (model No. PU-4180, Tokyo, Japan), a vacuum degasser, an autosampler (model No. AS-4150, Tokyo, Japan), a column oven (model No. CO-4062, Tokyo, Japan), and a fluorescence detector (model No. FP-4020, Tokyo, Japan) operating at the excitation and emission wavelengths of 378 nm and 492 nm, respectively. The detector signal was amplified tenfold. System control and data acquisition processes were performed using the ChromNAV2 software. Spectra Manager ver. 2 was used to analyze the spectra.

The samples (5 µL) were injected onto a 150 × 4.6 mm, 3.6 µm particle size Aeris™ WIDEPORE XB-C18 column (Phenomenex, CA, USA) using an autosampler. The mobile phase consisted of 0.1% TFA in water (A) and 0.1% TFA in ACN (B). All the analyses were performed at room temperature and the flow rate of the mobile phase was 1 mL min^{-1}. Chromatographic separation of HSA and low-molecular-weight thiols was achieved in

35 min with gradient elution as follows: 0–5 min, 5% B; 5–11 min, 5–8% B; 11–18 min, 8–15% B; 18–20 min, 15–30% B; 20–25 min, 30–40% B; 25–26 min, 40–50% B; 26–30 min, 50–70% B; 30–33 min, 70–5% B; 33–35 min, 5% B. Peaks were identified by comparing retention times and fluorescence spectra with corresponding data from the authentic standard.

3.3. Preparation of Stock and Buffer Solutions

Stock solutions of 0.01 mol L^{-1} GSSG, Hcy-Hcy, Cys-Gly and 0.1 mol L^{-1} Cys-Cys and NAC required for method development were prepared by dissolving appropriate amounts of the compounds in 100 µL of 1 mol L^{-1} HCl and diluting to 1 mL. A stock solution of 0.01 mol L^{-1} α-LA was prepared in 0.1 mol L^{-1} NaOH. The stock solution of HSA was prepared by dissolving 300 mg of the protein in 1 mL deionized water. These solutions were kept at 4 °C for one week without any noticeable change in the analyte content. The working solutions were prepared daily by appropriate dilutions with deionized water as needed and processed without delay.

The stock solution of the reducing agent was prepared daily by dissolving appropriate amounts of $NaBH_4$ in 500 µL of 0.1 mol L^{-1} NaOH to give a concentration of 6 mol L^{-1} and then preparing a mixture of $NaBH_4$ and DMSO in a volume ratio of 2:1.

The stock solution of 0.1 mol L^{-1} mBBr was prepared in ACN and stored at 4 °C in the dark for up to one month, without any noticeable change in the content.

The buffer solution of 0.2 mol L^{-1} Tris–HCl, pH 9.5, containing 0.2 mmol L^{-1} EDTA-Na_2 was prepared by dissolving 2.4228 g Tris base in 100 mL water and adjusted with 3 mol L^{-1} HCl to pH 9.5 by potentiometric titration. Then, the stock solution of 0.02 mol L^{-1} EDTA was prepared by dissolving 0.0074448 g of the compound in a 0.2 mol L^{-1} Tris–HCl buffer solution, pH 9.5. Finally, 900 µL EDTA stock solution was mixed with 89.1 mL buffer solution for a final volume of 90 mL.

3.4. Human Plasma Sample Collection and Storage

Blood (2 mL) was collected by venipuncture from 10 apparently healthy subjects of various ages into vacutainer tubes containing EDTA. The tubes were immediately placed on ice and centrifuged at 3500 rpm at 4 °C for 10 min. After centrifugation, the clear plasma supernatant was collected and stored at −80 °C until analysis.

3.5. Analytical Method Validation

The methods were validated according to the International Conference on Harmonization (ICH) guidelines for validation of analytical procedures [66] and the Food and Drug Administration (FDA) guidelines for analytical procedures and methods validation for drugs and biologics [65].

3.5.1. Preparation of Calibration Standards

To prepare calibration standards for the determination of total thiols in human plasma, 70 µL plasma from each of the apparently healthy individuals was placed in a test tube and spiked with 10 µL disulfide mixture to obtain the following concentrations: 1.76, 3.0, 5.0, 20.0, 25.0, 30.0 mg mL^{-1} plasma for HAS; 0.29, 0.5, 1.0, 2.0, 4.0, 5.0 nmol mL^{-1} plasma for α-LA; 1.16, 3.0, 5.0, 10.0, 20.0, 35.0 nmol mL^{-1} plasma for GSH; 9.83, 20.0, 35.0, 50.0, 250.0, 450.0 nmol mL^{-1} plasma for Cys; 0.55, 1.0, 3.0, 5.0, 20.0, 40.0 nmol mL^{-1} for Hcy; 0.34, 1.0, 5.0, 10.0, 30.0, 50.0 nmol mL^{-1} for NAC; and 1.45, 5.0, 10.0, 20.0, 30.0, 45.0 nmol mL^{-1} plasma for Cys-Gly. In the next step, 20 µL 1-octanol was added. Subsequently, the disulfide bonds were reduced for five minutes at room temperature by treatment with 30 µL of a mixture containing 6 mol L^{-1} $NaBH_4$ in DMSO (2:1; *v:v*) followed by adding 6 µL of 3 mol L^{-1} HCl. After the reduction reaction ended, 657 µL of 0.2 mol L^{-1} Tris–HCl (pH 9.5) with 0.2 mmol L^{-1} EDTA buffer were added. Eventually, the thiols were derivatized with 7 µL of 0.1 mol L^{-1} mBBr for 10 min at room temperature in the dark. To stop the derivatization reaction, 200 µL of 1 mol L^{-1} HCl were added. Of the final sample, 5 µL were injected into the chromatographic column. In all cases, the calibration standards were prepared in triplicate.

3.5.2. Calibration Curves

The calibration range was 1.76–30.0 mg mL^{-1} plasma for HSA, 0.29–5.0 nmol mL^{-1} plasma for α-LA, 1.16–35.0 nmol mL^{-1} plasma for GSH, 9.83–450.0 nmol mL^{-1} plasma for Cys, 0.55–40.0 nmol mL^{-1} for Hcy, 0.34–50.0 nmol mL^{-1} for NAC, and 1.45–45.0 nmol mL^{-1} plasma for Cys-Gly. The calibration curves were constructed using a linear least-squares regression model by plotting the peak area of the respective thiol mBBr derivative against the analyte concentration.

3.5.3. Limits of Detection and Quantification

LOD and LOQ were determined based on the standard deviation of the intercept and the slope of the calibration curve obtained for standard solutions of the analytes. The calibration standards and curves were prepared as follows: 70 µL proxy matrix (0.9% NaCl in 10 mmol L^{-1} phosphate buffer, pH 7.4) was placed in each test tube and spiked with 10 µL disulfide mixture to obtain the following concentrations: 1.0, 3.0, 5.0, 20.0, 25.0, and 30.0 mg mL^{-1} for HSA; 0.1, 0.3, 0.5, 2.0, 4.0, and 5.0 nmol mL^{-1} for α-LA; 0.3, 1.0, 5.0, 10.0, 20.0, and 35.0 nmol mL^{-1} for GSH; 5.0, 7.0, 10.0, 50.0, 250.0, and 450.0 nmol mL^{-1} for Cys; 0.3, 0.5, 1.0, 5.0, 20.0, and 40.0 nmol mL^{-1} for Hcy; 1.0, 3.0, 5.0, 7.0, 10.0, and 50.0 nmol mL^{-1} for NAC; 0.7, 5.0, 10.0, 20.0, 30.0, and 45.0 nmol mL^{-1} for Cys-Gly. The further procedure was the same as in Section 3.5.1.

3.5.4. Precision and Accuracy

Assay precision was determined intra-day and inter-day. Intra-day precision was assessed by assaying samples with the same concentration in triplicate and on the same day. Plasma samples were enriched with thiol standards at the following concentrations: 1.76, 5.0, 15.0, and 30.0 mg mL^{-1} for has; 0.29, 0.9, 1.75, and 4.0 nmol mL^{-1} for α-LA; 1.16, 4.0, 15.0, and 30.0 nmol mL^{-1} for GSH; 9.83, 30.0, 100.0, and 350.0 nmol mL^{-1} for Cys; 0.55, 1.7, 15.0, and 30 nmol mL^{-1} for Hcy; 0.34, 1.0, 20.0, and 40.0 nmol mL^{-1} for NAC; and 1.45, 4.5, 20.0, and 35.0 nmol mL^{-1} for Cys-Gly. The inter-day precision was investigated by comparing the assays on three different days. Three sample solutions with the same concentration as above were prepared and assayed in triplicate.

3.5.5. Matrix Effect Evaluation

The slope comparison method was used to evaluate the matrix effect. First, the slopes of the standard curves in plasma were compared with the slopes of the standard curves in a proxy matrix (0.9% NaCl in 10 mmol L^{-1} phosphate buffer, pH 7.4). In this case, the matrix effect was determined by spiking 70 µL of plasma or proxy matrix with 10 µL of mixed standard solutions to obtain low, medium, and high analyte concentrations as follows: 7.0, 50.0, and 250.0 nmol mL^{-1} for Cys; 2.0, 10.0, and 30.0 nmol mL^{-1} for Cys-Gly; 0.5, 5.0, and 20.0 nmol mL^{-1} for GSH and Hcy; 0.5, 2.0, and 4.0 nmol mL^{-1} for α-LA; 5.0, 10.0, and 150.0 nmol mL^{-1} for NAC; and 5.0, 10.0, and 25.0 mg mL^{-1} for HSA. The rest of the procedure was the same as in Section 3.5.1, and all the analyses were performed in triplicate. The next step was to evaluate the matrix effect in plasma samples from five different subjects at the same analyte concentrations and using the same protocol as above. The slopes of the calibration curves from the standard addition experiments were then compared for all the analytes.

3.6. Application to Real Samples

The content of total thiols in plasma samples from 10 potentially healthy volunteers was estimated. The method of single standard addition was used. For this purpose, measurements of the analytical signal were made for the plasma samples containing only the analyte and for the plasma samples with the addition of a known amount of the standard at concentrations of 50.0 nmol mL^{-1} for Cys and NAC, 20.0 nmol mL^{-1} for Cys-Gly, GSH, and Hcy, 4.0 nmol mL^{-1} for α-LA, and 10.0 mg mL^{-1} for HSA. The procedure was the same as presented in Section 3.5.1.

3.7. Statistical Analysis

All calculations, graphs, and statistical analyses were performed using Microsoft Excel 16.0 (Microsoft Corporation). Each value in the calibration charts represents the mean of three independent measurements with the standard deviations indicated. Unless otherwise noted, the graphs show a representative set of results from a plasma sample obtained from a single individual. All the results were presented as the means ± SD of three chromatographic runs. Linear regression was applied to develop an equation for predicting thiol concentration in plasma. Linear least-squares regression was used to calculate the linearity relationship between peak ranges and analyte concentrations.

4. Conclusions

In recent years, the detection of biothiols has attracted considerable interest because of their central role in a variety of physiological and pathological processes in the human body [52]. Continuous monitoring of the concentrations of endogenous thiols in biological fluids is extremely important to determine their content changes during disease development. Therefore, it can be a valuable source of information when assessing a patient's health status. HPLC-FLD is the most widely used method for the determination of endogenous thiols in biological samples due to its high sensitivity and robustness [14]. In this work, we propose a new, sensitive, and simple protocol for the simultaneous determination of HSA and six low-molecular-weight thiols in human plasma. The assay is based on the reduction of disulfide bonds with NaBH$_4$ followed by pre-column derivatization with mBBr. The thiol–bimane derivatives are then separated and quantified by RP-HPLC with fluorescence detection at the excitation and emission wavelengths of 378 nm and 492 nm, respectively. The major advantage of the presented method is that it allows the simultaneous determination of compounds with very different physicochemical properties in a complex matrix at the same analytical wavelengths and in a relatively short time of 35 min. The method was validated according to the FDA [65] and ICH [66] guidelines and the recovery, accuracy, and precision values meet the criteria for the analysis of biological samples. The devised method can be successfully used to determine HSA, α-LA, GSH, Cys, Hcy, Cys-Gly, and NAC concentrations in plasma samples derived from potentially healthy individuals.

Supplementary Materials: The following are available online, Figure S1: Dependence of the mean peak area on the pH of the buffer used for the derivatization reaction, n = 3, Figure S2: Dependence of the mean peak area on the volume of mBBr, n = 3, Figure S3: Dependence of the mean peak area on the time of the derivatization reaction, n = 3.

Author Contributions: Conceptualization, methodology, software, validation, formal analysis, investigation, data curation, writing—original draft preparation, resources, visualization, K.K.; writing—review and editing, supervision, project administration, funding acquisition, G.C. and R.G. All authors have read and agreed to the published version of the manuscript.

Funding: This research received no external funding.

Institutional Review Board Statement: The study was conducted according to the guidelines of the Declaration of Helsinki and approved by the Ethics Committee of the University of Lodz (protocol code 10/KBBN-UŁ/I/2020-21, approved on 15 December 2020).

Informed Consent Statement: Informed consent was obtained from all the subjects involved in the study.

Data Availability Statement: HPLC data are available from the authors.

Conflicts of Interest: The authors declare no conflict of interest.

Sample Availability: Plasma samples were damaged during the experiments and are not available from the authors.

Appendix A

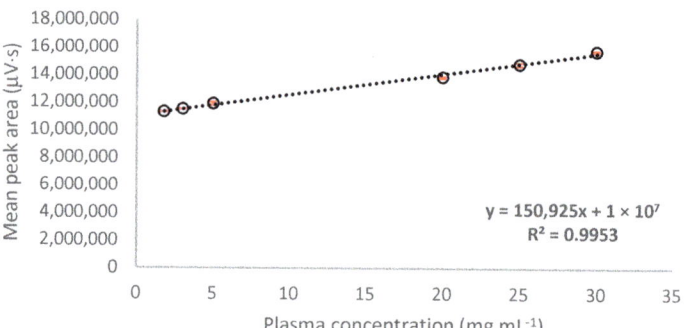

Figure A1. Calibration curve for human serum albumin.

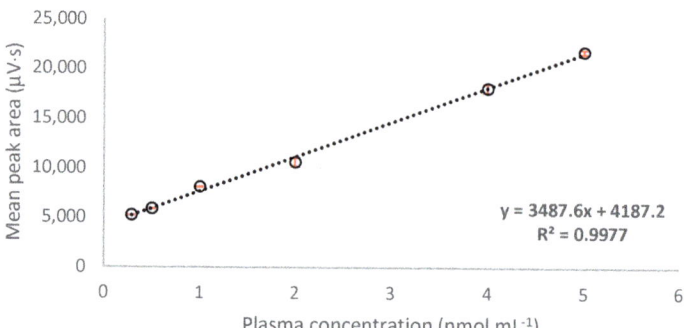

Figure A2. Calibration curve for α-lipoic acid.

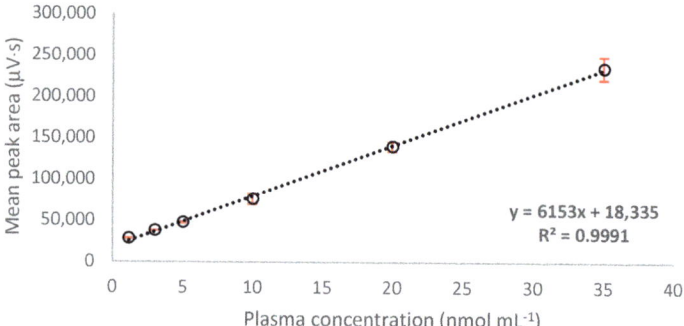

Figure A3. Calibration curve for glutathione.

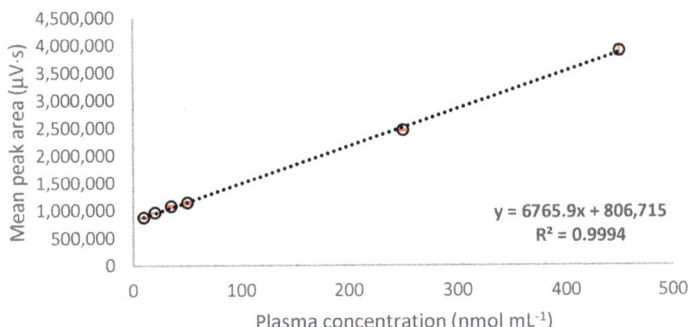

Figure A4. Calibration curve for cysteine.

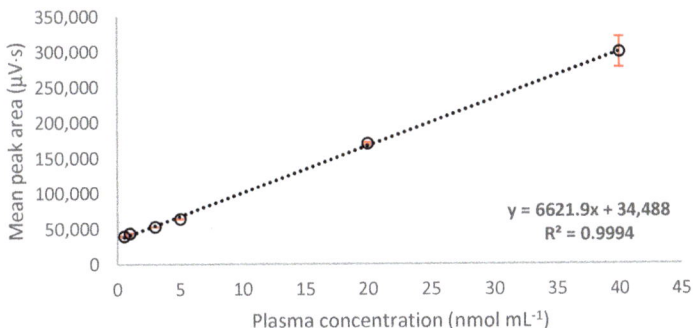

Figure A5. Calibration curve for homocysteine.

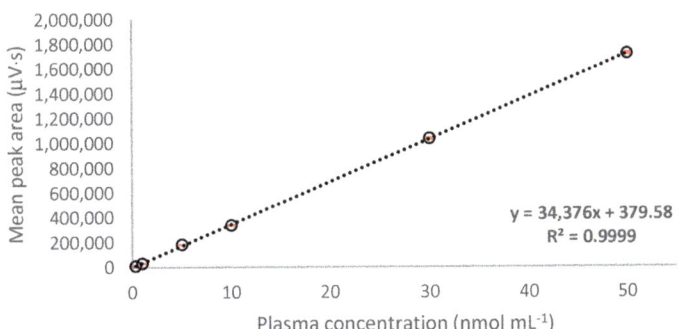

Figure A6. Calibration curve for *N*-acetyl-L-cysteine.

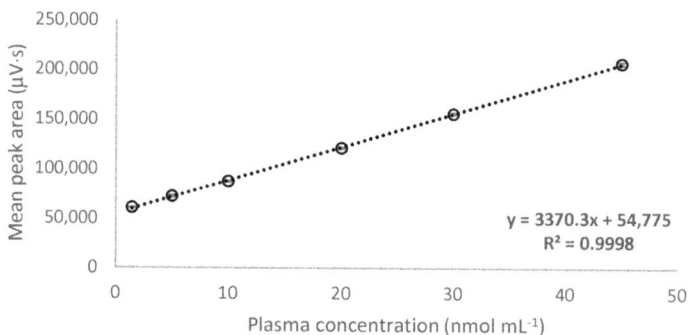

Figure A7. Calibration curve for cysteinylglycine.

References

1. Prakash, M.; Shetty, M.S.; Tilak, P.; Anwar, N. Total Thiols: Biomedical importance and their alteration in various disorders. *Online J. Health Allied Sci.* **2009**, *8*, 6664.
2. Onat, T.; Yalçın, S.; Kırmızı, D.A.; Başer, E.; Ercan, M.; Kara, M.; Esinler, D.; Yalvaç, E.S.; Çaltekin, M.D. The relationship between oxidative stress and preeclampsia. The serum Ischemia-modified albumin levels and thiol/disulfide homeostasis. *Turk. J. Obstet. Gynecol.* **2020**, *17*, 102–107. [CrossRef] [PubMed]
3. Pfaff, A.R.; Beltz, J.; King, E.; Ercal, N. Medicinal Thiols: Current Status and New Perspectives. *Mini Rev. Med. Chem.* **2019**, *20*, 513–529. [CrossRef] [PubMed]
4. Gowacki, R.; Bald, E. Determination of n-acetylcysteine and main endogenous thiols in human plasma by HPLC with ultraviolet detection in the form of their s-quinolinium derivatives. *J. Liq. Chromatogr. Relat. Technol.* **2009**, *32*, 2530–2544. [CrossRef]
5. Bald, E.; Chwatko, G.; Głowacki, R.; Kuśmierek, K. Analysis of plasma thiols by high-performance liquid chromatography with ultraviolet detection. *J. Chromatogr. A* **2004**, *1032*, 109–115. [CrossRef] [PubMed]
6. Chwatko, G.; Bald, E. Determination of cysteine in human plasma by high-performance liquid chromatography and ultraviolet detection after pre-column derivatization with 2-chloro-1-methylpyridinium iodide. *Talanta* **2000**, *52*, 509–515. [CrossRef]
7. Kono, Y.; Iizuka, H.; Isokawa, M.; Tsunoda, M.; Ichiba, H.; Sadamoto, K.; Fukushima, T. HPLC-fluorescence determination of thiol compounds in the serum of human male and female subjects using HILIC-mode column. *Biomed. Chromatogr.* **2014**, *28*, 589–593. [CrossRef]
8. Persichilli, S.; Gervasoni, J.; Castagnola, M.; Zuppi, C.; Zappacosta, B. A Reversed-Phase HPLC Fluorimetric Method for Simultaneous Determination of Homocysteine-Related Thiols in Different Body Fluids. *Lab. Med.* **2011**, *42*, 657–662. [CrossRef]
9. Townsend, D.M.; Tew, K.D.; Tapiero, H. Sulfur containing amino acids and human disease. *Biomed. Pharmacother.* **2004**, *58*, 47–55. [CrossRef]
10. Bicili, S.; Isik, M.; Alisik, M. Thiol and glutathione homeostasis parameters as plasma biomarkers of oxidative stress in age-related macular degeneration. *J. Clin. Exp. Ophthalmol.* **2020**, *11*, 1000862.
11. Turell, L.; Radi, R.; Alvarez, B. The thiol pool in human plasma: The central contribution of albumin to redox processes. *Free Radic. Biol. Med.* **2013**, *65*, 244–253. [CrossRef] [PubMed]
12. Erel, O.; Neselioglu, S. A novel and automated assay for thiol/disulphide homeostasis. *Clin. Biochem.* **2014**, *47*, 326–332. [CrossRef] [PubMed]
13. Isokawa, M.; Funatsu, T.; Tsunoda, M. Fast and simultaneous analysis of biothiols by high-performance liquid chromatography with fluorescence detection under hydrophilic interaction chromatography conditions. *Analyst* **2013**, *138*, 3802–3808. [CrossRef] [PubMed]
14. Toyo'oka, T. Recent advances in separation and detection methods for thiol compounds in biological samples. *J. Chromatogr. B* **2009**, *877*, 3318–3330. [CrossRef]
15. Tsikas, D.; Sandmann, J.; Ikic, M.; Fauler, J.; Stichtenoth, D.O.; Frölich, J.C. Analysis of cysteine and N-acetylcysteine in human plasma by high-performance liquid chromatography at the basal state and after oral administration of N-acetylcysteine. *J. Chromatogr. B Biomed. Sci. Appl.* **1998**, *708*, 55–60. [CrossRef]
16. Mokhtari, V.; Afsharian, P.; Shahhoseini, M.; Kalantar, S.M.; Moini, A. A review on various uses of N-acetylcysteine. *Cell J.* **2017**, *19*, 11–17. [PubMed]
17. Šalamon, Š.; Kramar, B.; Pirc Marolt, T.; Poljšak, B.; Milisav, I. Medical and Dietary Uses of N-Acetylcysteine. *Antioxidants* **2019**, *8*, 111. [CrossRef] [PubMed]
18. Fishbane, S. N-Acetylcysteine in the Prevention of Contrast-Induced Nephropathy. *Clin. J. Am. Soc. Nephrol.* **2008**, *3*, 281–287. [CrossRef] [PubMed]
19. Aldini, G.; Altomare, A.; Baron, G.; Vistoli, G.; Carini, M.; Borsani, L.; Sergio, F. N-Acetylcysteine as an antioxidant and disulphide breaking agent: The reasons why. *Free Radic. Res.* **2018**, *52*, 751–762. [CrossRef]

20. Moldéus, P.; Cotgreave, I.A. N-acetylcysteine. *Methods Enzymol.* **1994**, *234*, 482–492.
21. Wiklund, O.; Fager, G.; Andersson, A.; Lundstam, U.; Masson, P.; Hultberg, B. N-acetylcysteine treatment lowers plasma homocysteine but not serum lipoprotein(a) levels. *Atherosclerosis* **1996**, *119*, 99–106. [CrossRef]
22. Atmaca, G. Antioxidant effects of sulfur-containing amino acids. *Yonsei Med. J.* **2004**, *45*, 776–788. [CrossRef] [PubMed]
23. Biewenga, G.P.; Haenen, G.R.M.M.; Bast, A. The pharmacology of the antioxidant: Lipoic acid. *Gen. Pharmacol. Vasc. Syst.* **1997**, *29*, 315–331. [CrossRef]
24. Sharmilabanu, S. The Protective Role of Alpha Lipoic Acid on Organ Damages Induced By Oxidative Stress. *World J. Pharm. Pharm. Sci.* **2017**, *6*, 521–529. [CrossRef]
25. Zhang, S.J.; Ge, Q.F.; Guo, D.W.; Hu, W.X.; Liu, H.Z. Synthesis and anticancer evaluation of α-lipoic acid derivatives. *Bioorganic Med. Chem. Lett.* **2010**, *20*, 3078–3083. [CrossRef]
26. Malińska, D.; Winiarska, K. Kwas liponowy–charakterystyka i zastosowanie w terapii. *Postepy Hig. Med. Dosw.* **2005**, 535–543.
27. Skorupa, A.; Michałkiewicz, S. α-lipoic acid–antioxidant of antioxidants–properties and determination metods. *Wiad. Chem.* **2017**, *71*, 11–12.
28. Moini, H.; Packer, L.; Saris, N.E.L. Antioxidant and prooxidant activities of α-lipoic acid and dihydrolipoic acid. *Toxicol. Appl. Pharmacol.* **2002**, *182*, 84–90. [CrossRef]
29. Christodoulou, J.; Sadler, P.J.; Tucker, A. A New Structural Transition of Serum Albumin Dependent on the State of Cys34: Detection by 1H-NMR Spectroscopy. *Eur. J. Biochem.* **1994**, *225*, 363–368. [CrossRef]
30. Borowczyk, K.; Wyszczelska-Rokiel, M.; Kubalczyk, P.; Głowacki, R. Simultaneous determination of albumin and low-molecular-mass thiols in plasma by HPLC with UV detection. *J. Chromatogr. B* **2015**, *981–982*, 57–64. [CrossRef]
31. Ogasawara, Y.; Mukai, Y.; Togawa, T.; Suzuki, T.; Tanabe, S.; Ishii, K. Determination of plasma thiol bound to albumin using affinity chromatography and high-performance liquid chromatography with fluorescence detection: Ratio of cysteinyl albumin as a possible biomarker of oxidative stress. *J. Chromatogr. B* **2007**, *845*, 157–163. [CrossRef] [PubMed]
32. Ivanov, A.R.; Nazimov, I.V.; Baratova, L.A. Qualitative and quantitative determination of biologically active low-molecular-mass thiols in human blood by reversed-phase high-performance liquid chromatography with photometry and fluorescence detection. *J. Chromatogr. A* **2000**, *870*, 433–442. [CrossRef]
33. Go, Y.M.; Jones, D.P. Cysteine/cystine redox signaling in cardiovascular disease. *Free Radic. Biol. Med.* **2011**, *50*, 495–509. [CrossRef] [PubMed]
34. Matteucci, E.; Giampietro, O. Thiol signalling network with an eye to diabetes. *Molecules* **2010**, *15*, 8890–8903. [CrossRef] [PubMed]
35. Prabhu, A.; Sarcar, B.; Kahali, S.; Yuan, Z.; Johnson, J.J.; Adam, K.P.; Kensicki, E.; Chinnaiyan, P. Cysteine catabolism: A novel metabolic pathway contributing to glioblastoma growth. *Cancer Res.* **2014**, *74*, 787–796. [CrossRef] [PubMed]
36. Tetik, S.; Ahmad, S.; Alturfan, A.A.; Fresko, I.; Disbudak, M.; Sahin, Y.; Aksoy, H.; Yardimci, K.T. Determination of oxidant stress in plasma of rheumatoid arthritis and primary osteoarthritis patients. *Indian J. Biochem. Biophys.* **2010**, *47*, 353–358. [PubMed]
37. Rodrigues, S.D.; Batista, G.B.; Ingberman, M.; Pecoits-Filho, R.; Nakao, L.S. Plasma Cysteine/Cystine Reduction Potential Correlates with Plasma Creatinine Levels in Chronic Kidney Disease. *Blood Purif.* **2012**, *34*, 231–237. [CrossRef] [PubMed]
38. Steele, M.L.; Fuller, S.; MacZurek, A.E.; Kersaitis, C.; Ooi, L.; Münch, G. Chronic inflammation alters production and release of glutathione and related thiols in human U373 astroglial cells. *Cell. Mol. Neurobiol.* **2013**, *33*, 19–30. [CrossRef]
39. Borowczyk, K.; Olejarz, P.; Chwatko, G.; Szylberg, M.; Głowacki, R. A Simplified Method for Simultaneous Determination of α-Lipoic Acid and Low-Molecular-Mass Thiols in Human Plasma. *Int. J. Mol. Sci.* **2020**, *21*, 1049. [CrossRef] [PubMed]
40. Głowacki, R.; Bald, E. Fully automated method for simultaneous determination of total cysteine, cysteinylglycine, glutathione and homocysteine in plasma by HPLC with UV absorbance detection. *J. Chromatogr. B* **2009**, *877*, 3400–3404. [CrossRef]
41. Mansoor, M.A.; Svardal, A.M.; Ueland, P.M. Determination of the in vivo redox status of cysteine, cysteinylglycine, homocysteine, and glutathione in human plasma. *Anal. Biochem.* **1992**, *200*, 218–229. [CrossRef]
42. Fahey, R.C.; Newton, G.L. DeterMination of Low-Molecular-Weight Thiols Using Monobromobimane Fluorescent Labeling and High-Performance Liquid Chromatography. *Methods Enzymol.* **1987**, *143*, 85–96. [PubMed]
43. Fahey, R.C.; Newton, G.L.; Dorian, R.; Kosower, E.M. Analysis of biological thiols: Derivatization with monobromotrimethylammoniobimane and characterization by electrophoresis and chromatography. *Anal. Biochem.* **1980**, *107*, 1–10. [CrossRef]
44. Huang, K.J.; Han, C.H.; Li, J.; Wu, Z.W.; Liu, Y.M.; Wu, Y.Y. LC determination of thiols derivatized with 4-(aminosulfonyl)-7-fluoro-2,1, 3-benzoxadiazole after SPE. *Chromatographia* **2011**, *74*, 145–150. [CrossRef]
45. Satoh, S.; Toyo'oka, T.; Fukushima, T.; Inagaki, S. Simultaneous determination of α-lipoic acid and its reduced form by high-performance liquid chromatography with fluorescence detection. *J. Chromatogr. B* **2007**, *854*, 109–115. [CrossRef]
46. Cevasco, G.; Piatek, A.M.; Scapolla, C.; Thea, S. An improved method for simultaneous analysis of aminothiols in human plasma by high-performance liquid chromatography with fluorescence detection. *J. Chromatogr. A* **2010**, *1217*, 2158–2162. [CrossRef] [PubMed]
47. Ferin, R.; Pavão, M.L.; Baptista, J. Methodology for a rapid and simultaneous determination of total cysteine, homocysteine, cysteinylglycine and glutathione in plasma by isocratic RP-HPLC. *J. Chromatogr. B* **2012**, *911*, 15–20. [CrossRef] [PubMed]
48. Carru, C.; Deiana, L.; Sotgia, S.; Pes, G.M.; Zinellu, A. Plasma thiols redox status by laser-induced fluorescence capillary electrophoresis. *Electrophoresis* **2004**, *25*, 882–889. [CrossRef] [PubMed]

49. Caussé, E.; Malatray, P.; Calaf, R.; Charpiot, P.; Candito, M.; Bayle, C.; Valdiguié, P.; Salvayre, R.; Couderc, F. Plasma total homocysteine and other thiols analyzed by capillary electrophoresis/laser-induced fluorescence detection: Comparison with two other methods. *Electrophoresis* **2000**, *21*, 2074–2079. [CrossRef]
50. Chwatko, G. Spectrophotometric method for the determination of total thiols in human urine. *Ann. Clin. Lab. Sci.* **2013**, *43*, 424–428.
51. Özyürek, M.; Baki, S.; Güngör, N.; Çelik, S.E.; Güçlü, K.; Apak, R. Determination of biothiols by a novel on-line HPLC-DTNB assay with post-column detection. *Anal. Chim. Acta* **2012**, *750*, 173–181. [CrossRef]
52. Isokawa, M.; Kanamori, T.; Funatsu, T.; Tsunoda, M. Analytical methods involving separation techniques for determination of low-molecular-weight biothiols in human plasma and blood. *J. Chromatogr. B* **2014**, *964*, 103–115. [CrossRef] [PubMed]
53. Fiskerstrand, T.; Refsum, H.; Kvalheim, G.; Ueland, P.M. Homocysteine and other thiols in plasma and urine: Automated determination and sample stability. *Clin. Chem.* **1993**, *39*, 263–271. [CrossRef] [PubMed]
54. Svardal, A.M.; Mansoor, M.A.; Ueland, P.M. Determination of reduced, oxidized, and protein-bound glutathione in human plasma with precolumn derivatization with monobromobimane and liquid chromatography. *Anal. Biochem.* **1990**, *184*, 338–346. [CrossRef]
55. Witt, W.; Rüstow, B. Determination of lipoic acid by precolumn derivatization with monobromobimane and reversed-phase high-performance liquid chromatography. *J. Chromatogr. B Biomed. Sci. Appl.* **1998**, *705*, 127–131. [CrossRef]
56. Jacobsen, D.W.; Gatautis, V.J.; Green, R.; Robinson, K.; Savon, S.R.; Secic, M.; Ji, J.; Otto, J.M.; Taylor, L.M. Rapid HPLC determination of total homocysteine and other thiols in serum and plasma: Sex differences and correlation with cobalamin and folate concentrations in healthy subjects. *Clin. Chem.* **1994**, *40*, 873–881. [CrossRef]
57. Głowacki, R.; Stachniuk, J.; Borowczyk, K. A simple HPLC—UV method for simultaneous determination of cysteine and cysteinylglycine in biological fluids. *Acta Chromatogr.* **2016**, *28*, 333–346. [CrossRef]
58. Graham, D.E.; Harich, K.C.; White, R.H. Reductive dehalogenation of monobromobimane by tris(2-carboxyethyl)phosphine. *Anal. Biochem.* **2003**, *318*, 325–328. [CrossRef]
59. Borra, S.; Featherstone, D.E.; Shippy, S.A. Total cysteine and glutathione determination in hemolymph of individual adult D. melanogaster. *Anal. Chim. Acta* **2015**, *853*, 660–667. [CrossRef]
60. Monostori, P.; Wittmann, G.; Karg, E.; Túri, S. Determination of glutathione and glutathione disulfide in biological samples: An in-depth review. *J. Chromatogr. B* **2009**, *877*, 3331–3346. [CrossRef]
61. Sano, A.; Nakamura, H. Chemiluminescence Detection of Thiols by High-Performance Liquid Chromatography Using o-Phthalaldehyde and N-(4-Aminobutyl)-N-ethylisoluminol as Precolumn Derivatization Reagents. *Anal. Sci.* **1998**, *14*, 731–735. [CrossRef]
62. Concha-Herrera, V.; Torres-Lapasió, J.R.; García-Alvarez-Coque, M.C. Chromatographic determination of thiols after pre-column derivatization with o-phthalaldehyde and isoleucine. *J. Liq. Chromatogr. Relat. Technol.* **2004**, *27*, 1593–1609. [CrossRef]
63. Tsai, C.J.; Lin, Y.C.; Chen, Y.L.; Feng, C.H. Chemical derivatization combined with capillary LC or MALDI-TOF MS for trace determination of lipoic acid in cosmetics and integrated protein expression profiling in human keratinocytes. *Talanta* **2014**, *130*, 347–355. [CrossRef] [PubMed]
64. Li, H.; Kong, Y.; Chang, L.; Feng, Z.; Chang, N.; Liu, J.; Long, J. Determination of lipoic acid in biological samples with acetonitrile-salt stacking method in CE. *Chromatographia* **2014**, *77*, 145–150. [CrossRef]
65. U.S. Department of Health and Human Services, FDA. *Bioanalytical Method Validation Guidance for Industry*; FDA: 2013-D-1020; U.S. Department of Health and Human Services: Washington, DC, USA, 2018. Available online: https://www.fda.gov/media/70858/download (accessed on 30 May 2021).
66. ICH M10. *Bioanalytical Method Validation*; International Council for Harmonisation of Technical Requirements for Pharmaceuticals for Human Use: Geneva, Switzerland, 2019; Available online: https://database.ich.org/sites/default/files/M10_EWG_Draft_Guideline.pdf (accessed on 30 May 2021).
67. Cuadros-Rodríguez, L.; Gámiz-Gracia, L.; Almansa-López, E.M.; Bosque-Sendra, J.M. Calibration in chemical measurement processes. II. A methodological approach. *TrAC Trends Anal. Chem.* **2001**, *20*, 620–636. [CrossRef]

Article

GC-MS Studies on Derivatization of Creatinine and Creatine by BSTFA and Their Measurement in Human Urine

Olga Begou [†], Kathrin Weber [†], Bibiana Beckmann and Dimitrios Tsikas *

Institute of Toxicology, Core Unit Proteomics, Hannover Medical School, 30625 Hannover, Germany; mpegolga@chem.auth.gr (O.B.); kathrin.weber89@gmx.net (K.W.); beckmann.bibiana@mh-hannover.de (B.B.)
* Correspondence: tsikas.dimitros@mh-hannover.de
† These authors contributed equally to this work.

Abstract: In consideration of its relatively constant urinary excretion rate, creatinine (2-amino-1-methyl-5H-imidazol-4-one, MW 113.1) in urine is a useful endogenous biochemical parameter to correct the urinary excretion rate of numerous endogenous and exogenous substances. Reliable measurement of creatinine by gas chromatography (GC)-based methods requires derivatization of its amine and keto groups. Creatinine exists in equilibrium with its open form creatine (methylguanidoacetic acid, MW 131.1), which has a guanidine and a carboxylic group. Trimethylsilylation and trifluoroacetylation of creatinine and creatine are the oldest reported derivatization methods for their GC-mass spectrometry (MS) analysis in human serum using flame- or electron-ionization. We performed GC-MS studies on the derivatization of creatinine (d_0-creatinine), [*methylo*-$^2H_3$]creatinine (d_3-creatinine, internal standard) and creatine (d_0-creatine) with *N,O-bis*(trimethylsilyl)trifluoroacetamide (BSTFA) using standard derivatization conditions (60 min, 60 °C), yet in the absence of any base. Reaction products were characterized both in the negative-ion chemical ionization (NICI) and in the positive-ion chemical ionization (PICI) mode. Creatinine and creatine reacted with BSTFA to form several derivatives. Their early eluting *N,N,O-tris*(trimethylsilyl) derivatives (8.9 min) were found to be useful for the precise and accurate measurement of the sum of creatinine and creatine in human urine (10 μL, up to 20 mM) by selected-ion monitoring (SIM) of m/z 271 (d_0-creatinine/d_0-creatine) and m/z 274 (d_3-creatinine) in the NICI mode. In the PICI mode, SIM of m/z 256, m/z 259, m/z 272 and m/z 275 was performed. BSTFA derivatization of d_0-creatine from a freshly prepared solution in distilled water resulted in formation of two lMate-eluting derivatives (14.08 min, 14.72 min), presumably creatinyl-creatinine, with the creatininyl residue existing in its enol form (14.08 min) and keto form (14.72 min). Our results suggest that BSTFA derivatization does not allow specific analysis of creatine and creatinine by GC-MS. Preliminary analyses suggest that pentafluoropropionic anhydride (PFPA) is also not useful for the measurement of creatinine in the presence of creatine. Both BSTFA and PFPA facilitate the conversion of creatine to creatinine. Specific measurement of creatinine in urine is possible by using pentafluorobenzyl bromide in aqueous acetone.

Keywords: BSTFA; creatine; creatinine; derivatization; quantification; silylation; TMS; validation

1. Introduction

Creatinine (2-amino-1,5-dihydro-1-methyl-4H-imidazol-4-one, MW 113.12; see Scheme 1) is the end-product of creatine catabolism. Creatinine is excreted in the urine with a fairly constant rate and is generally used for the correction of renal excretion rates of endogenous and exogenous substances. This correction is indispensable in clinical studies when urine specimens from spontaneous micturition must be analyzed [1]. The mean concentration of creatinine in urine samples of healthy adults is approximately 12–13 mM, with men excreting higher amounts of creatinine than women [1]. Besides the spectrophotometric method based on the famous Jaffé reaction [2] many different analytical methods are currently available for creatinine. They include spectrophotometric, enzymatic and instrumental methods based on HPLC, GC-MS, LC-MS and LC-MS/MS [3–19]. Lawson [20] and Siekmann [21] demonstrated by electron ionization (EI) that creatinine reacts with silylation reagents to form its N,N,O-*tris*(trimethylsilyl) derivative. Björkhem and colleagues used trifluoroacetic anhydride for the derivatization of creatinine and its GC-MS analysis [22]. Trimethylsilylation derivatization reactions used in MS-based methods were found to be associated with interferences due to formation of several derivatives [23]. To our knowledge, the GC-MS measurement of urinary creatinine as N,N,O-*tris*(trimethylsilyl) derivative by negative-ion chemical ionization (NICI) or positive-ion chemical ionization (PICI) has not been reported thus far.

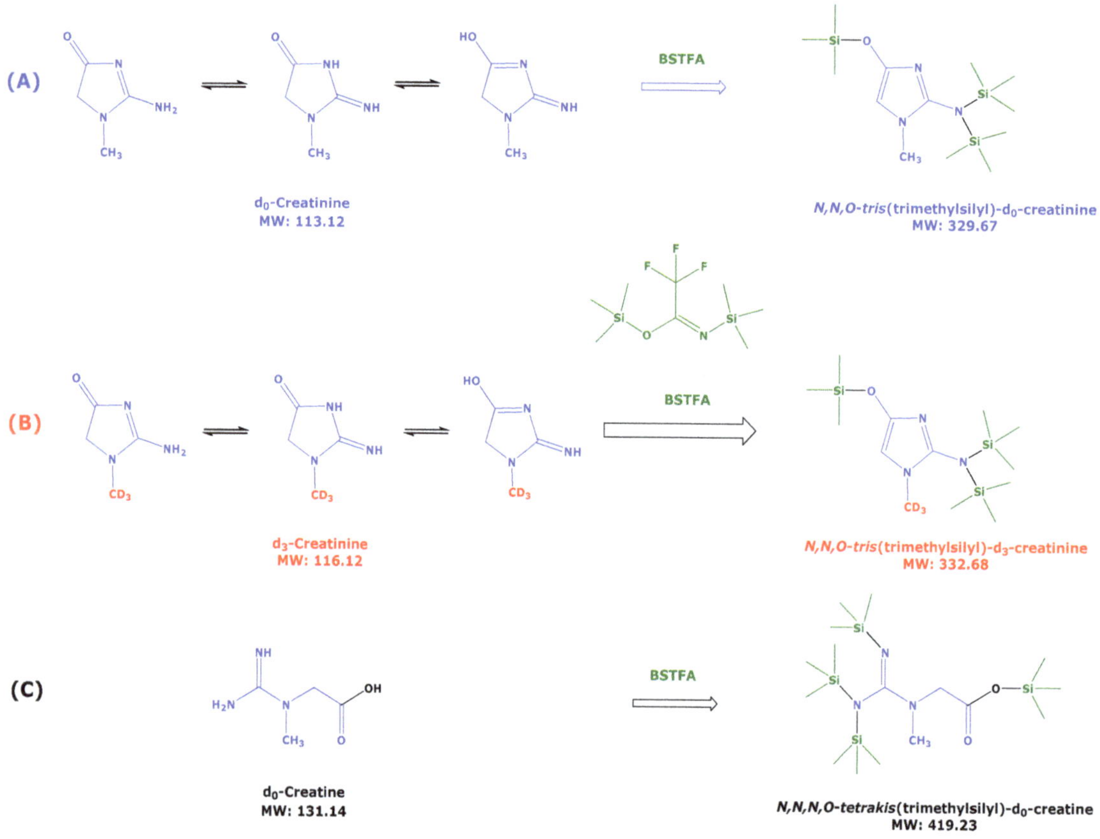

Scheme 1. Schematic of the expected derivatization reaction and products of (**A**) unlabeled creatinine (d_0-creatinine), (**B**) deuterium-labelled creatinine ([methylo-^{2}H$_3$]creatinine, d_3-creatinine) and (**C**) unlabeled (d_0-creatine) and with N,O-*tris*(trimethylsilyl)trifluoroacetamide (BSTFA) to form their N,N,O-*tris*(trimethylsilyl) creatinine derivatives (**A**,**B**) and N,N,N',O-tetrakis(trimethylsilyl creatine derivative (**C**).

In the present study, we investigated in detail the derivatization of unlabelled creatinine (d_0-creatinine), commercially available [methylo-^{2}H$_3$]creatinine (d_3-creatinine) and unlabelled creatine (d_0-creatine) by N,O-*bis*(trimethylsilyl)trifluoroacetamide (BSTFA), one of the oldest trimethylsilylation reagents for amino acids [24]. It is well known that BSTFA reacts with many functional groups, notably hydroxyl, carboxyl and amine groups. Based on this knowledge we expected that BSTFA will react with creatinine and creatine to form N- and O-derivatives (Scheme 1).

In our study, we used GC-MS in the NICI and in the PICI mode, confirmed the formation of the expected N,N,O-$tris$(trimethylsilyl) derivatives and identified several derivatives of creatinine and creatine that have not reported thus far. As creatinine and creatine are in a pH-dependent equilibrium and inter-convertible, our results suggest that BSTFA and GC-MS are not specific for creatinine and creatine but allow measurement of their sum. Using d_3-creatinine as the internal standard we demonstrate that creatinine can be reliably quantitated in 10-µL aliquots of human urine by GC-MS as N,N,O-$tris$(trimethylsilyl) derivative with minimum labour.

2. Materials and Methods

2.1. Chemicals and Materials

Unlabeled creatine (d_0-creatine), unlabeled creatine phosphate $4 \times H_2O$, unlabeled creatinine (d_0-creatinine) and trideuterocreatinine, i.e., [methyl-$^2H_3$]creatinine (d_3-creatinine; declared isotopic purity of >99 atom% ^2H) were obtained from Aldrich (Steinheim, Germany). Stock solutions of d_0-creatine, d_0-creatinine and d_3-creatinine (each 20 mM) were prepared in deionized water and stored in a refrigerator at 8 °C. BSTFA was obtained from Macherey-Nagel (Düren, Germany). Glassware for GC–MS (i.e., 1.5 mL autosampler glass vials and 0.2 mL microvials) and a fused-silica capillary column Optima 17 (15 m × 0.25 mm I.D., 0.25 µm film thickness) were purchased from Macherey-Nagel.

2.2. Derivatization Procedure for Creatinine in Human Urine Samples

Urine samples used in method development and validation were obtained from healthy volunteers being members of the researcher group and authors of this manuscript. Urine samples (1-mL aliquots) were kept frozen at −18 °C until analysis. Prior to sample derivatization, urine samples were thawed and centrifuged (5800× g, 5 min). Urine (10 µL) and synthetic creatinine-containing samples (usually 10 µL) were evaporated to complete dryness under a stream of nitrogen. Subsequently, the samples were treated with 100 µL absolute ethanol and the solvents were evaporated to dryness by a stream nitrogen gas to remove remaining water. Then, the residues were reconstituted with pure BSTFA (100 µL), the glass vials were tightly closed and heated for 60 min at 60 °C in a thermostat. After cooling to room temperature, aliquots (about 90 µL) were transferred into 1.8-mL autosampler glass vials equipped with 200-µL microinserts. Aqueous solutions (usually 10 µL aliquots) of creatinine and creatine were derivatized as described above.

2.3. GC–MS Conditions

In this work, we used a GC-MS method previously used in our group for amino acid derivatives [25]. GC-MS analyses were performed on a single-quadrupole mass spectrometer model ISQ directly interfaced with a Trace 1310 series gas chromatograph equipped with an autosampler AS 1310 from ThermoFisher (Dreieich, Germany). The following oven temperature program was used with helium as the carrier gas at a constant flow rate of 1 mL/min: 0.5 min at 40 °C, then increased to 210 °C at a rate of 15 °C/min and to 320 °C at a rate of 35 °C/min, respectively, and held at 320 °C for 1 min. Interface, injector and ion-source were kept at 300 °C, 280 °C and 250 °C, respectively. Electron energy was set to 70 eV and electron current to 50 µA. Methane (2.4 mL/min) was used as the reagent gas for NICI and PICI. Aliquots (1 µL from derivatization mixtures) were injected in the splitless mode by means of the autosampler using a 10-µL Hamilton needle, which was cleaned automatically three times with toluene (5 µL) after each injection. Quantitative analyses were performed in the selected-ion monitoring (SIM) mode. The peak area (PA) values of d_0-creatinine and d_3-creatinine were calculated automatically by the GC–MS software (Xcalibur and Quan Browser). The concentration of d_0-creatinine was calculated by multiplying the peak area ratio (PAR) of d_0-creatinine to d_3-creatinine with the concentration of d_3-creatinine added to the sample. Statistical analyses and graphs were performed and prepared by GraphPad Prism 7 (San Diego, CA, USA).

2.4. HPLC Analysis of Creatine, Creatinine and Creatine-Phosphate in HCl Solutions

We used a HPLC method previously reported by our group for creatinine measurement in human urine [16]. HPLC analyses were carried out on the an HPLC system consisting of an Agilent 1100 Series binary pump G1312A, an Agilent 1100 Series Degaser G1322A, an Agilent 1100 Series oven column Colcom G1316A, an Agilent 1100 Series VWD detector (all Agilent, Waldbronn, Germany, and an MP3 autosampler (Gerstel, Mülheim, Germany), ChemStation for LC-Systems Rev.B.0402SP1 (212) and Gerstel Maestro Version 1.3.20.41.13.5 were used to control the HPLC system and evaluate the analyses. HPLC analyses were performed on a Kinetex 5 µm EVO C18 100 Å column (250 × 4.6 mm) from Phenomenex (Aschaffenburg, Germany) at a fixed column temperature of 20 °C. The mobile phase was 100 mM sodium acetate, pH 7.5, 10 vol% methanol and was pumped at a flow rate of 1.0 mL/min. Samples (20 µL) were injected by means of the autosampler. The effluent was monitored at 210 nm. The analysis time was 5 min. The retention time was 2.073 ± 0.018 min for creatine-phosphate, 2.252 ± 0.007 min for creatine and 2.547 ± 0.007 min for creatinine.

3. Results

3.1. Generation of GC-MS Spectra and Characterization of Derivatization Products of d_0-Creatine and of d_3-Creatinine

Each 100 nmol of d_0-creatinine and d_3-creatinine taken from their aqueous solutions were combined, the solvent was evaporated to dryness and derivatized with 100 µL BSTFA as described above. Derivatization resulted in a yellow-colored clear solution. The sample was analyzed by GC-MS in the PICI and NICI mode consecutively by injecting 1-µL aliquots of the BSTFA solutions corresponding each to 1 nmol of d_0-creatinine and d_3-creatinine (assuming quantitative derivatization). GC-MS spectra were generated by scanning the quadrupole in the mass-to-charge (m/z) ratio range of 50–650 and 50–1000 (1 scan per s). We observed two chromatographic peaks with the retention time of 8.6 min and 8.9 min (major peak) and their GC-MS spectra contained paired m/z values differing by 3 Da due to the three deuterium atoms in methyl group of d_3-creatinine (Figure 1).

The most intense anions in the NICI mass spectrum (Figure 1A) of the GC peak eluting at 8.9 min were m/z 271 and m/z 274 (base peaks). Less intense anions were found at m/z 199 and m/z 202, and very weak ions (intensity < 1%) were m/z 326 and m/z 329, presumably due to molecular anions of the derivatives (i.e., [M]$^-$). These data indicate the presence of the unlabeled methyl group in d_0-creatinine and of the deuterium-labeled methyl group of d_3-creatinine in this peak (Figure 1A). The NICI spectrum of this GC peak also contained weak anions at m/z 144 and m/z 186 that do not carry the original methyl group of creatinine (Figure 1A). The PICI mass spectrum of the GC peak eluting at 8.9 min contained intense cations at m/z 272, m/z 275, m/z 256 and m/z 259, less intense cations at m/z 300 and m/z 303, weaks ions at m/z 312 and m/z 315, and very weak ions (intensity < 1%) at m/z 330 and m/z 333, presumably due to the protonated molecules of the derivatives (i.e., [M+H]$^+$) (Figure 1B).

These data indicate the presence of the unlabeled methyl group in d_0-creatinine and the deuterium-labeled methyl group of d_3-creatinine in this peak (Figure 1). Comparison of the total ion intensity in the NICI and PICI mass spectra (1.85×10^6 versus 9.92×10^5, Figure 1) suggests that NICI may allow for a more sensitive detection of creatinine than PICI. Proposed fragmentation mechanisms of the *N,N,O*-trimethylsilyl derivatives in the PICI are shown in Scheme 2.

The smaller GC peak eluting at 8.6 min had closely comparable NICI and PICI mass spectra to those of the *N,N,O-tris*(trimethylsilyl) derivative (data not shown). These observations suggest that the GC peak with the retention time of 8.6 min is an isomer of the *N,N,O-tris*(trimethylsilyl) derivative of creatinine.

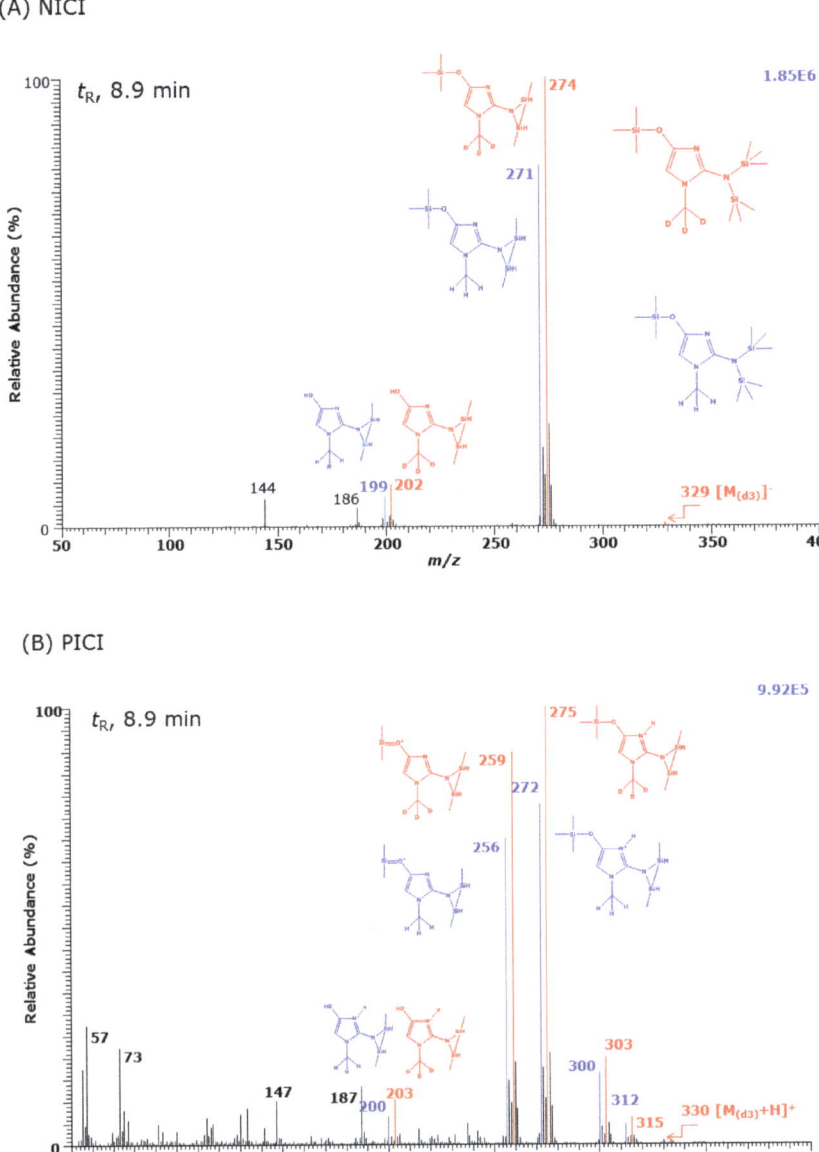

Figure 1. (**A**) Negative-ion chemical ionization (NICI) and (**B**) positive-ion chemical ionization (PICI) GC-MS spectra generated from a mixture of d_0-creatinine (blue) and d_3-creatinine (red) after derivatization with *N,O-tris*(trimethylsilyl)trifluoroacetamide (BSTFA) at 60 °C for 60 min (each 1 nmol injected). The retention time (t_R) of the GC-MS peak was 8.9 min. Insets indicate the proposed structures of the derivatives and ions. See Scheme 2.

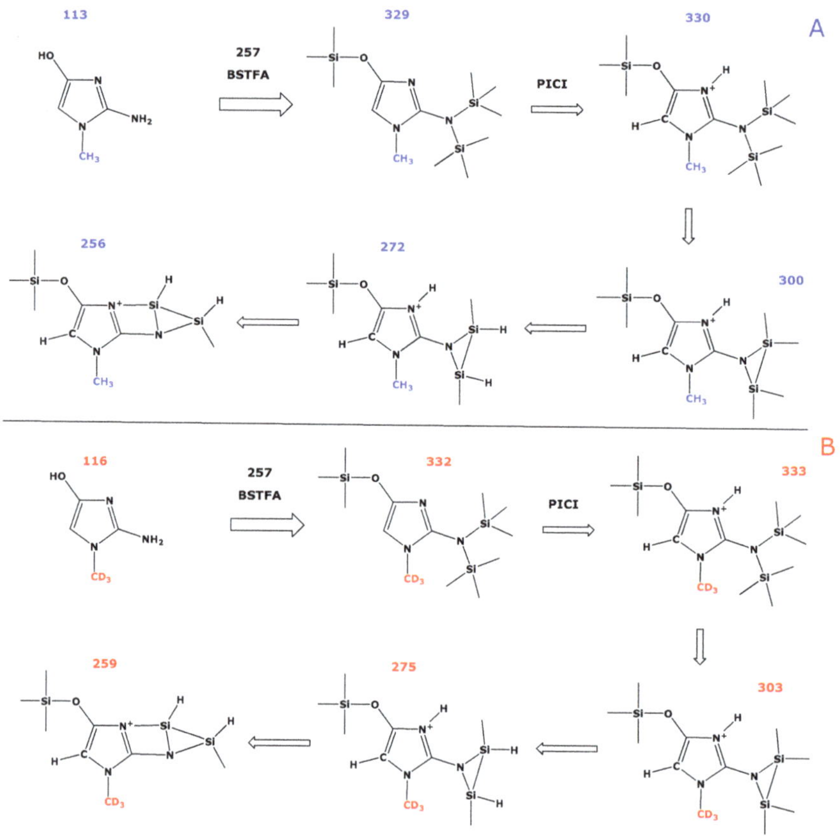

Scheme 2. Proposed fragmentation mechanisms for the N,N,O-trimethylsilyl derivatives of d_0-creatinine (**A**, blue) and d_3-creatinine (**B**, red) of the GC-MS peak with the retention time of 8.9 min in the PICI mode. The numbers give the molecular weight of the neutral substances and the cations. See Figure 1B.

The GC peak with the retention time of 8.7 min was only detectable in the PICI mode. The PICI mass spectrum of this peak contained three pairs of cations differing by 3 Da due to the presence of d_3-creatinine, i.e., m/z 314 and m/z 317 (base peaks), m/z 330 and m/z 333 ($[M+H]^+$), and m/z 358 and m/z 361 ($[M+C_2H_4+H]^+$) (Figure 2). Adducts such as C_2H_4 (28 Da) are common in PICI of amines such as dimethyl amine and derive from the reactant gas methane [26]. Presumably, the adduct is on the non-ring amine group. These observations suggest the GC peak eluting at 8.7 min is a creatinine derivative with three trimethylsilyl (TMS) groups, most likely the N^2,N^3,O^4-tris(trimethylsilyl) derivative. It cannot ionize in the NICI mode, presumably because of the inability to form anions by loss of an H atom or by capturing an electron due to the lack of electron-capturing atoms and functional groups in the derivative. The cations with m/z 314 and m/z 317 seem to be very stable and do not fragment. The cations m/z 55, m/z 57, m/z 73 and m/z 147 are shared by d_0-creatinine and d_3-creatinine and are likely to be associated with the TMS groups (see also [23]) of the derivatives (see also Figure 1B).

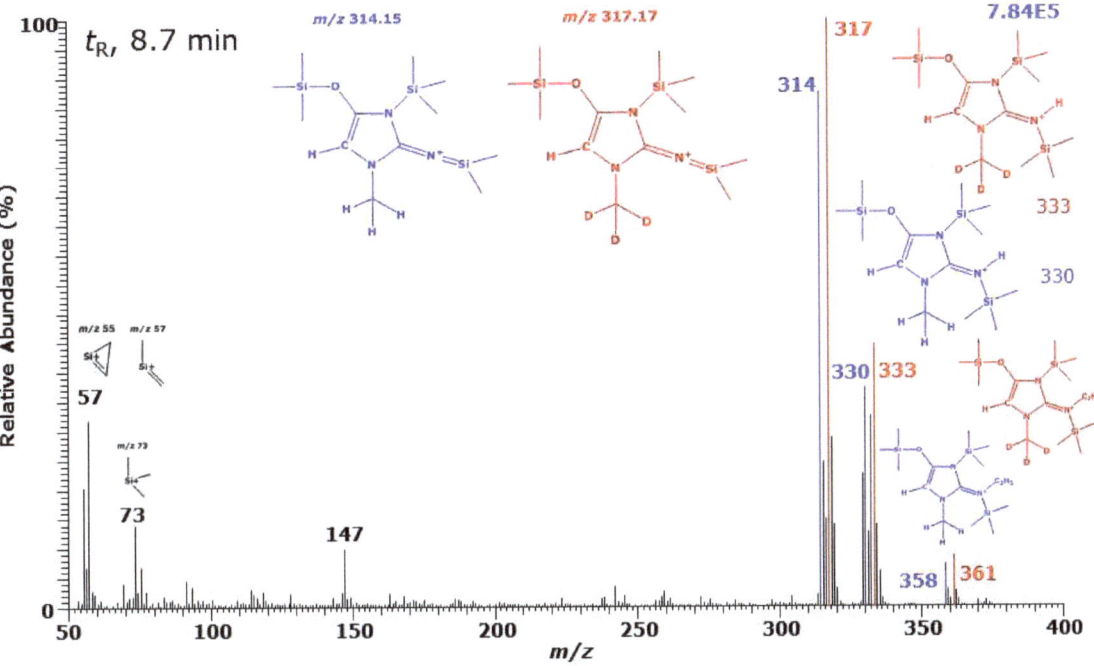

Figure 2. PICI GC-MS spectrum generated from a mixture of d_0-creatinine (blue) and d_3-creatinine (red) after derivatization with BSTFA at 60 °C for 60 min. The retention time (t_R) of the GC-MS peak was 8.7 min. Insets indicate the proposed structures for the mass fragments. See also Figure 1B.

3.2. Generation of GC-MS Spectra and Characterization of Derivatization Products of d_0-Creatine

Derivatization of d_0-creatine with BSTFA (60 °C, 60 min) resulted in the formation of three GC-MS peaks with the retention times of the d_0-creatinine. The NICI and PICI mass spectra of these derivatives were virtually identical with those of the d_0-creatinine derivatives (data not shown). In order to investigate the potential formation of additional derivatives of d_0-creatine we extended the upper m/z scanning range to 1000 and the acquisition time to 16 min. We observed two GC-MS eluting at 14.08 min (minor peaks) and 14.72 min (major peaks) in the NICI and PICI mode. The corresponding GC-MS spectra of these d_0-creatine derivatives and the relatively high difference in their long retention times suggest that these peaks are not derivatives of d_0-creatine or d_0-creatinine (Figure 3). A possible explanation could be the formation of a creatinyl-derivative by the reaction of two creatine molecules and/or by the reaction of a creatine molecule and with a molecule of creatinine formed from creatine during the derivatization. The peak with shorter retention time could be due to its TMS ether functionality compared to the keto group.

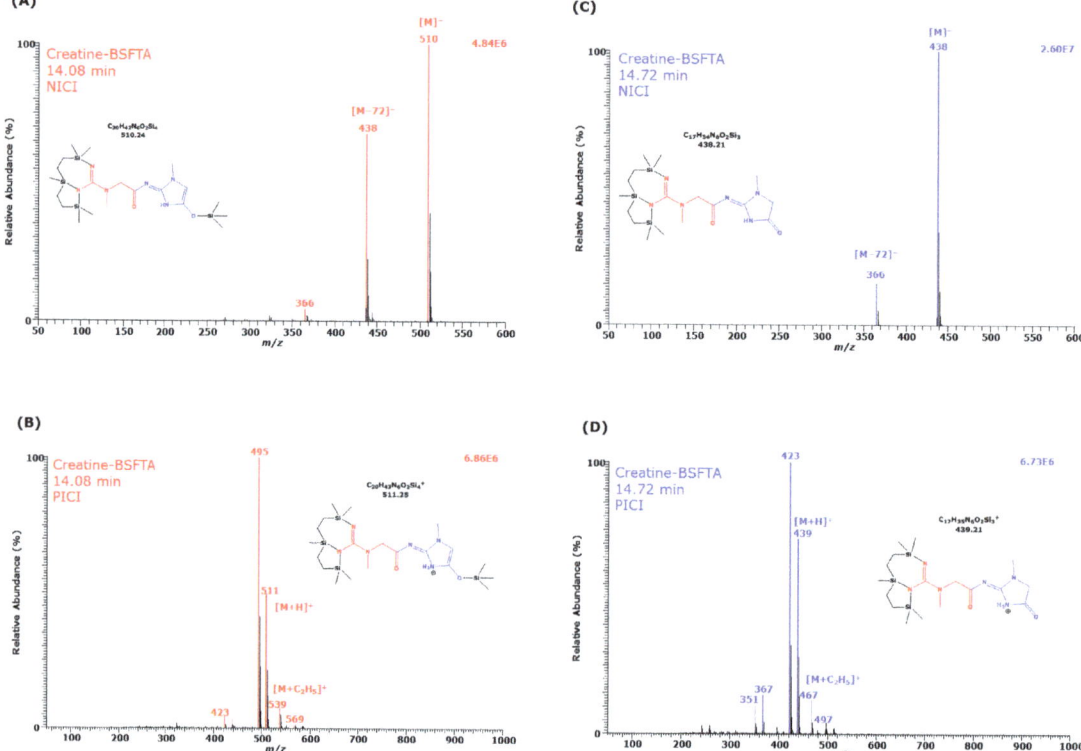

Figure 3. NICI (**A,C**) and PICI (**B,D**) GC-MS spectra generated from the BSTFA derivatization (60 °C, 60 min) of a freshly prepared solution of d_0-creatine in deionized water upon its evaporation to complete dryness. The retention times (t_R) of the GC-MS peak were 14.08 min (minor peak, red) and 14.72 min (major peak, blue). Insets indicate the proposed structures for the mass fragments. The same oven column temperature program was used as in Figures 1 and 2.

3.3. Standardization of [methylo-$^2H_3$]Creatinine

The isotopic purity of stable isotope-labelled analogs is of particular importance in quantitative analyses [27]. The isotopic purity of the commercially available [methylo-$^2H_3$]creatinine was verified as follows.

Nine separate samples containing each 100 nmol of d_0-creatinine and d_3-creatinine were derivatized with BSTFA (100 µL) and 1-µL aliquots of their solutions were analyzed by SIM of m/z 256, m/z 259, m/z 271, and m/z 274 (peak with retention time 8.9 min). Analysis of the sample containing d_0-creatinine generated a mean PAR of 0.02223 ± 0.00329 (RSD, 15%; $n = 9$) for m/z 259 to m/z 256, and a PAR of 0.01302 ± 0.00209 (RSD, 16%; $n = 9$) for m/z 274 to m/z 271. Analysis of the samples that contained d_3-creatinine produced a mean PAR of 0.000588 ± 0.0001411 (RSD, 24%; $n = 9$) for m/z 256 to m/z 259 and a mean PAR of 0.01108 ± 0.0002861 (RSD, 2.6%; $n = 9$) for m/z 271 to m/z 274. These observations indicate the presence of only very low amounts of d_0-creatinine in the commercial [methylo-$^2H_3$]creatinine and confirm its declared isotopic purity (>99 atom% ^2H).

3.4. Method Linearity, Precision and Accuracy

For quantitative analyses of creatinine, we selected the N,N,O-tris(trimethylsilyl) derivative of creatinine with the retention of 8.9 min. The structure of this creatinine derivative is most likely N^2,N^2,O^4-tris(trimethylsilyl). The structure with the ring-N^3 atom of creatinine, which is not derivatized, allows both PICI and NICI. In the NICI mode, SIM of m/z 271 for d_0-creatinine and m/z 274 for d_3-creatinine was performed. A representative GC-MS chromatogram is shown in Figure 4 and indicates peaks with closely comparable intensity (2.66×10^6 versus 2.56×10^6) due to injection of nominally 1 nmol of each analyte. In the PICI mode, SIM of m/z 256 and m/z 272 for d_0-creatinine and of m/z 259 and m/z 275 for d_3-creatinine was performed. The dwell-time was 108 ms for all ions and the electron multiplier voltage was set to 2025 V.

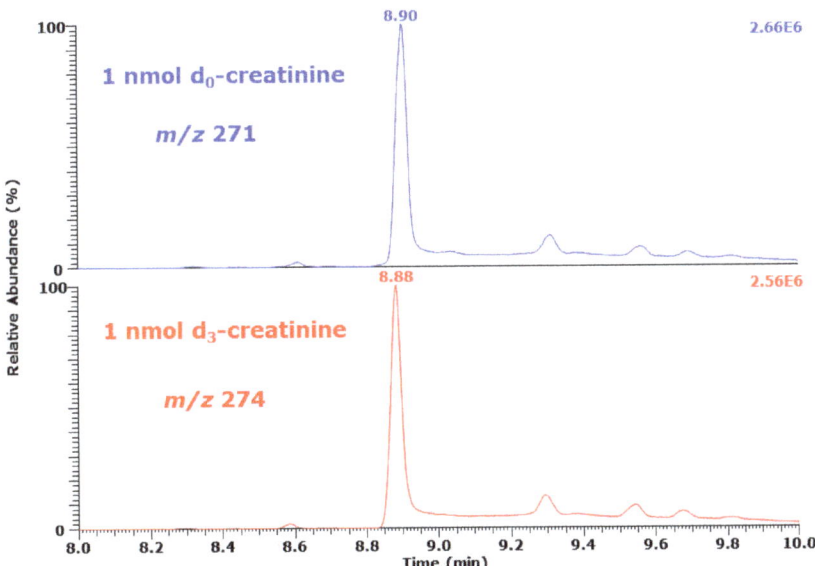

Figure 4. Partial GC-MS chromatograms from the analysis of an equimolar mixture of d_0-creatinine (blue) and d_3-creatinine (red) after derivatization with BSTFA at 60 °C for 60 min (each 1 nmol injected). SIM of m/z 271 for d_0-creatinine and m/z 274 for d_3-creatinine was performed in the NICI mode.

Stock solutions (each 20 mM) of d_0-creatinine and d_3-creatinine were freshly prepared in Ampuwa deionized water. Dilutions of the stock solution of d_0-creatinine were prepared using Ampuwa water providing d_0-creatinine concentrations of 0, 2, 4, 6, 8, 10, 14 and 20 mM. Each 10-µL aliquots of these solutions were combined with each 5-µL aliquots of the 20 mM d_3-creatinine stock solution. After evaporation to dryness under a stream of nitrogen gas, reconstitution of the residue in absolute ethanol and renewed evaporation to dryness, derivatization with 100 µL BSTFA each was performed (60 min, 60 °C). Then. 1-µL aliquots of the samples were injected in the splitless mode and analyzed in the PICI mode by SIM of m/z 256, m/z 259, m/z 272 and m/z 275. The amounts injected were 1 nmol for d_3-creatinine in each sample and varying amounts of d_0-creatinine (i.e., 0.0, 0.2, 0.4, 0.6, 0.8, 1.0, 1.4, 2 nmol). These analyses were performed by three persons in triplicate for each concentration. The precision (relative standard deviation, RSD) ranged between 0.1% and 8.4%. Linear regression analysis between the PAR m/z 256 to m/z 259 (y) or the PAR m/z 272 to m/z 275 (y) and the amount of d_0-creatinine (nmol) (x) for all data resulted in straight lines with the regression equations $y = 0.033 + 0.0087x$ ($r^2 = 0.9945$) and $y = 0.001 + 0.0098x$ ($r^2 = 0.9937$), respectively (Figure 5). The reciprocal values of slopes of the straight lines were 115 nmol and 102 nmol and correspond to the nominal amount of d_3-creatinine of 100 nmol used in the linearity experiment. Thus, SIM of m/z 272 and m/z 275 yields a higher mean accuracy than SIM of m/z 256 and m/z 259 (87% vs. 98%) (Figure 5).

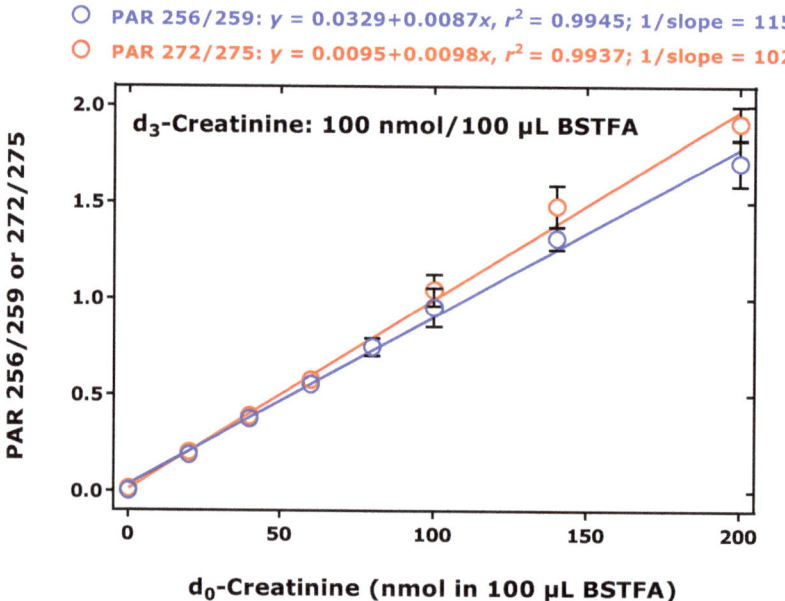

Figure 5. Linear relationships between the peak area ratio (PAR) values and d_0-creatinine amounts obtained by SIM of m/z 256 and m/z 272 for d_0-creatinine and of m/z 259 and m/z 275 for d_3-creatinine in the PICI mode. The indicated d_0-creatinine amounts and each 100 nmol d_3-creatinine were derivatized with BSTFA (100 µL) at 60 °C for 60 min and 1 µL aliquots of the reaction mixture were injected in the splitless mode. Data are shown as mean ± standard deviation ($n = 9$). These analyses were performed by three persons in triplicate for each d_0-creatinine amount. For more details, see the text.

3.5. Measurement of Creatinine in Human Urine in the NICI Mode

The method was validated in human urine samples in the NICI mode by the three persons who performed the experiment described above. Three healthy volunteers (#1, #2, #3) donated urine samples by spontaneous micturition. To 10-µL urine aliquots, d_0-creatinine was added to reach final added concentrations of 2, 4, 6, 10, 14 and 20 mM. d_3-Creatinine was also added to these samples to reach a fixed concentration of 10 mM in each urine sample. After evaporation to dryness under a stream on nitrogen gas, reconstitution of the residues in 100 µL aliquots of absolute ethanol and renewed evaporation to dryness, derivatization each with 100 µL BSTFA was performed (60 min, 60 °C) and 1-µL aliquots were injected and analyzed by SIM of m/z 271 and m/z 274 in the NICI mode. Linear regression analysis between the PAR of m/z 271 to m/z 274 (y) and the concentration of d_0-creatinine (mM) (x) resulted in straight lines (Figure 6). The reciprocal slope values of the straight lines were 10.9 mM for urine #1, 11.0 mM for urine #2, and 10.2 mM for urine #3. Based on the nominal concentration of d_3-creatinine of 10 mM in the urine samples, the mean accuracy is calculated to be 109%, 110% and 102% in the three human urine samples in the concentration range investigated. The y axis intercept values indicate mean basal creatinine+creatine concentrations of 1.7, 1.8 and 1 mM, respectively.

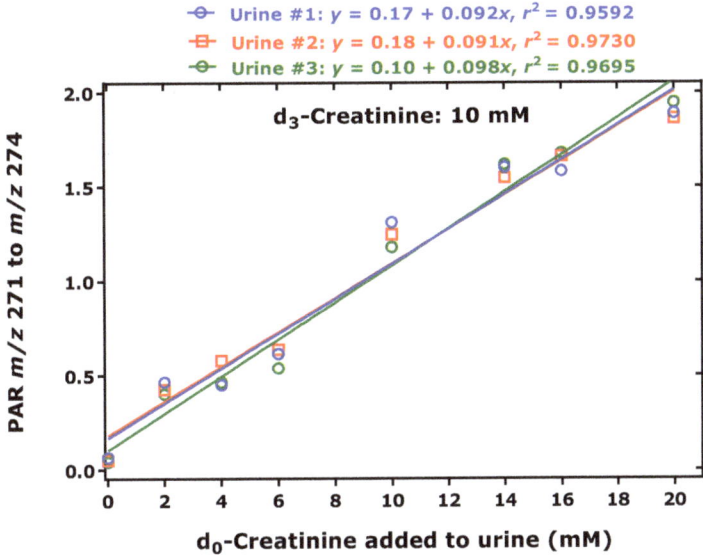

Figure 6. Linear relationships between the peak area ratio (PAR) values and the varying d_0-creatinine concentrations added to human urine samples donated by three healthy volunteers and regression equations. d_3-Creatinine was added at the fixed concentration of 10 mM and served as the internal standard. SIM of m/z 271 and m/z 274 for d_0-creatinine and d_3-creatinine was performed in the NICI mode, respectively. The analyses were performed by the three persons who performed the analyses shown in Figure 5. For more details, see the text.

3.6. HPLC Analysis of Creatinine in HCl Solutions of Creatine

The aim of these analyses was to estimate the extent of formation of creatinine from creatine and creatine-phosphate in hydrohloric acid solutions of varying molarity and incubation time at room. Linear relationships between the response (y), i.e., peak area, mAU×min at 210 nm, and the creatinine concentration in µM (x) was observed: $y = 7.4 + 4.92\, x$, ($r^2 = 0.9999$) (range, 0–1000 µM). This regression equation was used to measure the concentration of creatinine in creatine solutions in hydrochloric acid (Figure 7). The concentration of creatinine in freshly prepared 5000 µM creatine solutions ranged between 2 and 7 µM and increased with increasing HCl molarity and incubation time up to 56 µM at 1 M HCl and 360 min (Figure 7A) and up to 2500–3000 µM after 63 days in 250–1000 mM HCl solutions (Figure 7B). The sigmoidal creatinine-incubation time profile in a 5000 µM solution of creatine in 25 mM HCl is shown in (Figure 7C). The highest creatinine concentration was determined to be 2150 µM after 69 days. Creatine-phosphate was found to be stable in deionized water. Similar experiments with HCl-solutions of creatine-phosphate did not result in formation of considerable amounts of creatinine (data not shown). Except for creatine and creatinine we did not detect appearance of additional peaks within HPLC run time of 5 min and UV absorbance detection at 210, 232 and 250 nm.

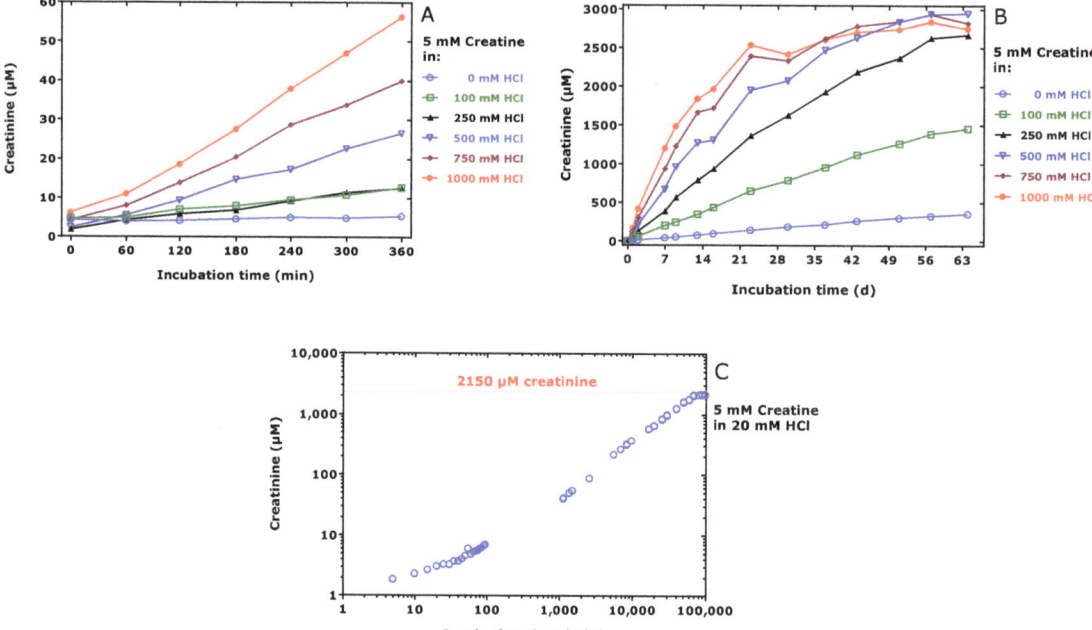

Figure 7. (**A–C**) Creatinine formed upon incubation of 5 mM creatinine in deionized water and in the indicated HCl solutions for the indicated times at room temperature (22–25 °C). Analyses were performed by HPLC with UV absorbance detection at 210 nm. Note the double decadic logarithmic scale in panel (**C**).

4. Discussion

Silylation is one of the most widely used derivatization reaction in analytical chemistry, notably in GC-based methods. Silylation reagents such as BSTFA and MSTFA are not specific, but react with different functionalities of organic compounds, especially of hydroxyl and amine groups, to form *O*- and *N*-trimethylsilyl derivatives [23]. Such derivatives are volatile and thermally stable in non-aqueous systems, best properties in GC-based analytical methods.

Creatinine, 2-amino-1-methyl-5*H*-imidazol-4-one (Scheme 1), is an endogenous substance, the final metabolite of creatine catabolism. Creatinine can be formed chemically from creatine by acid-catalyzed cyclization (Scheme 1). The most significant field of interest in creatinine is Clinical Chemistry. Serum creatinine serves as an indicator of kidney function. Urinary creatinine is of particular importance in clinical, pharmacological and epidemiological studies, where biomarkers must be measured in urine collected from spontaneous micturition, i.e., when the urine volume and the time between two urine collections are unknown. This particular importance is because creatinine is excreted in the urine with a relatively constant rate primarily via glomerular filtration mainly depending on age and gender. The great interest in creatinine in various disciplines led to the development of many analytical methodologies based on different principles. As an organic amine, derivatization of creatinine improves its physicochemical properties so that its analysis becomes feasible by GC also coupled with mass spectrometry (MS) [3–23]. Thus, GC-MS was used several decades ago for the quantitative measurement of creatinine in biological samples including serum and urine using stable isotope-labelled analogs of creatinine [20–22].

Using trimethylsilylation (no conditions reported), Lawson found by GC-MS and EI that creatinine is converted to a single derivative, which was identified as the *N,N,O*-TMS derivative [20]. The EI mass spectrum of this derivative contained two ions at *m/z* 329, which is the the molecular radical cation [M]$^{•+}$, and m/z 314 due to the loss of methyl radical ([M-CH$_3$]$^+$) from one of the three TMS groups [20]. This derivative obviously corresponds to the derivatives of d$_0$-creatinine of d$_3$-creatinine in our study with the retention time of 8.9 min. Siekmann extracted creatinine from human serum samples by cation-exchange resin Ag 50W-X2, derivatized by N-methyl-N-trimethylsilyl-trifluoroacetamide (MSTFA) in anhydrous pyridine (1:1, v/v) by heating (40 min, 60 °C) [21]. Siekmann reported on the formation of a single GC-MS peak (by SIM), of which the EI spectrum was very similar to that reported by Lawson [20], supporting the formation of a N,N,O-TMS derivative of creatinine. Neither Lawson nor Siekmann reported in their papers analogous analyses with creatine.

Our observations strongly suggest that derivatization of creatinine (60 min, 60 °C) with pure BSTFA, i.e., in the absence of any solvents such as pyridine generates at least three derivatives. The derivative eluting at 8.9 min is most likely N^2,N^2,O^4-*tris*(trimethylsilyl)-creatinine, identical with that proposed by Lawson [20] and Siekmann [21]. The second major derivative formed under the same derivatization conditions is most likely N^2,N^3,O^4-*tris*(trimethylsilyl)-creatinine with the retention time of 8.7 min (Scheme 3). This derivative has not been reported thus far. N^2,N^3,O^4-*tris*(Trimethylsilyl)-creatinine elutes in front of N^2,N^2,O^4-*tris*(trimethylsilyl)-creatinine presumably because all derivatizable N atoms of creatinine are derivatized.

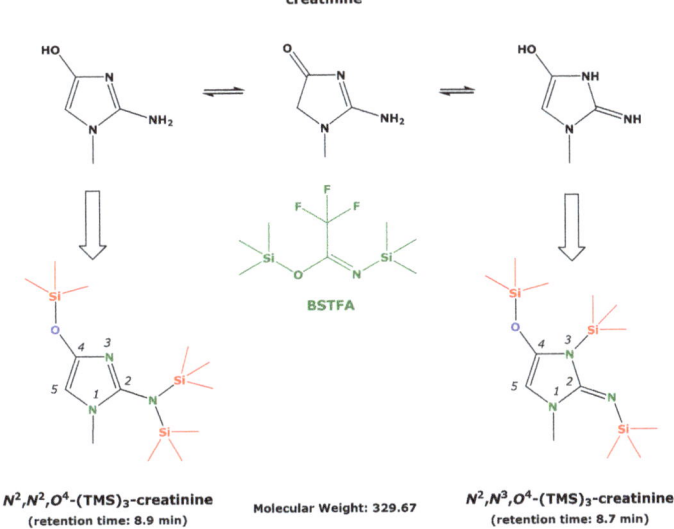

Scheme 3. Proposed chemical structures for two derivatives of creatinine formed by its reaction with N,O-*tris*(trimethylsilyl)trifluoroacetamide (BSTFA) at 60 °C for 60 min in pure BSTFA.

Our study also strongly suggests that derivatization of creatine with pure BSTFA under same conditions (60 min, 60 °C) generates the same two derivatives eluting at 8.7 min and 8.9 min. The results from HPLC analyses of creatine solutions in deionized water and hydrochloric acid solutions suggest that creatine cyclizes to form creatinine, yet a very low extent. One may therefore assume that the derivatives N^2,N^3,O^4-*tris*(trimethylsilyl) and N^2,N^2,O^4-*tris*(trimethylsilyl) are formed during the BSTFA derivatization step.

To the best of our knowledge, the present study is the first to demonstrate the formation of two new derivatives from creatine via BSTFA derivatization (60 min, 60 °C). These derivatives emerge from the column 5 to 6 min later than the above-mentioned derivatives of creatinine and creatine. Our study strongly suggests that both late-eluting TMS derivatives stem from a creatine-creatinine adduct. As no creatinine was initially present in the creatine sample, the detected creatine-creatinine is likely to have been formed by alternative mechanisms. One possible mechanism could involve formation of the O-TMS ester of creatine (Scheme 4), which is likely to be formed more rapidly and to a higher extent than the N-TMS [24]. Subsequently, free amine groups may attack the chemically activated carboxylic group to make the creatinine residues. Thus far, only one group has reported on the synthesis of creatinyl-amino acids derivatives such as creatinyl-glycine, which has been reported to be neuroprotective [28]. In the NICI and PICI mass spectra of these derivatives we obtained mass fragments being each by 4 Da (see Figure 3). An explanation for this finding could be loss of 4 H atoms in total on the three trimethylsilyl groups of the terminal guanidine group. We do not know whether this results from the derivatization or ionization irrespective of the ionization mode. Such a phenomenon has not been reported thus far. Yet, there is an indication that this may occur in N,N-di-trimethylsilyl derivatives [29–31]. Thus, in the EI mass spectra of the per-trimethylsilylated 1-phosphono-2-amino-ethane (MW 413) and O-phosphorylethanolamine (MW 429) the cation m/z 174 was observed, which was assigned to $[CH_2=N(Si(CH_3)_3)_2]^+$. These spectra also contained m/z 172 with intensity ratio of 2:1. A possible structure for m/z 172 could be $[CH_2=N(Si(CH_3)_2CH_2)_2]^+$.

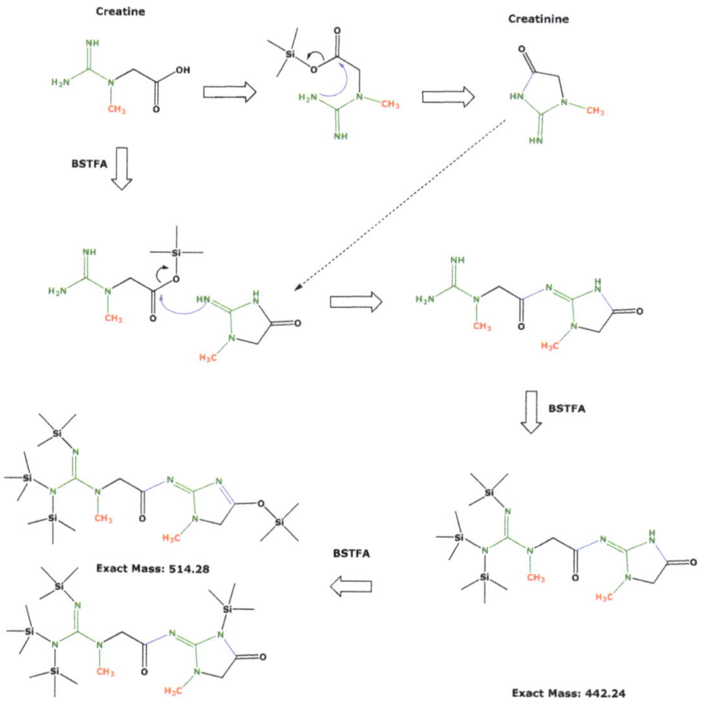

Scheme 4. Proposed chemical structures for the formation of creatinyl-creatinine derivatives from the derivatization of creatine with pure N,O-tris(trimethylsilyl)trifluoroacetamide (BSTFA) at 60 °C for 60 min. See the NICI and PICI mass spectra of these derivatives in Figure 3.

Silylation of this compound with a perdeuterated silylation reagent shifted these cations to m/z 192 and m/z 188 [30], strongly supporting the bridging of two methyl groups of the neighboring TMS groups on the amine group. BSTFA and other silylation agents can react with various functionalities [23], including acetamide groups such as that of acetaminophen (paracetamol) to generate its O,O-di-TMS derivative [31], and their derivatives undergo multiple fragmentations and rearrangements during ionization such as EI [32].

Pentafluoropropionic anhydride (PFPA) is another useful derivatization reagent in GC-MS. Like BSTFA, PFPA also reacts with amine, hydroxylic and carboxylic groups for instance of amino acids [33]. The N-pentafluoropropionyl derivatives are considerably more stable than the O-pentafluoropropionyl derivatives [34]. As BSTFA derivatization does not allow discrimination between creatinine and creatine, we tested the utility of PFPA. Under conditions previously reported for amino acids [25,32], i.e., heating the analytes in PFPA-ethyl acetate (1:4, v/v; 65 °C, 30 min), we observed each only one peak from creatine, d_0-creatinine and d_3-creatinine. The GC-NICI-MS spectra of creatine and d_0-creatinine derivatives were virtually identical: m/z 221 (6 %; $[M-HF-H_2O]^-$), 239 (100 %; $[M-HF]^-$) and m/z 259 (6 %; $[M]^-$; $C_7H_6F_5N_3O_2$); the GC-MS spectrum of the d_3-creatinine derivative eluted a few seconds earlier: m/z 224 (6 %), 242 (100 %) and m/z 262 (6 %). These results indicate the formation presumably of N^2-pentafluoropropionyl from both, creatine and creatinine. These observations suggest that PFPA reacts with the carboxylic group of creatine to form the mixed anhydride. Subsequently, the N^2-imine group attacks intramolecularly the carboxylic group, with pentafluoropropionic acid leaving the molecule, analogous to the BSTFA derivative of creatine.

As far we are informed, creatinine has not be measured by GC-MS in human urine after derivatization with BSTFA. Our study indicates that creatinine can be quantified precisely and accurately by GC-MS in only 10-μL aliquots of human urine using d_3-creatinine as internal standard in relevant concentration ranges. The method does not require any organic solvent or base like pyridine for derivatization and/or extraction for GC-MS analysis. Excess BSTFA serves as a solvent, in which the TMS derivatives are readily soluble, yet no other charged endogenous constituents present in urine. As BSTFA is highly reactive towards numerous substances [23], it is possible that many endogenous substances also form volatile TMS derivatives that do not accumulate in the GC column.

Currently available data suggest that specific measurement of creatinine in urine and other biological samples is possible by using 2,3,4,5,6-pentafluorobenzyl (PFB) bromide (PFB-Br) in aqueous acetone (60 min, 50 °C) [34]. Creatinine reacts with PFB-Br to form a single derivative, i.e., N^2-PFB-creatinine. Interestingly, we found that the N^2-PFB-derivatives of d_0-creatinine and d_3-creatinine react with PFPA (65 °C, 30 min) to form their N^2-PFB,N^3-PFP derivatives (MW=439.21, $C_{14}H_7F_{10}N_3O_2$; MW=442.23, $C_{14}H_4D_3F_{10}N_3O_2$, respectively) with relative retention time of 1.29 with respect to the N^2-PFB-derivatives.

5. Conclusions

BSTFA is known for many decades as a useful derivatization reagent for the GC-MS analysis of creatinine, but its utility to measure creatinine in human urine has not been reported thus far. This study investigated the derivatization of creatinine and its precursor creatine with BSTFA. Both substances react with BSTFA (60 °C, 60 min) to form three derivatives of virtually identical structures. Creatinine and creatine were found to react with PFPA (65 °C, 30 min) to form a single N-pentafluoropropionyl derivative. These observations indicate that BSTFA and PFPA is much more effective in the conversion of creatine to creatinine than lowering the pH by inorganic acids such as hydrochloric acid. Our findings suggest that BSTFA and PFPA are not useful for the simultaneous measurement of creatinine and creatine. The N,N,O-trimethylsilyl derivative of creatinine and creatine with the retention time of 8.9 min is useful for their quantitative measurement in human urine both in the NICI and PICI mode using trideuteromethyl creatinine as internal standard. Under the same derivatization conditions, creatine reacts with BSTFA and forms two creatinyl-creatine derivatives with retention times of 14.07 min and 14.72 min, suggesting intermediate formation of creatinine and its conjugation with creatine. Our results show that methyl groups of the TMS residues react to form -CH_2-CH_2-bridges and are supported by previous reports on alkyl amines. Such a cyclization reaction is more likely to occur during EI, PICI and NICI, rather than during the derivatization with BSTFA. The possibility that creatinine, but not creatine, reacts with PFB-Br and the N^2-PFB-creatinine derivative reacts with PFPA to form N^2-PFB,N^3-PFP-creatinine offers the possibility to measure biological creatinine and creatine simultaneously by GC-MS. This could be of particular importance in the area of Clinical Chemistry and in clinical trials.

Author Contributions: Conceptualization, D.T.; methodology, O.B., K.W. and B.B.; software, O.B. and D.T.; validation, O.B., K.W. and D.T.; formal analysis, D.T.; investigation, O.B., K.W. and B.B.; resources, D.T.; data curation, O.B., K.W., B.B. and D.T.; writing—original draft preparation, D.T., O.B. and K.W.; writing, O.B., K.W. and D.T.; visualization, O.B. and D.T.; supervision, D.T.; project administration, D.T.; funding acquisition, D.T. All authors have read and agreed to the published version of the manuscript.

Funding: This research received no external funding.

Institutional Review Board Statement: Ethical review and approval were waived for this study, due to the use of spot urine samples donated by three volunteers being authors of this article.

Informed Consent Statement: Patient consent was waived due to the use of spot urine samples donated by three volunteers being authors of this article.

Data Availability Statement: The study did not report any data.

Conflicts of Interest: The authors declare no conflict of interest.

Sample Availability: Not available.

References

1. Szadkowski, D.; Jörgensen, A.; Essing, H.G.; Schaller, K.H. Creatinine elimination rate as reference value for analysis of urine samples. I. Effect of daily urine volume and circadian rhythm on creatinine excretion. *Z. Klin. Chem. Klin. Biochem.* **1970**, *8*, 529–533.
2. Jaffé, M. Ueber den Niederschlag welchen Pikrinsäure in normalen Harn erzeugt und über eine neue reaction des Kreatinins. *Z. Physiol. Chem.* **1886**, *10*, 391–400.
3. Yatzidis, H. New method for direct determination of "true" creatinine. *Clin. Chem.* **1974**, *20*, 1131–1134. [CrossRef]
4. Bergam, A.; Ohman, G. Effect of detergent on kinetic Jaffé-method assay of creatinine. *Clin. Chem.* **1980**, *26*, 1729–1732.
5. Welch, M.J.; Cohen, A.; Hertz, H.S.; Ng, K.J.; Schaffer, R.; Van der Lijn, P.; White, E., 5th. Determination of serum creatinine by isotope dilution mass spectrometry as a candidate definitive method. *Anal. Chem.* **1986**, *58*, 1681–1685. [CrossRef]
6. Paroni, R.; Arcelloni, C.; Fermo, I.; Bonini, P.A. Determination of creatinine in serum and urine by a rapid liquid-chromatographic method. *Clin. Chem.* **1990**, *36*, 830–836. [CrossRef] [PubMed]
7. Ekelund, S.; Paby, P. High-performance liquid chromatographic determination of creatinine. *Scand. J. Clin. Lab. Investig.* **1991**, *51*, 67–71. [CrossRef]
8. Linnet, K.; Bruunshuus, I. HPLC with enzymatic detection as a candidate reference method for serum creatinine. *Clin. Chem.* **1991**, *37*, 1669–1675. [CrossRef]
9. Sugita, O.; Uchiyama, K.; Yamada, T.; Sato, T.; Okada, M.; Takeuchi, K. Reference values of serum and urine creatinine, and of creatinine clearance by a new enzymatic method. *Ann. Clin. Biochem.* **1992**, *29*, 523–528. [CrossRef]
10. Takatsu, A.; Nishi, S. Determination of serum creatinine using isotope dilution method using discharge-assisted thermospray liquid chromatography/mass spectrometry. *Biol. Mass Spectrom.* **1993**, *22*, 643–646. [CrossRef]
11. Thienpont, L.M.; van Landuyt, K.G.; Stockl, D.; de Leenheer, A.P. Candidate reference method for determining serum creatinine by isocratic HPLC: Validation with isotope dilution gas chromatography-mass spectrometry and application for accuracy assessment of routine test kits. *Clin. Chem.* **1995**, *41*, 995–1003. [CrossRef]
12. Carobene, A.; Ferrero, C.; Ceriotti, F.; Modenese, A.; Besozzi, M.; De Giorgi, E.; Franzin, M.; Franzini, C.; Kienle, M.G.; Magni, F. Creatinine measurement proficiency testing: Assignment of matrix-adjusted ID GC-MS target values. *Clin. Chem.* **1997**, *43*, 1342–1347. [CrossRef]
13. Yasuda, M.; Sugahara, K.; Zhang, J.; Ageta, T.; Nakayama Shuin, T.; Kodama, H. Simultaneous determination of creatinine, creatine, and guanidinoacetic acid in human serum and urine using liquid chromatography-atmospheric pressure chemical ionization mass spectrometry. *Anal. Biochem.* **1997**, *253*, 231–235. [CrossRef] [PubMed]
14. Smith-Palmer, T. Separation methods applicable to urinary creatine and creatinine. *J. Chromatogr. B* **2002**, *781*, 93–106. [CrossRef]
15. Stokes, P.; O'Connor, G. Development of a liquid chromatography-mass spectrometry method for the high-accuracy determination of creatinine in serum. *J. Chromatogr. B* **2003**, *794*, 125–136. [CrossRef]
16. Tsikas, D.; Wolf, A.; Frölich, J.C. Simplified HPLC method for urinary and circulating creatinine. *Clin. Chem.* **2004**, *50*, 201–203. [CrossRef] [PubMed]
17. Owen, L.J.; Wear, J.E.; Keevil, B.G. Validation of a liquid chromatography tandem mass spectrometry assay for serum creatinine and comparison with enzymatic and Jaffe methods. *Ann. Clin. Biochem.* **2006**, *43*, 118–123. [CrossRef] [PubMed]
18. Takahashi, N.; Boysen, G.; Li, F.; Li, Y.; Swenberg, J.A. Tandem mass spectrometry measurements of creatinine in mouse plasma and urine for determining glomerular filtration rate. *Kidney Int.* **2007**, *71*, 266–271. [CrossRef]

19. Park, E.K.; Watanabe, T.; Gee, S.J.; Schenker, M.B.; Hammock, B.D. Creatinine measurements in 24 h urine by liquid chromatography—Tandem Mass Spectrometry. *J. Agric. Food Chem.* **2008**, *56*, 333–336. [CrossRef] [PubMed]
20. Lawson, A.M. Prospects for mass spectrometry in clinical chemistry. *Ann. Clin. Biochem.* **1975**, *12*, 51–57. [CrossRef] [PubMed]
21. Siekmann, L. Determination of creatinine in human serum by isotope dilution-mass spectrometry. *J. Clin. Chem. Clin. Biochem.* **1985**, *23*, 137–144. [PubMed]
22. Björkhem, I.; Blomstrand, R.; Ohman, G. Mass fragmentography of creatinine proposed as a reference method. *Clin. Chem.* **1977**, *23*, 2114–2121. [CrossRef]
23. Little, J.L. Artifacts in trimethylsilyl derivatization reactions and ways to avoid them. *J. Chromatogr. A* **1999**, *844*, 1–22. [CrossRef]
24. Stalling, D.L.; Gehrke, C.W.; Zumwalt, R.W. A new silylation reagent for amino acids bis(trimethylsilyl)trifluoroacetamide (BSTFA). *Biochem. Biophys. Res. Commun.* **1968**, *31*, 616–622. [CrossRef]
25. Baskal, S.; Bollenbach, A.; Tsikas, D. GC-MS Discrimination of Citrulline from Ornithine and Homocitrulline from Lysine by Chemical Derivatization: Evidence of Formation of N^5-Carboxy-ornithine and N^6-Carboxy-lysine. *Molecules* **2021**, *26*, 2301. [CrossRef]
26. Chobanyan, K.; Mitschke, A.; Gutzki, F.M.; Stichtenoth, D.O.; Tsikas, D. Accurate quantification of dimethylamine (DMA) in human plasma and serum by GC-MS and GC-tandem MS as pentafluorobenzamide derivative in the positive-ion chemical ionization mode. *J. Chromatogr. B* **2007**, *851*, 240–249. [CrossRef]
27. Begou, O.; Drabert, K.; Theodoridis, G.; Tsikas, D. GC-NICI-MS analysis of acetazolamide and other sulfonamide (R-SO_2-NH_2) drugs as pentafluorobenzyl derivatives [R-SO_2-N(PFB)$_2$] and quantification of pharmacological acetazolamide in human urine. *J. Pharm. Anal.* **2020**, *10*, 49–59. [CrossRef]
28. Burov, S.; Leko, M.; Dorosh, M.; Dobrodumov, A.; Veselkina, O. Creatinyl amino acids: New hybrid compounds with neuroprotective activity. *J. Pept. Sci.* **2011**, *17*, 620–626. [CrossRef]
29. Karlsson, K.A. Analysis of compounds containing phosphate and phosphonate by gas-liquid chromatography and mass spectrometry. *Biochem. Biophys. Res. Commun.* **1970**, *39*, 847–851. [CrossRef]
30. Shimojo, T.; Schroepfer, G.J., Jr. Sphinganine-1-phosphate lyase: Identification of ethanolamine 1-phosphate as product. *Biochim. Biophys. Acta* **1976**, *43*, 433–446.
31. Caban, M.; Stepnowski, P. Silylation of acetaminophen by trifluoroacetamide-based silylation agents. *J. Pharm. Biomed. Anal.* **2018**, *154*, 433–437. [CrossRef] [PubMed]
32. Harvey, D.J.; Vouros, P. Mass Spectrometric Fragmentation of Trimethylsilyl and Related Alkylsilyl Derivatives. *Mass Spectrom. Rev.* **2020**, *39*, 105–211. [CrossRef] [PubMed]
33. Bollenbach, A.; Hanff, E.; Beckmann, B.; Kruger, R.; Tsikas, D. GC-MS quantification of urinary symmetric dimethylarginine (SDMA), a whole-body symmetric L-arginine methylation index. *Anal. Biochem.* **2018**, *556*, 40–44. [CrossRef] [PubMed]
34. Tsikas, D.; Wolf, A.; Mitschke, A.; Gutzki, F.M.; Will, W.; Bader, M. GC-MS determination of creatinine in human biological fluids as pentafluorobenzyl derivative in clinical studies and biomonitoring: Inter-laboratory comparison in urine with Jaffe, HPLC and enzymatic assays. *J. Chromatogr. B Anal. Technol. Biomed. Life Sci.* **2010**, *878*, 2582–2592. [CrossRef] [PubMed]

Article

GC-MS Discrimination of Citrulline from Ornithine and Homocitrulline from Lysine by Chemical Derivatization: Evidence of Formation of N^5-Carboxy-ornithine and N^6-Carboxy-lysine

Svetlana Baskal, Alexander Bollenbach and Dimitrios Tsikas *

Core Unit Proteomics, Institute of Toxicology, Hannover Medical School, 30625 Hannover, Germany; baskal.svetlana@mh-hannover.de (S.B.); bollenbach.alex@gmail.com (A.B.)
* Correspondence: Tsikas.dimitros@mh-hannover.de

Citation: Baskal, S.; Bollenbach, A.; Tsikas, D. GC-MS Discrimination of Citrulline from Ornithine and Homocitrulline from Lysine by Chemical Derivatization: Evidence of Formation of N^5-Carboxy-ornithine and N^6-Carboxy-lysine. *Molecules* **2021**, *26*, 2301. https://doi.org/10.3390/molecules26082301

Academic Editor: Paraskevas D. Tzanavaras

Received: 24 March 2021
Accepted: 13 April 2021
Published: 15 April 2021

Publisher's Note: MDPI stays neutral with regard to jurisdictional claims in published maps and institutional affiliations.

Copyright: © 2021 by the authors. Licensee MDPI, Basel, Switzerland. This article is an open access article distributed under the terms and conditions of the Creative Commons Attribution (CC BY) license (https://creativecommons.org/licenses/by/4.0/).

Abstract: Derivatization of amino acids by 2 M HCl/CH$_3$OH (60 min, 80 °C) followed by derivatization of the intermediate methyl esters with pentafluoropropionic anhydride (PFPA) in ethyl acetate (30 min, 65 °C) is a useful two-step derivatization procedure (procedure A) for their quantitative measurement in biological samples by gas chromatography-mass spectrometry (GC-MS) as methyl ester pentafluoropropionic (PFP) derivatives, (Me)$_m$-(PFP)$_n$. This procedure allows in situ preparation of trideutero-methyl esters PFP derivatives, (d$_3$Me)$_m$-(PFP)$_n$, from synthetic amino acids and 2 M HCl/CD$_3$OD for use as internal standards. However, procedure A converts citrulline (Cit) to ornithine (Orn) and homocitrulline (hCit) to lysine (Lys) due to the instability of their carbamide groups under the acidic conditions of the esterification step. In the present study, we investigated whether reversing the order of the two-step derivatization may allow discrimination and simultaneous analysis of these amino acids. Pentafluoropropionylation (30 min, 65 °C) and subsequent methyl esterification (30 min, 80 °C), i.e., procedure B, of Cit resulted in the formation of six open and cyclic reaction products. The most abundant product is likely to be N^5-Carboxy-Orn. The second most abundant product was confirmed to be Orn. The most abundant reaction product of hCit was confirmed to be Lys, with the minor reaction product likely being N^6-Carboxy-Lys. Mechanisms are proposed for the formation of the reaction products of Cit and hCit via procedure B. It is assumed that at the first derivatization step, amino acids form (N,O)-PFP derivatives including mixed anhydrides. At the second derivatization step, the Cit-(PFP)$_4$ and hCit-(PFP)$_4$ are esterified on their C^1-Carboxylic groups and on their activated N^{ureido} groups. Procedure B also allows in situ preparation of (d$_3$Me)$_m$-(PFP)$_n$ from synthetic amino acids for use as internal standards. It is demonstrated that the derivatization procedure B enables discrimination between Cit and Orn, and between hCit and Lys. The utility of procedure B to measure simultaneously these amino acids in biological samples such as plasma and urine remains to be demonstrated. Further work is required to optimize the derivatization conditions of procedure B for biological amino acids.

Keywords: amino acids; derivatization; esterification; GC-MS; pentafluoropropionic anhydride; ureide

1. Introduction

Analysis of amino acids, dipeptides, and tripeptides such as glutathione by gas chromatography-mass spectrometry (GC-MS) requires suitable derivatization reactions to convert them into volatile and thermally stable derivatives [1–9]. Derivatization of amino acids with 2 M HCl in methanol (CH$_3$OH) (60 min, 80 °C) yields their mono- and di-methyl esters. Subsequent reaction with pentafluoropropionic anhydride (PFPA) in ethyl acetate (30 min, 65 °C) generates the N- and O-pentafluoropropionyl (PFP) derivatives. The methyl ester (Me) PFP derivatives ((Me)$_m$-(PFP)$_n$) obtained by this procedure (here designated as procedure A) are useful for the quantitative measurement of biological amino acids by GC-MS [7]. However, the carbamoyl-amino acids citrulline (Cit) and homocitrulline (hCit) (Figure 1) are converted under these reaction conditions into the methyl esters of ornithine (Orn) and lysine (Lys), respectively [7]. Analogously, glutamine (Gln) and asparagine (Asn) are converted into glutamate (Glu) and aspartate (Asp), respectively [7]. For not yet fully understood reasons, the derivatization procedure A was found to be not useful for the GC-MS analysis of N^G,N'^G-dimethylarginine (symmetric dimethylarginine, SDMA), in contrast to its structural isomer N^G,N^G-dimethylarginine (asymmetric dimethylarginine, ADMA) and to their precursor arginine. This difficulty was in part overcome by using a single derivatization reaction with PFPA, which most likely generates the *tetrakis*(pentafluoropropionyl) derivative of SDMA, i.e., SDMA-(PFP)$_4$ [8]. This derivatization reaction, i.e., (N,O)-pentafluoroprionylation, enables quantitative measurement of SDMA in human urine, but requires the use of commercially available stable-isotope labelled SDMA analogue such as [N^G,N'^G-^{2}H$_6$]dimethylarginine [8] and is less sensitive compared to the GC-MS analysis of ADMA as Me-PFP derivative. Interestingly, the tripeptides glutathione and its analogue ophthalmic acid were also found to react with PFPA under the same derivatization conditions, which enabled their GC-MS analysis [9].

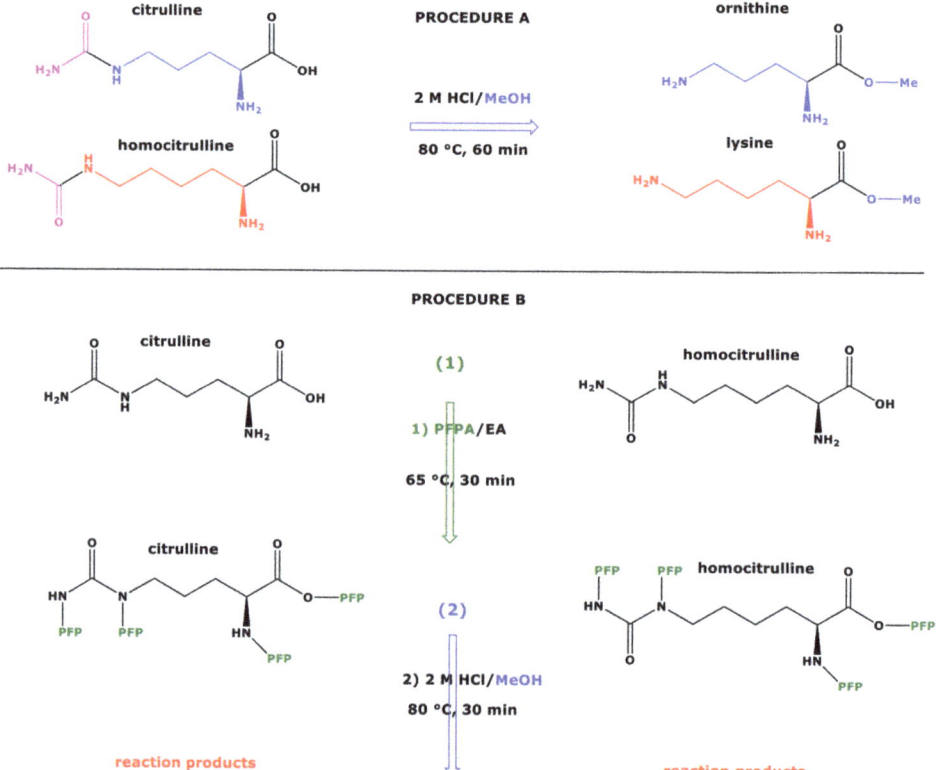

Figure 1. Upper panel, procedure (**A**) Schematic of the reactions of citrulline and homocitrulline with 2 M HCl/MeOH forming the methyl esters of ornithine and lysine, respectively. Lower panel, procedure (**B**) Schematic of the two-step derivatization of citrulline (left) and homocitrulline (right) first with PFPA/EA to form their PFP derivatives with the proposed formulas Cit-(PFP)$_4$ and hCit-(PFP)$_4$, respectively. Subsequently, these derivatives react with 2 M HCl/MeOH (procedure B) to form reaction products that were characterized structurally by GC-MS in the present study. Cit, citrulline; hCit, homocitrulline; MeOH, methanol; PFPA, pentafluoropropionic anhydride; PFP, pentafluoropropionyl residue; EA, ethyl acetate.

The aim of the present study was to find derivatization conditions that would allow discrimination of Cit from Orn, and of hCit from Lys. Our previous observations that SDMA can be measured in human urine by GC-MS by using PFPA/EA as the first derivatization step [8] prompt us to investigate whether the derivatization of Cit and hCit with PFPA/EA as the first step may also be useful for their GC-MS analysis and for their discrimination from Orn and Lys, respectively. Analogous to SDMA, we assumed intermediate formation of Cit-(PFP)$_4$ and hCit-(PFP)$_4$ (Figure 1). In order to investigate potential reactions of the putative intermediates, we coupled the PFPA/EA derivatization with the classical esterification with 2 M HCl/CH$_3$OH and with 2 M HCl/CD$_3$OD to prepare stable-isotope-labelled analogs of Cit and hCit. De facto, this resulted in a reversed order of the original two-step derivatization procedure A, which is specified as procedure B in the present work (Figure 1). In most investigations using derivatization procedure B, we used experimental conditions previously found to be optimum for the derivatization and GC-MS analysis of amino acids and the tripeptides glutathione and ophthalmic acid [7–9].

2. Materials and Methods

2.1. Chemicals, Materials and Reagents

All amino acids (chemical purity, 95 to 98%) were obtained from Sigma-Aldrich. Tetradeuterated methanol (CD$_3$OD, 99% at ^{2}H) and pentafluoropropionic anhydride were supplied by Aldrich (Steinheim, Germany). Methanol was obtained from Chemsolute (Renningen, Germany). Hydrochloric acid (37 wt%) was purchased from Baker (Deventer, The Netherlands). Ethyl acetate was obtained from Merck (Darmstadt, Germany). Glassware for GC-MS (1.5 mL autosampler glass vials and 0.2 mL microvials) and the fused-silica capillary column Optima 17 (15 M × 0.25 mm I.D., 0.25 μM film thickness) were purchased from Macherey–Nagel (Düren, Germany). Separate stock solutions of amino acids were prepared by dissolving accurately weighed amounts of commercially available amino acids in deionized water. Stock solutions were diluted with deionized water as appropriate.

For the preparation of unlabelled methyl esters and deuterium-labelled methyl esters of amino acids, two derivatization reagents were prepared. To 80 mL ice-cold CH$_3$OH were added 16 mL of 37 wt% HCl slowly under gentle mixing. Analogously, to 80 mL ice-cold CD$_3$OD, 16 mL of 37 wt% HCl were added slowly under gentle mixing. The concentration of HCl in these methanolic solutions was each 2 M. In the present article, these solutions are denoted as 2 M HCl/CH$_3$OH and 2 M HCl/CD$_3$OD, respectively. The PFPA-ethyl acetate reagent (PFPA/EA) was prepared daily by diluting pure PFPA in ethyl acetate (EA) (1:4, v/v).

2.2. Derivatization Procedures A and B for Amino Acids and Generation of GC-MS Spectra

Procedure A. Solid amino acids were derivatized first with 2 M HCl/CH$_3$OH or 2 M HCl/CD$_3$OD and then with PFPA/EA in autosampler glass vials. Briefly, residues were reconstituted in 100 μL aliquots of a 2 M HCl/CH$_3$OH or 2 M HCl/CD$_3$OD solution and the glass vials were tightly sealed. Esterification was performed by heating the samples for 60 min at 80 °C. After cooling the samples of the esterification reaction to room temperature, solvents and reagents were evaporated to dryness under a stream of nitrogen. Aliquots (100 μL) of the PFPA/EA solution were added, and the glass vials were tightly sealed and heated for 30 min at 65 °C to prepare N-pentafluoropropionic amides of the methyl esters. Then, residues were treated first with 200 μL aliquots of 400 mM borate buffer, pH 8.5, and immediately thereafter with 200 μL aliquots of toluene, followed by immediate vortex-mixing for 60 s and centrifugation (4000× g, 5 min, 18 °C). Aliquots (150 μL) of the upper organic phase were transferred into autosampler glass vials equipped with microinserts, and the samples were sealed and subjected to GC-MS analysis.

Procedure B. Solid amino acids were derivatized first with PFPA/EA (30 min, 65 °C) and then with 2 M HCl/CH$_3$OH or 2 M HCl/CD$_3$OD (30 min, 80 °C). Briefly, aliquots (100 μL) of a freshly prepared PFPA/EA solution were added, the glass vials were tightly sealed and heated for 30 min at 65 °C to prepare N-pentafluoropropionic amides of the methyl esters. After cooling the samples to room temperature, solvents and reagents were evaporated to dryness under a stream of nitrogen. Then, residues were reconstituted in 100 µL aliquots of a 2 M HCl/CH$_3$OH or 2 M HCl/CD$_3$OD solution and the glass vials were tightly sealed. Esterification was performed by heating the samples for 30 min at 80 °C. After cooling to room temperature, solvents and reagents were evaporated to dryness under a stream of nitrogen. Residues were treated directly with toluene (200 µL), shortly vortex-mixed, aliquots (150 µL) of the upper organic phase were transferred into autosampler glass vials equipped with microinserts, and the samples were sealed and subjected to GC-MS analysis.

2.3. Generation of GC-MS Spectra

GC-MS spectra were obtained using negative-ion chemical ionization (NICI) after separate derivatization of 5 nmol of each amino acid using both derivatization procedures as described above. The derivatives were extracted with toluene (1 mL), 1 µL aliquots containing 5 pmol of each analyte (assuming quantitative yield) were injected in the splitless mode, and mass spectra were generated in the scan mode in the mass-to-charge (m/z) range 50 to 650 (1 s per scan). The GC-MS software Xcalibur and Quan Browser were used. ChemDrawProfessional 15.0 was used to draw chemical structures and to convert structures into names. GraphPad Prism 7.0 (San Diego, CA, USA) was used in statistical analyses and to prepare graphs.

2.4. GC-MS Conditions

All analyses were performed on a GC-MS apparatus consisting of a single quadrupole mass spectrometer model ISQ, a Trace 1210 series gas chromatograph, and an AS1310 autosampler from ThermoFisher (Dreieich, Germany). The injector temperature was kept at 280 °C. Helium was used as the carrier gas at a constant flow rate of 1.0 mL/min. The oven temperature was held at 40 °C for 0.5 min and ramped to 210 °C at a rate of 15 °C/min and then to 320 °C at a rate 35 °C/min. Interface and ion-source temperatures were set to 300 °C and 250 °C, respectively. Electron energy was 70 eV and electron current 50 µA. Methane was used as the reagent gas for NICI at a constant flow rate of 2.4 mL/min. In quantitative analyses, the dwell time was 100 ms for each ion in the selected-ion monitoring (SIM) mode and the electron multiplier voltage was set to 1400 V.

3. Results

3.1. Derivatization of Citrulline and Structural Characterization of Its Reaction Products by GC-MS

Scanning of the Cit samples derivatized by procedure B resulted in the elution each of six GC-MS peaks using CH$_3$OH (Supplementary Materials Figure S1A) and CD$_3$OD (Figure S1B). In the latter case, the peaks I, II, V, and VI eluted a few seconds in front of the peaks of the Cit sample derivatized with CH$_3$OH, indicating the presence of deuterium atoms in these peaks [7] (see Table 1). The almost identical retention times of the minor peaks III (retention time, 9.36 min) and peaks IV (retention time, 9.67 min) suggest that they are not methyl esters, but rather cyclic compounds.

The mass spectra of the peaks I (retention time, 8.36 min, 8.33 min) contained four corresponding ions that differed by 3 Da each, suggesting the presence of a single methylated carboxylic group (Figure S1(A1,B1)) (Table 1). A tentative structure of this molecule could be (S)-3-amino-2-oxopiperidine-1-Carboxylic acid (non-derivatized).

Table 1. GC-MS retention times (t_R, min) and most intense ions in the mass spectra of the six reaction products of citrulline derivatized with procedure B (first PFPA/EA then 2 M HCl/CH$_3$OH or 2 M HCl/CD$_3$OD). For comparison, synthetic ornithine standard (Orn-Std) was also derivatized with procedure B. See also Figure S1.

Peak No.	t_R	Spectrum Intensity	m/z (Intensity, %)
Peak I	8.36	1.0×10^7	235 (2), 278 (5), **298** [a] (100), 318 (2)
Peak I	8.36	1.3×10^7	238 (2), 281 (4), **301** (100), 321 (1)
Peak II	8.67	1.5×10^7	275 (5), 398 (18), **418** (100)
Peak II	8.65	1.2×10^7	278 (5), 401 (20), **421** (100)
Peak III	9.36	1.2×10^6	218 (92), **238** (100)
Peak III	9.36	1.5×10^6	218 (90), **238** (100)
Peak IV	9.68	2.8×10^6	220 (20), **240** (100), 258 (5)
Peak IV	9.67	3.0×10^6	220 (15), **240** (100), 258 (5)
Peak V	10.57	2.5×10^6	162 (43), 278 (15), **298** (100), 486 (2)
Peak V	10.52	2.7×10^6	162 (25), 281 (12), **301** (100), 489 (6)
Peak VI	10.75	2.5×10^7	290 (5), 298 (27), **330** (100), 349 (3)
Peak VI	10.71	2.9×10^7	296 (5), 301 (25), **336** (100), 355 (3)
Orn-Std	8.67	1.1×10^8	275 (3), 398 (20), **418** (100), 437 (8)
Orn-Std	8.63	1.2×10^8	278 (2), 401 (20), **421** (100), 440 (10)

[a] Bold numbers indicate mass fragments with the highest intensity in the mass spectrum (i.e., base peaks).

Derivatization of Cit by procedure B resulted in the formation of the peaks II (retention time, 8.67 min, 8.63 min) (Supplementary Materials: Figure S1(A1,B1)) (Table 1). Peaks II had virtually the same mass spectra as the unlabelled Me-PFP (d$_0$Me-PFP) and the labelled Me-PFP (d$_3$Me-PFP) derivatives of Orn (Figure S1(C1,C2)), indicating conversion of Cit to Orn by both procedures as observed previously using procedure A [7].

The mass spectra in combination with the retention times of the peaks III and the peaks IV suggest that the peak III corresponds to (S)-3-amino-4,5-dihydropyridin-2(3H)-one (Figure S1(A3,B3)) and peak IV corresponds to (S)-3-aminopiperidin-2-one (Figure S1(A4,B4)) (Table 1).

The mass spectra of the minor peaks V (retention time, 10.57 min, 10.52 min) contained corresponding ions that did not differ (m/z 162) or did differ by 3 Da each (m/z, 301/298; m/z, 489/486) suggesting the presence of an intact methylated carboxylic group and presumably a fragmented methyl ester (Figure S1(A5,B5)) (Table 1). A tentative structure of this molecule could be (S)-2-amino-5-(Carboxyamino)pentanoic acid, which could be trivially named N^5-Carboxy-ornithine.

The most intense GC-MS peaks of Cit derivatized by procedure B were the peaks VI, which eluted at 10.75 min (using CH$_3$OH) and 10.71 min (using CD$_3$OD) (Figure S1(A6,B6)). The GC-MS spectra of these peaks contained several corresponding mass fragments that differed by 3 Da (m/z 298/301) or 6 Da (m/z 296/290, m/z 336/330, m/z 355/349) suggesting the presence of two carboxylic groups in these ions (Table 1).

3.2. Effects of the PFPA/EA Derivatization Time in Procedure B on the Reaction Products

The derivatization conditions used in procedure A in the present study were found to be optimal in previous studies [7,8]. In this experiment, we investigated the effect of the derivatization time of the pentafluoropropionylation reaction of Cit in procedure B. For this, two sets of 10 µL aliquots of Cit samples in distilled water (50, 100, 150, 200, 250 µM) were derivatized first with PFPA/EA at 65 °C for 30 min and subsequently with 2 M HCl/CH$_3$OH for 10, 20, 30, 40, and 60 min at 80 °C. A 100 µM Cit sample in distilled water served as internal standard and was derivatized in parallel under the same conditions using 2 M HCl/CD$_3$OD. After toluene extraction, GC-MS analysis was performed by SIM of m/z 298 and m/z 301 for peak I, m/z 418 and m/z 421 for peak II (i.e., Orn), m/z 298 and m/z 301 for peak V, and m/z 330 and m/z 336 for peak VI (see Table 1).

The results of this experiment are illustrated in Figures 2–4. The peak area of the internal standards varied between 11 and 16% (m/z 301, Peak I), between 7 and 14% (m/z 421,

Peak II), between 10 and 21% (m/z 301, Peak V), and between 13 and 17% (m/z 336, Peak VI). There were significant time effects for all monitored internal standards ($P < 0.0001$, two-way ANOVA) and a Cit concentration effect for Peak II ($P = 0.043$, two-way ANOVA). With the exception of the Peak I, the peak areas of the Peaks II, V, and VI had a minimum at the esterification time of 30 min. When combining all data of the incubation times of the esterification, the peak area ratios (y) of all peaks depended linearly upon the Cit concentration (in µM) (x): $y = -0.223 + 0.011\,x$, $r^2 = 0.9943$ for Peak I, $y = 0.241 + 0.0098\,x$, $r^2 = 0.9712$ for Peak II, $y = -0.469 + 0.012\,x$, $r^2 = 0.9944$ for Peak V, and $y = -0.415 + 0.013\,x$, $r^2 = 0.9923$ for Peak VI. Linear relationships were observed between the peak area ratio (PAR) values (y) of the individual peaks and the derivatized Cit concentration (x) resulted in straight lines for all derivatization times of the esterification step (Figure 4). Linear regression analysis of the mean PAR of all peaks vs. The concentrations of derivatized Cit resulted in the regression equation $y = -0.217 + 0.0117\,x$, $r^2 = 0.9973$. The reciprocal of the slope value of this regression equation indicates a mean concentration of 85.3 µM for the sum of internal standards (nominal concentration, 100 µM).

Taken together, these results demonstrate the principle applicability of the procedure B for the quantitative GC-MS analysis of Cit in aqueous solutions.

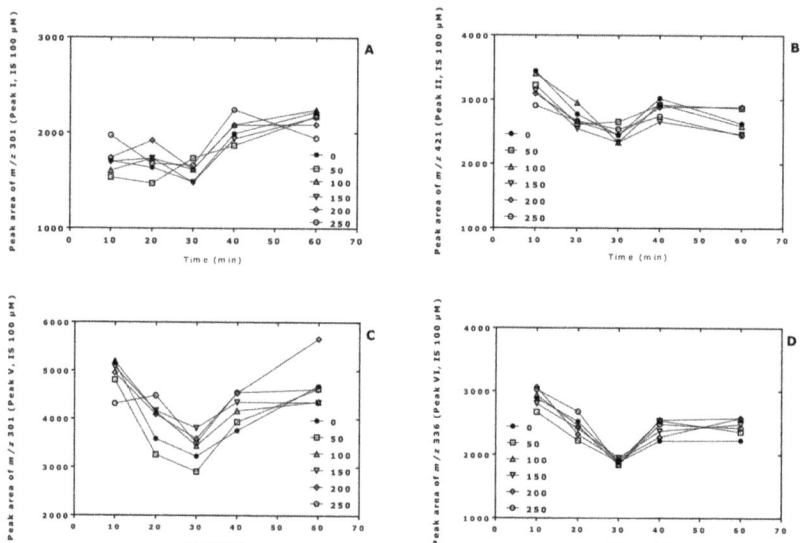

Figure 2. Time profiles of the peak areas of the internal standards (generated from 100 µM Cit) upon derivatization of aqueous Cit (0, 50, 100, 150, 200, 250 µM; see insert) using procedure B, i.e., first with PFPA/EA for a fixed time of 30 min at 65 °C and subsequently with 2 M HCl/CH$_3$OH (2 M HCl/CD$_3$OD for the internal standards) at 80 °C for the indicated times. (**A**) Peak I; (**B**) Peak II; (**C**) Peak V; (**D**) Peak VI. See also Table 1 and Figure 3.

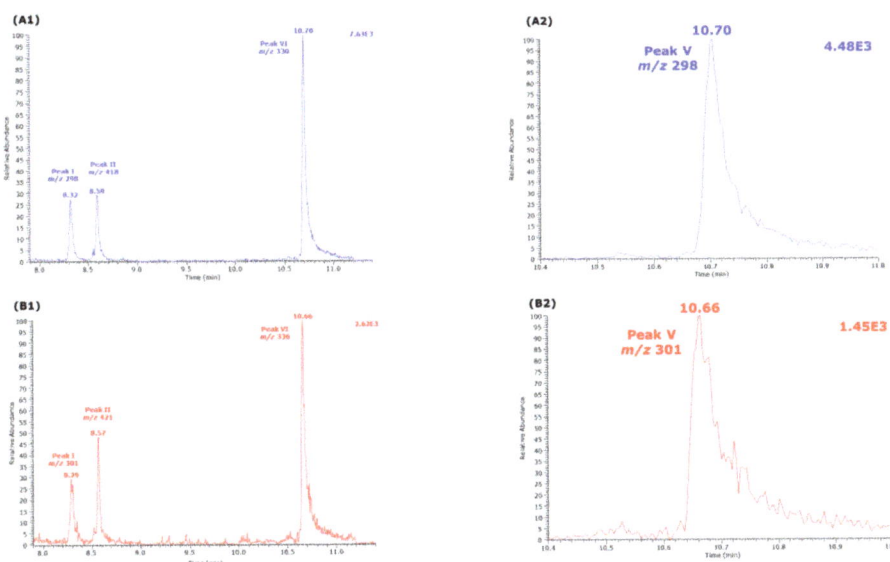

Figure 3. Partial GC-MS chromatograms from the analysis of an aqueous citrulline sample (250 µM) derivatized by procedure B, i.e., first with PFPA/EA (30 min, 65 °C) followed with 2 M HCl in CH$_3$OH or CD$_3$OD (60 min, 80 °C). Selected ion monitoring (SIM, 100 ms) of (**A1,A2**) m/z 298 and m/z 301 for Peak I, m/z 418 and m/z 421 for Peak II, and m/z 330 and m/z 336 for Peak VI was performed. (**B1,B2**) SIM (100 ms) of m/z 298 and m/z 301 for Peak V. See also Table 1 and Figure 2.

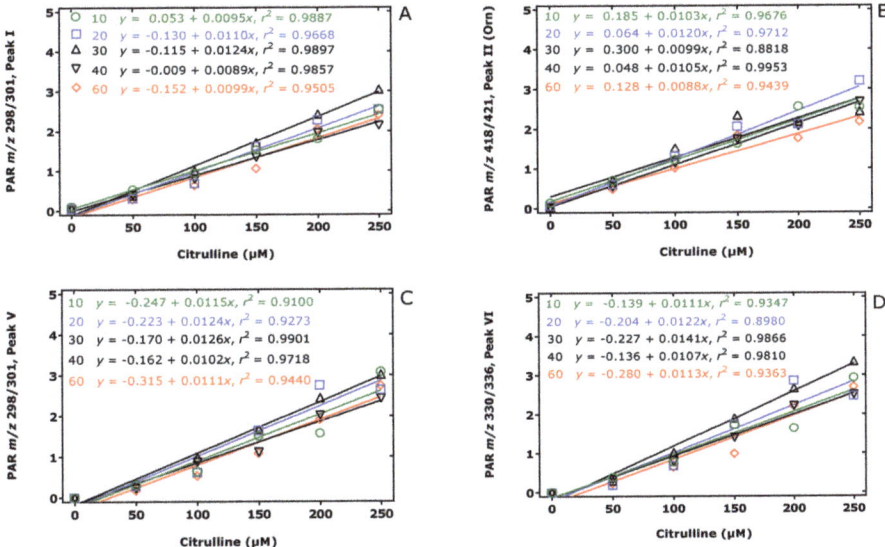

Figure 4. Linear regression analysis between the peak area ratio (PAR) values (y) for (**A**) Peak I, (**B**) Peak II, (**C**) Peak V, and (**D**) Peak VI to the corresponding internal standards (generated from 100 µM Cit) and the Cit concentration (x) upon derivatization of aqueous Cit (0, 50, 100, 150, 200, 250 µM) first with PFPA/EA for a fixed time of 30 min at 65 °C and subsequently with 2 M HCl/CH$_3$OH (2 M HCl/CD$_3$OD for the internal standard) at 80 °C for 10 (green circles), 20 (blue squares), 30 (black upper triangles), 40 (black lower triangles), 60 (red diamonds) min. Insets indicate the regression equations. Note that Peak II corresponds to the Orn derivative. See also Table 1 and Figure 3.

3.3. Derivatization of Homocitrulline and Structural Characterization of Its Reaction Products by GC-MS

Scanning of the hCit samples derivatized by procedure B resulted in the elution each of an intense GC-MS peak using CH_3OH (Figure S1D) and CD_3OD (Figure S1E) and two minor peaks (Table 2).

Table 2. GC-MS retention times (t_R, min) and most intense ions in the mass spectra of the three reaction products of homocitrulline derivatized with procedure B (PFPA/EA then 2 M HCl/CH_3OH or 2 M HCl/CD_3OD). For comparison, synthetic lysine standard (Lys-Std) was also derivatized with procedure B. See also Figure S1.

Peak No.	t_R	Spectrum Intensity	*m/z* (Intensity, %)
Peak I	9.21	4.7×10^5	272 (9), 292 (5), **312** [a] (100), 338 (3)
Peak I	9.18	5.9×10^5	272 (8), 295 (3), **315** (100), 338 (3)
Peak II	9.54	2.5×10^7	289 (11), 392 (8), 412 (27), **432** (100)
Peak II	9.52	1.1×10^7	292 (10), 395 (5), 415 (27), **435** (100)
Peak III	11.47	4.0×10^5	**312** (100), 344 (42)
Peak III	11.42	7.7×10^5	**315** (100), 350 (48)
Lys-Std	9.54	5.6×10^6	289 (15), 392 (16), 412 (50), **432** (100), 451 (5)
Lys-Std	9.52	5.6×10^6	292 (16), 395 (14), 415 (30), **435** (100), 454 (4)

[a] Bold numbers indicate mass fragments with the highest intensity in the mass spectrum (i.e., base peaks).

The major GC-MS peaks eluted at 9.54 min and 9.52 min, respectively. The mass spectra of these peaks are very similar to those obtained from the derivatization of hCit by procedure B, as well as to those of the d_0Me-PFP and d_3Me-PFP derivatives of synthetic Lys confirming previous observations of the conversion of hCit to Lys [7] (Figure 1). The minor GC-MS peaks eluting at 9.21 min and 9.18 min could correspond to (S)-3-amino-2-oxoazepane-1-Carboxylic acid (Figure S1(D1,E1)). The minor GC-MS peaks eluting at 11.47 min and 11.42 min could correspond to N^6-Carboxy-lysine (Figure S1(D3,E3)), analogous to N^5-Carboxy-ornithine obtained from Cit using procedure B. The small differences in the retention times is indicative of the presence of deuterium atoms in the earlier eluting peaks.

4. Discussion

Procedure A allows for the reliable quantitative determination of amino acids and their metabolites in biological samples by GC-MS [7,10]. During the first esterification step, however, Cit and hCit undergo almost complete conversion to the methyl esters of Orn and Lys, respectively. The same happens to Gln and Asn, which are converted to the methyl esters of Glu and Asp, respectively [7]. These observations strongly indicate that the carbamide groups of Cit, hCit, Gln, and Asn are labile under the strong esterification conditions. This circumstance prevents simultaneous measurement of Cit, Orn, hCit, Lys, Gln, Glu, Asn, and Asp [7]. We have hypothesized that reversing the order of the derivatization procedure A may present a way to prevent the abovementioned conversions. In the present study, we investigated this possibility for Cit and hCit using procedure B, i.e., first pentafluoropropionylation and subsequently esterification, using previously optimized derivatization conditions [7]. Cit and hCit reacted to form five and three reaction products, respectively. The tentative chemical structures of these reaction products are illustrated in Figure 5.

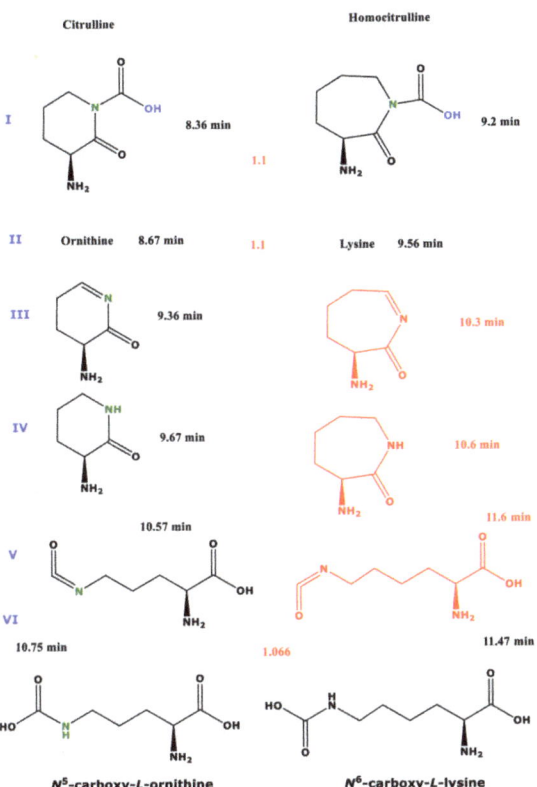

Figure 5. Proposed structures of the reaction products of citrulline (left panel) and homocitrulline (right panel) using procedure B (first PFPA/EA then 2 M HCl/MeOH). The red-marked structures were not found to be derivatization products of homocitrulline. The red numbers between the structures are the relative retention times of the homocitrulline products to the corresponding citrulline reaction products.

One major reaction product of Cit was identified as Orn. This observation suggests that pentafluoropropionylation prevents conversion of Cit to Orn, albeit not entirely. The major reaction product of hCit was identified as Lys. The reaction products of hCit corresponding to the Cit-derived peaks III, IV, and V were not observed (Figure 5). These observations suggest that pentafluoropropionylation prevents conversion of hCit to Lys to only a minor extent. The conversion of Cit to four reaction products in addition to Orn suggest that pentafluoropropionylation of Cit enables additional reactions during the second reaction step of procedure B. The different reaction behaviour of Cit and hCit could be due to the longer side chain of these homologue amino acids: 3 vs. 4 CH_2 groups. It is assumed that this structural difference plays a major role in the formation of cyclic reaction products (Figure 5). Interestingly, procedure B resulted in the formation of N^5-Carboxy-Orn from Cit as a major reaction product and N^6-Carboxy-Lys from hCit as a minor reaction product. Because of the commercially unavailability of synthetic standards of N^5-Carboxy-L-Orn and N^6-Carboxy-L-Lys, we were not able to unequivocally demonstrate the formation of these reaction products. Nevertheless, these putative reaction products enable discrimination of Cit from Orn, and of hCit from Lys, respectively. It is interesting to note that the physiological occurrence and the biological significance of the free amino acids N^5-Carboxy-Orn and N^6-Carboxy-Lys (Chemical Entities of Biological Interest (ChEBI):43575) have not been reported thus far. However, a N^6-Carboxy-Lys residue was found to be present in the active site of class D β-lactamases and to play a significant role in the hydrolysis of β-lactam antibiotics [11,12]. Our study provides useful information for forthcoming studies on these uncommon amino acids.

Based on the results of our study, we propose potential mechanisms that may explain the reaction products of Cit and hCit during the derivatization procedure B. Being a highly reactive derivatization reagent, PFPA is likely to react with all functional groups of free amino acids and those in tripeptides [8,9]. We therefore assume that PFPA/EA reacts with all functional groups of Cit to form its N,N,N,O-(PFP)$_4$ derivative (Figure 6). An intact Cit-(PFP)$_4$ derivative was not observed in our study. An explanation could be that the remaining Cit-(PFP)$_4$ extracted into toluene decomposed during the injection in the hot injector (280 °C). This is more likely to happen to the O-PFP residue, as N-PFP residues of derivatized amino acids are considerably stable [7]. A more plausible explanation for our observations is that the O-PFP residue of the Cit derivative is a mixed anhydride of PFPA and the carboxylic group Cit. As such, the Cit-(PFP)$_4$ derivative is likely to undergo several reactions with 2 M HCl/CH$_3$OH (Figure 6). The reaction of the Cit-(PFP)$_4$ derivative with 2 M HCl/CH$_3$OH will always generate its C^1-Carboxy-methyl ester. Analogously, the reaction of the Cit-(PFP)$_4$ derivative with 2 M HCl/CD$_3$OD will generate the C^1-Carboxy-trideutero-methyl ester. This provides a way to prepare deuterium-labelled internal standards for quantitative analyses. Especially the N-PFP residue on the carbamide functionality of the Cit-(PFP)$_4$ derivative opens ways for additional reactions, which leads to the formation of open reaction products including N^5-Carboxy-Orn from Cit and N^6-Carboxy-Lys from hCit and several cyclic reaction products that can be utilized both in analytical and organic preparative chemistry (Figure 6).

The reaction time of the esterification reaction performed at 80 °C has an effect on the yield of individual reaction products. In a proof-of-principle experiment, we found that procedure B is useful for the quantitative analysis of Cit in aqueous solution for several esterification times. Yet, the quantitative determination of Cit, Orn, hCit, and Lys in biological samples by GC-MS using procedure B remains to be optimized and validated. Our preliminary studies suggest that the derivatization procedure B can be extended to Gln and Asn, which are converted into Glu and Asp, respectively. The derivatization procedure B possess the potential to simultaneously quantitate a large number of biological amino acids and their metabolites by GC-MS using in situ prepared (d$_3$Me)$_m$-(PFP)$_n$ or commercially available stable-isotope labelled amino acids as internal standards.

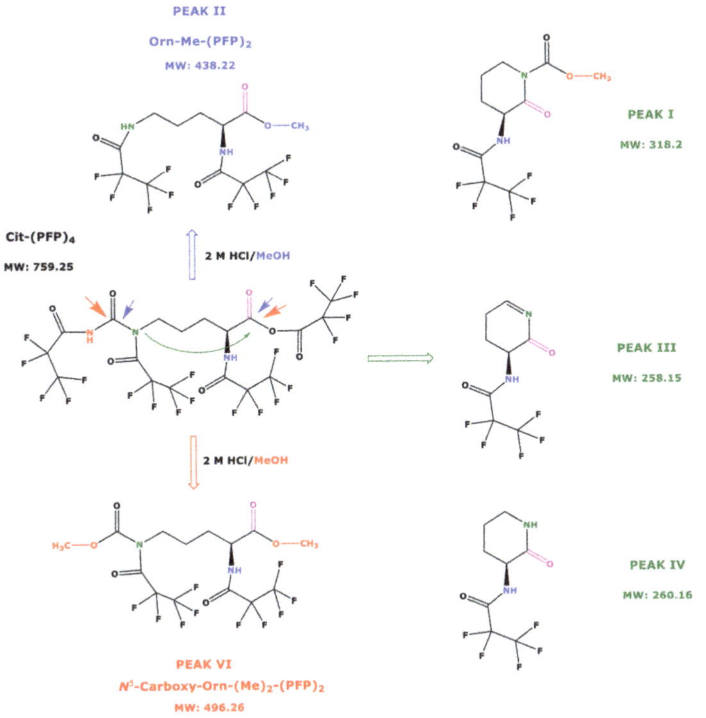

Figure 6. Proposed reaction products of the very first, not yet identified *N,N,N,O*-(PFP)₄ derivative of Cit using procedure B. Blue and red arrows indicate the carbonyl moieties, which are attacked by methanol, and the green arrow indicates the intramolecular attack of N^5 on C^1 carbonyl group.

Supplementary Materials: The following are available online, Figure S1: Separate derivatization of citrulline and homocitrulline (5 nmol each) by procedure B, i.e., first with PFPA/EA and then 2 M HCl/CH₃OH or 2 M HCl/CD₃OD, and structural characterization of their reaction products by GC-MS.

Author Contributions: Conceptualization, D.T.; methodology, S.B.; software, S.B. and D.T.; validation, S.B., A.B. and D.T.; formal analysis, D.T.; investigation, S.B. and A.B.; resources, D.T.; data curation, S.B. and D.T. writing—original draft preparation, D.T., S.B. and A.B.; writing, S.B., A.B. and D.T.; visualization, S.B. and D.T.; supervision, D.T.; project administration, D.T.; funding acquisition, D.T. All authors have read and agreed to the published version of the manuscript.

Funding: This research received no external funding.

Institutional Review Board Statement: Ethical review and approval were waived for this study, due to the use of human urine samples originally collected in previously ethically approved study.

Informed Consent Statement: Subject consent was waived due to the use of human urine samples originally collected in previously ethically approved study.

Data Availability Statement: The study did not report any data.

Acknowledgments: We thank Bibiana Beckmann for administrative assistance.

Conflicts of Interest: The authors declare no conflict of interest.

Sample Availability: Samples of the compounds are not available from the authors.

References

1. Hušek, P.; Macek, K. Gas chromatography of amino acids. *J. Chromatogr. A* **1975**, *113*, 139–230. [CrossRef]
2. Hušek, P.; Švagera, Z.; Hanzlíková, D.; Řimnáčová, L.; Zahradníčková, H.; Opekarová, I.; Šimek, P. Profiling of urinary aminocarboxylic metabolites by in-situ heptafluorobutyl chloroformate mediated sample preparation and gas-mass spectrometry. *J. Chromatogr. A* **2016**, *1443*, 211–232. [CrossRef] [PubMed]
3. Ferré, S.; González-Ruiz, V.; Guillarme, D.; Rudaz, S. Analytical strategies for the determination of amino acids: Past, present and future trends. *J. Chromatogr. B Analyt. Technol. Biomed. Life Sci.* **2019**, *1132*, 121819. [CrossRef] [PubMed]
4. Xu, W.; Zhong, C.; Zou, C.; Wang, B.; Zhang, N. Analytical methods for amino acid determination in organisms. *Amino Acids* **2020**, *52*, 1071–1088. [CrossRef] [PubMed]
5. Zhao, L.; Ni, Y.; Su, M.; Li, H.; Dong, F.; Chen, W.; Wei, R.; Zhang, L.; Guiraud, S.P.; Martin, F.P.; et al. High Throughput and Quantitative Measurement of Microbial Metabolome by Gas Chromatography/Mass Spectrometry Using Automated Alkyl Chloroformate Derivatization. *Anal. Chem.* **2017**, *89*, 5565–5577. [CrossRef] [PubMed]
6. Baskal, S.; Bollenbach, A.; Tsikas, D. Two-step derivatization of amino acids for stable-isotope dilution GC–MS analysis: Long-term stability of methyl ester-pentafluoropropionic derivatives in toluene extracts. *Molecules* **2021**, *26*, 1726. [CrossRef] [PubMed]
7. Hanff, E.; Ruben, S.; Kreuzer, M.; Bollenbach, A.; Kayacelebi, A.A.; Das, A.M.; von Versen-Höynck, F.; von Kaisenberg, C.; Haffner, D.; Ückert, S.; et al. Development and validation of GC-MS methods for the comprehensive analysis of amino acids in plasma and urine and applications to the HELLP syndrome and pediatric kidney transplantation: Evidence of altered methylation, transamidination, and arginase activity. *Amino Acids* **2019**, *51*, 529–547. [CrossRef] [PubMed]
8. Bollenbach, A.; Hanff, E.; Beckmann, B.; Kruger, R.; Tsikas, D. GC-MS quantification of urinary symmetric dimethylarginine (SDMA), a whole-body symmetric L-arginine methylation index. *Anal. Biochem.* **2018**, *556*, 40–44. [CrossRef] [PubMed]
9. Bollenbach, A.; Tsikas, D. Measurement of the tripeptides glutathione and ophthalmic acid by gas chromatography-mass spectrometry. *Anal. Biochem.* **2020**, 113841. [CrossRef] [PubMed]
10. Tsikas, D. De novo synthesis of trideuteromethyl esters of amino acids for use in GC-MS and GC-tandem MS exemplified for ADMA in human plasma and urine: Standardization, validation, comparison and proof of evidence for their aptitude as internal standards. *J. Chromatogr. B* **2009**, *877*, 2308–2320. [CrossRef] [PubMed]
11. Schneider, K.D.; Bethel, C.R.; Distler, A.M.; Hujer, A.M.; Bonomo, R.A.; Leonard, D.A. Mutation of the active site carboxy-lysine (K70) of OXA-1 beta-lactamase results in a deacylation-deficient enzyme. *Biochemistry* **2009**, *48*, 6136–6145. [CrossRef] [PubMed]
12. Leonard, D.A.; Bonomo, R.A.; Powers, R.A. Class D β-lactamases: A reappraisal after five decades. *Acc. Chem. Res.* **2013**, *46*, 2407–2415. [CrossRef] [PubMed]

Article

Quantitative GC–MS Analysis of Artificially Aged Paints with Variable Pigment and Linseed Oil Ratios

Eliise Tammekivi *[ ], Signe Vahur, Martin Vilbaste and Ivo Leito [ ]

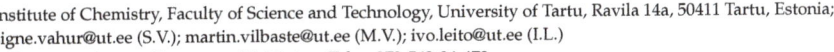

Institute of Chemistry, Faculty of Science and Technology, University of Tartu, Ravila 14a, 50411 Tartu, Estonia; signe.vahur@ut.ee (S.V.); martin.vilbaste@ut.ee (M.V.); ivo.leito@ut.ee (I.L.)
* Correspondence: eliise.tammekivi@ut.ee; Tel.: +372-562-34-473

Abstract: In this study, quantitative gas chromatography–mass spectrometry (GC–MS) analysis was used to evaluate the influence of pigment concentration on the drying of oil paints. Seven sets of artificially aged self-made paints with different pigments (yellow ochre, red ochre, natural cinnabar, zinc white, Prussian blue, chrome oxide green, hematite + kaolinite) and linseed oil mixtures were analysed. In the pigment + linseed oil mixtures, linseed oil concentration varied in the range of 10 to 95 g/100 g. The results demonstrate that the commonly used palmitic acid to stearic acid ratio (P/S) to distinguish between drying oils varied in a vast range (from especially low 0.6 to a common 1.6) even though the paints contained the same linseed oil. Therefore, the P/S ratio is an unreliable parameter, and other criteria should be included for confirmation. The pigment concentration had a substantial effect on the values used to characterise the degree of drying (azelaic acid to palmitic acid ratio (A/P) and the relative content of dicarboxylic acids (∑D)). The absolute quantification showed that almost all oil paint mock-ups were influenced by pigment concentration. Therefore, pigment concentration needs to be considered as another factor when characterising oil-based paint samples based on the lipid profile.

Keywords: GC–MS; pigment; linseed oil; derivatisation; quantification; P/S ratio; A/P ratio; ∑D

1. Introduction

Paints are complex mixtures that may consist of various organic and inorganic compounds (pigments, binders, fillers, and additives). During the drying/curing of the liquid mixture, the binding material goes through different chemical reactions (oxidation, crosslinking, hydrolysis of ester bonds, polymerisation), by which a solid paint layer is formed. Therefore, identifying the binder may be very challenging due to the loss of original compounds and the appearance of new ones [1–3].

One type of binding material whose identification still puzzles scientists are drying oils [4–6]. For example, the most common drying oils (linseed, poppy, and walnut oil) consist of the same fatty acids (palmitic, stearic, oleic, linoleic, and linolenic acid) bound together in triglyceride molecules. The only difference is the percentages of the beforementioned fatty acids [7]. It was discovered by Mills in the 1960s that the ratio of the contents of two saturated fatty acids—palmitic acid to stearic acid (P/S) ratio—is approximately stable during the drying process [8]. Since then, the P/S ratio has been used as one of the main criteria to differentiate between drying oils. Over the years, the P/S value for linseed oil has been reported to be around 1.4–2.4, for walnut oil, 2–4.5, and poppy oil, 3–8 [9]. However, some studies have questioned the stability of the P/S ratio [3,10,11]. For example, Schilling et al. [11] demonstrated that palmitic acid is around four times more prone to evaporate than stearic acid from a drying oil-based paint, which could lead to a decrease in the P/S ratio.

In a study by Keune et al. [10], the P/S ratio was monitored in oil paint mock-ups (artificially created paint samples) made with linseed oil mixed with different pigments.

They observed that depending on the pigment, the P/S ratio ranged from 0.8 (paints containing vine black or vermilion) to 1.7 (Naples yellow paints). However, this wide range of the observed P/S ratios could have been caused by the experimental conditions applied for the artificial ageing (light and high relative humidity) because, uncommonly, the non-pigmented linseed oil was reliquified during ageing.

Together with the P/S ratio, ratios of other fatty acids or dicarboxylic acids have been used to characterise the binding material or its degree of oxidation. One is the ratio of azelaic acid to palmitic acid (A/P). The higher the value, the more oxidised/dried is the oil because azelaic acid is one of the main dicarboxylic acids produced from the autoxidation of the unsaturated fatty acids present in the fresh oils [10,12]. Similar to the A/P ratio is the sum of the relative content of dicarboxylic acids to other fatty acids ($\sum$D), which again is higher for more oxidised oil [2]. Both values are used to differentiate drying oils from other lipids such as semi-drying (e.g., canola oil) and non-drying oils (e.g., castor oil) or egg (yolk, white, or whole egg) [13–15]. In rough terms, A/P > 1, together with $\sum$D > 40%, suggests that the binder is a drying oil. In contrast, A/P < 0.3 and $\sum$D < 15% indicate that the binder is an egg. In the case of a drying oil and egg mixture, the values should be in between [13]. However, numerous studies show that different factors such as the type of the pigment, the origin and pre-treatment of the oil, and environmental conditions affect the fatty acid composition, which in turn influence the P/S, A/P, and $\sum$D values and make the identification challenging [3,16–19]. Sometimes, even additives with a similar composition to the fatty acids (such as metal stearates) that have been added to paints affect the previously mentioned values [3,20].

However, one aspect that has been studied less is the effect of pigment concentration on the drying of the oil. To our knowledge, previous research has thoroughly addressed the influence of different pigments [21–28] but neglected the effect of the amount of pigment. The impact of pigment concentration on proteinaceous binders [29] has been studied but not on lipids.

In this study, we have addressed the questions of P/S stability and pigment concentration effect by analysing paint mock-ups in which the pigments and their percentages varied over a wide range (the narrowest pigment range was 25–70 g/100 g and the widest 5–90 g/100 g). Seven sets of aged paint samples were analysed, each containing different commercially available natural or synthetic pigment (chrome oxide green, natural cinnabar, yellow or red ochre, Prussian blue, zinc white, or a mixture of hematite and kaolinite) mixed with linseed oil. For the analysis, gas chromatography combined with mass spectrometry (GC–MS) was chosen because of its reputation as a standard method for identifying and characterising fatty acids in oil-based paints. All the mentioned paint mock-ups were derivatised with acid-catalysed methylation, and the sample preparation was modified to enable both absolute and relative quantification. The values of the most common ratios (P/S, A/P, $\sum$D) were examined, and the absolute quantification of palmitic, stearic, and oleic acid was performed based on ref [30].

To the best of our knowledge, this is the first time the absolute quantification method has been applied to characterise fatty acids in aged oil paints. Additionally, attenuated total reflection Fourier transform infrared (ATR–FT–IR) spectroscopy was used to support the GC–MS findings.

2. Materials and Methods

2.1. Materials

Clarified linseed oil was a product of Lefranc & Bourgeois, Paris, France. The pigments zinc white, Prussian blue LUX (45202), natural cinnabar (10620), and chrome oxide green (44200) were obtained from Kremer Pigmente GmbH & Co KG, Aichstetten, BW, Germany. Yellow ochre and red ochre are the product of Kreidezeit Naturfarben GmbH, Lamspringe, NI, Germany (purchased from Safran OÜ, Tartu, Estonia). Kaolinite was obtained from Bang & Bonsomer Group Oy, Helsinki, Finland and hematite from Reakhim.

Methanol (purity ≥ 99.9%), hexane (purity ≥ 97.0%), and toluene (purity ≥ 99.9%) were purchased from Honeywell (Charlotte, NC, USA). Concentrated sulfuric acid (purity 98%) was from VWR Chemicals (Radnor, PA, USA), K_2CO_3 (purity 99.5%) from Reakhim, glass wool from Supelco (Bellefonte, PA, Unites States), and hexadecane (purity ≥ 99%) from Sigma-Aldrich (St. Louis, MO, USA). For the absolute quantification, a standard mixture of fatty acid methyl esters (FAME) was purchased from Sigma-Aldrich. The concentrations of FAMEs used in this study were the following: methyl palmitate 9.9%, methyl stearate 5.95%, methyl oleate + methyl elaidate (Z + E) 34.9%, and their purities were in the range of 99.0–100.0%.

2.2. Preparation and Ageing of the Pigment and Linseed Oil Mixtures

This investigation is part of a larger research project related to the quantitative analysis of different painting materials (S. Vahur's grant PUT1521). All these self-made oil paint mixtures have been made during the framework of this project.

Six sets of pigment (yellow ochre, red ochre, natural cinnabar, zinc white, Prussian blue, or chrome oxide green) + linseed oil mixtures and one set of hematite + kaolinite + linseed oil mixture with different mass ratios were prepared on Petri dishes by weighing components and mixing thoroughly. A diverse set of pigments was chosen. They include both synthetic (Prussian blue, zinc white, chrome oxide green) and natural pigments (natural cinnabar, both ochres, hematite), as well as one that is known to produce metal soaps (zinc white). Among them are slow driers (cinnabar, zinc white), medium driers (chrome oxide green, ochres), and a fast drier (Prussian blue). The red ochre and hematite + kaolinite mixtures can both be called red ochres, but these were analysed to see how much additives affect the drying processes. Each set contained 10 to 16 paint mock-ups in which the approximate concentration of oil varied in the range of 10 to 95 g/100 g. The total initial weight for zinc white + linseed oil and yellow ochre + linseed oil mixtures was 3 g, and for the rest of the mock-ups, 1 g. The exact mass ratios of all the paint mock-ups are presented in the Supplementary Materials.

All the oil paint mock-ups were artificially aged eight to ten months. The Prussian blue, red ochre, chrome oxide green, hematite + kaolinite, and natural cinnabar oil paint mock-ups were artificially aged in a specially made chamber with relative humidity (RH) of 35 ± 10% and temperature of 72 ± 5 °C for six months and 62 ± 5 °C for four months. Zinc white + linseed oil and yellow ochre + linseed oil mixtures were aged eight months at 80 ± 2 °C in a drying oven (Heraeus, Thermo Scientific, Waltham, MA, USA).

After ageing, the paints were pulverised with a ball mill (Mini-mill Pulverisette 23, Fritsch) to obtain a homogeneous mixture. The pulverised paint samples were transferred into 4 mL vials and stored at room temperature. After about nine months, these self-made oil paint samples were analysed with GC–MS, and ATR–FT–IR spectra of the paints with the pigment concentration of 50 g/100 g were recorded.

2.3. Derivatisation of the Pigment and Linseed Oil Mixtures

GC–MS analysis of aged oil paints requires derivatisation, during which the dried polymeric structure consisting of polar compounds is converted into less polar and more volatile molecules. Depending on the limitations and question at hand, various derivatisation methods may be preferred [1,30,31]. In this work, acid-catalysed methylation (a technique that uses low-cost reagents and is suitable for the analysis of aged paint) was chosen because of the large sample amount and the high number of paint samples used in these experiments. Additionally, a derivatisation method with m-(trifluoromethyl)phenyltrimethylammonium hydroxide (TMTFTH) reagent was tested because of its suitability for the quantitative GC analysis of fresh oils [30]. However, complications occurred with this reagent when the absolute quantification of fatty acids in 10 mg of dried paint samples was attempted. Even when 0.5 mL (compared to the commonly used 15–50 µL [12,30,32]) was used for the derivatisation, the absolute quantities of fatty acids were lower than the values obtained with the acid-catalysed methylation.

One by one, each pigment + linseed oil set was derivatised and analysed with GC–MS in one series. The used derivatisation method was based on the procedure presented in ref [30], with slight modifications. To make the absolute quantification reliable, 10 to 12 mg of paint sample was weighed into a 15 mL glass vial. As some paint samples (especially chrome oxide green + linseed oil mixtures) remained visually heterogeneous even after pulverisation, pieces as different as possible were carefully selected for the derivatisation to enable the best overview of the whole paint. A total of 2 mL of methanol was added to the weighted portion, and the vial was sonicated for 15 min. Then, 0.4 mL of concentrated H_2SO_4 was added carefully to the methanolic solution. The vial with the derivatisation mixture was heated for 3 h at 80 °C in an oven (Heraeus). After that, the solution was allowed to cool to room temperature and then extracted with hexane (3 × 2 mL). The obtained hexane solution containing the methylated analytes was pipetted through a glass pipette filled with a layer of K_2CO_3 on top of a glass wool layer into another 15 mL glass vial. The combined hexane extracts were evaporated to dryness under a N_2 evaporator. Then, 2 mL of toluene was added to the residue and weighed on the analytical balance. The solution was stirred vigorously on a VWR vortex mixer to redissolve the analytes. From these stock solutions, dilutions in toluene were made into 1.5 mL Eppendorf® Safe-Lock PCR clean tubes. If the sample contained at least 70 g of oil per 100 g of the mixture, then 0.16 mL of stock solution and 0.4 mL of toluene were weighed; if the oil content was between 40g to 70 g of oil per 100 g, then 0.32 mL of stock solution and 0.24 mL of toluene were weighed; if the oil content was below 40 g/100 g, then 0.64 mL of stock solution was weighed. To all the solutions, 0.1 mL of internal standard solution (0.136 mg·g^{-1} hexadecane in toluene) was added and weighed. The solutions were mixed, and 50 µL of the solution was pipetted into a chromatographic vial with a conical insert (100 µL). To estimate the reproducibility of the applied derivatisation method, one of the most homogeneous samples (yellow ochre + linseed oil, oil concentration 50 g/100 g) was analysed on five analysis days, together with the other paint sets.

2.4. Preparation of the Calibration Solutions

For the absolute quantification, an internal standard method presented in ref [30] was used. To construct the calibration curves, eight calibration solutions from the FAME mixture were made in toluene, and the internal standard solution (0.363 mg·g^{-1} hexadecane in toluene) was added. All the calibration solutions were made by weighing. The concentrations of the compounds necessary for this study were as follows: 0.004 to 0.252 mg·g^{-1} for methyl palmitate, 0.002 to 0.151 mg·g^{-1} for methyl stearate, 0.013 to 0.888 mg·g^{-1} for methyl oleate, and around 0.036 mg·g^{-1} for the internal standard. From the calibration solution, 50 µL was pipetted into a chromatographic vial with a conical insert (100 µL). The calibration solutions were measured in the same GC–MS run with the derivatised pigment + linseed oil mixtures in random order.

2.5. Instrumentation

For preparing the paint samples, an analytical balance (Sartorius, Göttigen, NI, Germany, resolution 0.1 mg) had been used to weigh all the paint components. For the weighing the aged paint samples, another analytical balance (Precisa, Dietikon, Switzerland, resolution 0.01 mg) was used.

An Agilent (Santa Clara, CA, USA) 5975C inert XL MSD with a triple-axis detector, connected to an Agilent 7890A GC system with an Agilent G4513A autosampler, was used for the GC–MS analysis. The column was a 30 m × 0.25 mm in diameter, 0.25 µm film thickness Agilent DB-225MS capillary column (50% cyanopropylphenyl/50% methylpolysiloxane stationary phase). The injection volume was 0.5 µL. The temperature of the ion source and mass spectrometer were 230 °C and 280 °C, respectively. The inlet temperature was 300 °C, the splitless mode was used, and the split was opened after 2 min. The oven temperature program was as follows: initial temperature 80 °C, isothermal for 2 min, increased at 20 °C min^{-1} to 200 °C, isothermal for 4 min, increased 5 °C min^{-1} to 220 °C, isothermal for

5 min, and finally 10 °C min^{-1} to 230 °C, isothermal for 12 min, with a total run time of 34 min. The solvent delay was 5.6 min, electron ionisation (EI) with 70 eV was used, and helium 6.0 with a flow rate of 1.5 mL min^{-1} was used as the carrier gas. The analysis was performed in two modes—scan mode, where a total ion chromatogram (TIC) was recorded, and selected ion monitoring (SIM) mode. The scan mode was used for the qualitative analysis, and the mass range of 27–400 m/z was scanned. For quantitative analysis, SIM mode was used. At the start of the run, the signal corresponding to m/z values of 57 and 71 were measured to detect internal standard (hexadecane), and after 9.5 min, the signal of m/z values 55, 74, and 81 were measured (corresponding to the most intense fragments of methylated palmitic, stearic, and oleic acid). The measured chromatograms were analysed with Agilent MSD ChemStation and the mass spectra with NIST (National Institute of Standards and Technology) Mass Spectral Library Search 2.0.

For the ATR–FT–IR analysis, a Thermo Scientific (Waltham, MA, USA) Nicolet 6700 FT-IR spectrometer with a Smart Orbit diamond micro-ATR accessory (refractive index is 2.4 and the diameter of the active sample area is 1.5 mm) was used. The FT-IR spectrometer was equipped with a DLaTGS detector, Vectra aluminium interferometer, and CsI beamsplitter. The recorded wavenumber range was 4000–225 cm^{-1}, with a resolution of 4 cm^{-1}, and the number of scans 128. Constant purging with dry air was used to protect the spectrometer from atmospheric moisture. Thermo Electron's OMNIC 9 software was used to collect and process the spectra.

3. Results and Discussion

The GC–MS analysis of the seven artificially aged paint sets showed that all the samples contained some original fatty acids of linseed oil—palmitic (P), stearic (S), and oleic (O) acid—and the degradation products of the unsaturated fatty acids—azelaic (A), sebacic (Se), suberic (Su), and pimelic acid (Pi). Some of the paint sets contained linoleic and linolenic acid in small quantities (zinc white + linseed oil mixture) and/or other degradation products besides the above-mentioned dicarboxylic acids—9-oxononanoic acid, 10-oxoctadecanoic acid, and undecanedioic acid. Two representative chromatograms — zinc white + linseed oil (A) and Prussian blue + linseed oil (B) — measured in SIM mode are presented in Figure 1.

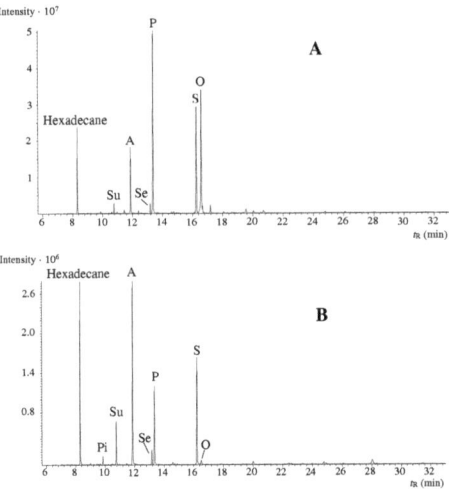

Figure 1. Representative selected ion monitoring (SIM) chromatograms of aged oil paint mock-ups, pigment concentration of 50 g/100 g. (**A**) zinc white and linseed oil and (**B**) Prussian blue and linseed oil. Hexadecane was used as the internal standard, and the abbreviations of other peaks are explained in the text.

3.1. Relative Quantification

The most common values (P/S, A/P, and $\sum$D) used to characterise a drying oil were found through relative quantification. The results are presented in Table 1. These values were also found for the same fresh linseed oil that was used to prepare the paint samples: P/S = 1.5, A/P = 0, and $\sum$D = 0%. The other ratios determined for all paint mock-ups (O/S, A/Su, A/Se) are presented in the Supplementary Materials.

Table 1. Palmitic acid to stearic acid ratio (P/S), azelaic acid to palmitic acid ratio (A/P), and the sum of the relative content of dicarboxylic acids ($\sum$D) in % of the studied pigment and linseed oil mixtures. The name of the pigment represents the studied pigment and linseed oil mixture.

Oil Concentration in g/100 g (ca. [a])	Chrome Oxide Green			Natural Cinnabar			Red Ochre			Prussian Blue			Hematite + Kaolinite			Yellow Ochre			Zinc White		
	P/S	A/P	$\sum$D	P/S	A/P	$\sum$D	P/S	A/P	$\sum$D	P/S	A/P	$\sum$D	P/S	A/P	$\sum$D	P/S	A/P	$\sum$D	P/S	A/P	$\sum$D
10				1.0	3.0	65.4															
15	0.8	2.3	55.6																1.5	0.4	15.5
20				0.9	2.7	54.5															
25	0.8	2.7	58.4				0.8	6.9	79.8							1.5	1.7	57.5	1.5	0.3	12.4
30	0.7	2.6	57.4	0.8	2.6	52.8	1.0	4.4	72.3	0.7	2.8	58.7	0.9	6.7	74.6	1.4	1.1	46.0	1.5	0.3	12.2
35	0.7	2.8	58.9	0.9	2.3	51.6	0.8	4.8	70.0	0.7	1.7	46.6	1.1	5.6	72.6	1.6	1.2	48.4	1.5	0.3	13.4
39							0.8	5.0	74.4												
40	0.8	2.3	55.6	0.8	2.2	50.4				0.7	2.0	49.3	1.0	3.1	59.0	1.4	1.2	47.0	1.5	0.3	13.1
42							0.9	5.7	75.1												
45	0.8	2.7	59.2	0.8	2.2	50.0	0.8	5.5	73.4	0.7	1.7	46.1	1.0	5.8	73.1	1.4	1.0	41.9	1.5	0.3	13.2
50	0.7	2.7	58.8	0.8	1.5	39.0	0.7	6.8	75.4	0.7	2.1	50.1	0.9	4.2	65.4	1.6	1.0	44.2	1.6	0.3	13.6
55	0.7	2.7	57.2	0.8	1.6	41.9	0.9	6.5	77.1	0.7	2.0	47.0	0.9	3.9	67.1	1.6	0.9	41.1	1.5	0.3	12.6
60	0.8	2.4	56.5	0.9	1.3	36.7	0.8	4.6	69.4	0.7	1.7	44.8	0.9	6.1	73.2	1.6	1.0	42.3	1.5	0.3	12.4
62										0.7	2.6	52.8									
65	0.9	1.7	50.6	0.9	1.2	33.7	0.9	3.1	62.5	0.7	2.7	56.0	1.0	2.0	48.3	1.6	1.0	42.5	1.5	0.3	13.2
70	0.8	2.0	52.5	0.8	1.7	42.4	0.8	3.7	66.4	0.6	2.7	53.8	1.0	4.5	68.8	1.6	0.8	39.7	1.5	0.3	12.3
75	0.9	1.5	46.2	1.0	0.9	28.6	0.9	2.9	62.5	0.7	3.1	58.2	1.0	4.9	70.4	1.6	0.8	37.6	1.5	0.3	12.5
80	1.0	1.2	39.3				0.9	2.5	58.7	0.6	2.9	57.3				1.6	0.7	35.2	1.5	0.3	13.3
85				1.0	0.9	28.4				0.7	2.4	54.1				1.6	0.7	34.4	1.6	0.3	14.4
90	1.0	1.1	36.2				0.9	1.9	49.4	0.7	2.7	57.1				1.6	0.6	30.0	1.5	0.3	14.1
95	1.1	0.8	31.0	1.2	0.5	20.0	0.9	2.0	51.9	0.8	2.4	54.2				1.6	0.6	27.8	1.5	0.3	14.5

[a]—These values are rounded. The exact values are presented in the Supplementary Materials.

In Table 1, the value of the P/S ratio ranges from 0.6 (Prussian blue + linseed oil) to 1.6 (yellow ochre + linseed oil and zinc white + linseed oil). The P/S ratio had significantly decreased (≤ 1.2) when Prussian blue, chrome oxide green, natural cinnabar, hematite + kaolinite, or red ochre were mixed with the linseed oil. This suggests that with these pigments, palmitic acid is more prone than stearic acid to evaporate during the drying process, which leads to a decrease in the P/S ratio. These results confirm the observations presented in several publications that the P/S ratio may not be the most reliable differentiator [2,3]. Here, we can also conclude that a low P/S ratio does not always imply that additives (e.g., metal stearates) have been added to the paint, as has been sometimes suggested [33]. Further investigation must be conducted to answer whether these pigments could have the same effect on walnut oil (P/S value of fresh oil is 2–4.5) by lowering the P/S value and complicating the differentiation from linseed oil. It is interesting to note that only a few studies see the decrease of the P/S value during paint drying [10,34], and others do not [21,22,26].

In yellow ochre and zinc white containing oil paint samples, the P/S ratio (average value of 1.6 and 1.5, respectively) is almost the same as the P/S ratio for fresh linseed oil (1.5). The ATR–FT–IR analysis also confirmed that the oil composition of these pigment and linseed oil mixtures differs from the other aged paints. For yellow ochre + linseed oil and zinc white + linseed oil sets, the ATR–FT–IR spectra (Figure 2) differ from the IR spectra of linseed oil mixtures with red ochre (Figure 2), natural cinnabar, Prussian blue, hematite + kaolinite, and chrome oxide green in the wavenumber range of 1530–1750 cm^{-1} (see Figures S1–S4 in the Supplementary Materials). The IR spectra of all the paint samples have an absorption band around 1730–1740 cm^{-1} that belongs to the C=O stretching vibration of the ester group in the triglyceride molecule. However, in the IR spectra of oil paint samples with yellow ochre and zinc white, a C=O stretching band near 1705 cm^{-1} is absent that is present in the spectra of all the other pigment and linseed oil pairs. This absorption has been assigned to carboxylic acids formed during the oxidation of the paint [25]. In the case of zinc white + linseed oil mixture, also absorptions corresponding to zinc carboxylates were observed—1587 cm^{-1} (amorphous structure) and 1539 cm^{-1} (crystalline structure) [35]. The formation of metal carboxylates (or metal soaps) from the metal cation and free carboxylic acids from the hydrolysis of triglycerides is a known phenomenon observed with some pigments, including zinc white [2,24,36]. Importantly, carboxylic acids in the anionic form are not prone to evaporation. Additionally, the absence of 1705 cm^{-1} absorbance in the spectrum of yellow ochre and linseed oil paint shows that also with this pigment, less free carboxylic acids are present. These results imply that the formation of free carboxylic acids (as opposed to carboxylate salts) may lead to the decrease of the P/S value. However, this correlation should be investigated further.

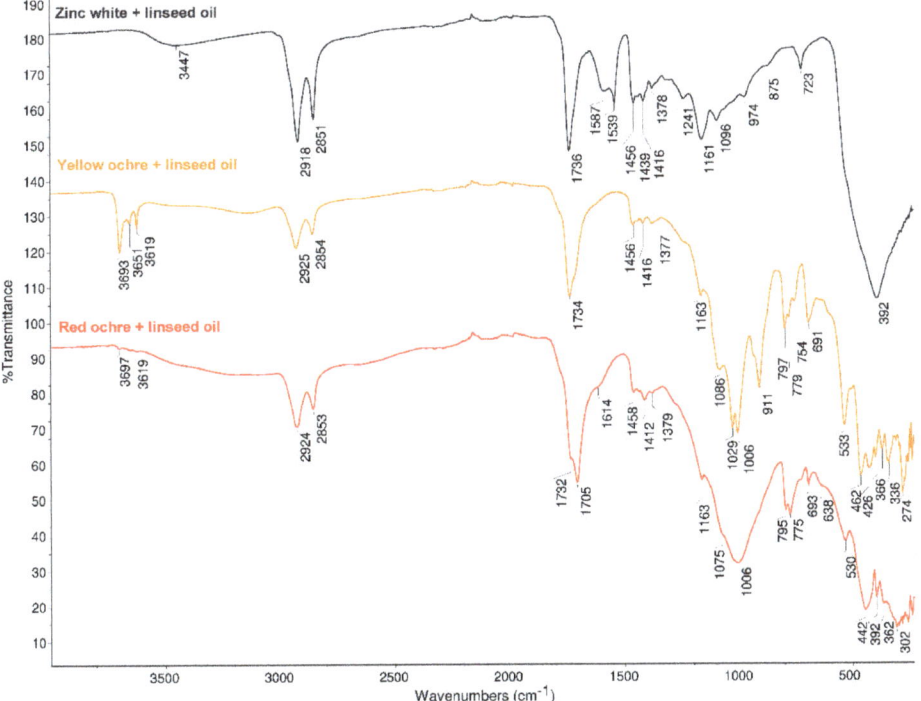

Figure 2. Attenuated total reflection Fourier transform infrared (ATR–FT–IR) spectra of aged paint mock-ups, pigment concentration 50 g/100 g. The spectra of the mixtures containing other pigments (pigment concentration 50 g/100 g) are presented in the Supplementary Materials.

Observing the A/P and $\sum$D values in Table 1, the highest values are found for red ochre and hematite + kaolinite containing oil paint mock-ups, implying that with these mixtures, the paints are the most oxidised/dried. In most cases, a trend can be observed in the range of one pigment and oil set. The only exception is the hematite + kaolinite + linseed oil paints, in which these values change more hectically over the scope of the set. This could imply that the powdered hematite + kaolinite paint mock-ups were less homogeneous than the other samples.

When observing the results of chrome oxide green, natural cinnabar, yellow ochre, and red ochre containing oil paints, the A/P and $\sum$D values increase when the linseed oil content decreases. This implies that besides the type of pigment, pigment concentration also influences the drying of the oil. With these pigments, the higher the pigment concentration, the more 'dried' is the oil—even though considering one set of pigment and oil mixtures, the samples have been dried over the same period. However, with Prussian blue + linseed oil and zinc white + linseed oil mixtures, these correlations cannot be made, which implies that the pigment concentration does not influence the degree of oil oxidation with these pigments. Therefore, interestingly, the pigment concentration effect is not in correlation with the siccative nature of the pigment. Even though Prussian blue is a known fast drier, the higher concentration of Prussian blue pigment in the paint mixture did not accelerate the drying processes. Likewise, cinnabar is known more as a slow drier; however, the higher concentration of cinnabar pigment accelerated the drying of the paint mock-ups. The fact that with some pigments the A/P and $\sum$D values increase together with the increase of the pigment concentration, but with other pigments, this correlation is absent leads to another conclusion: the increase in the A/P and $\sum$D values does indeed come from the higher pigment concentration, not from the fact that the samples contained less oil (one might think that at lower concentration the oil dries better).

As another interesting fact, the A/P and $\sum$D values are especially low for zinc white + linseed oil mixtures (average values of 0.3 and 13, respectively)—this again may confirm the presence of zinc carboxylates, which hinders the loss of palmitic, stearic, and oleic acid. Therefore, these results show that the A/P ratio does not straightforwardly indicate if the oil has been pre-polymerised (A/P < 0.5 for the zinc white set) or not (A/P > 1 for all the other pigment sets), even though these criteria have been suggested previously [28]. Additionally, here it can be seen that although the A/P ratio is below 0.3 and $\sum$D is below 15%, the binding material is not an egg. Therefore, in similar cases, those values would lead to incorrect identification of the binding material. These results show that both A/P and $\sum$D values are affected by the pigment type and, interestingly, in most cases, also by the pigment percentage in the paint mock-up. Therefore, these ratios should only be used cautiously to characterise the state of the dried oil binder.

3.2. Absolute Quantification of Fatty Acids

3.2.1. Intermediate Precision Estimation

For estimating the intermediate precision (within-lab reproducibility) of the analysis, one of the yellow ochre and linseed oil mixtures (concentration of oil 50 g/100 g) was analysed on five days, spread over two months (see Table 2). This sample was chosen because the yellow ochre + linseed oil mixtures were one of the most homogeneous, and therefore, the contribution to scatter resulting from sampling should be the lowest. The fatty acids present in the FAME calibration mixture and also in the dried paint samples were methylated palmitic, stearic, and oleic acid, which were quantified with the absolute quantification method.

Table 2. Results of absolute quantification of fatty acids in dried yellow ochre and linseed oil mixture (concentration of oil 50 g/100 g) [a].

g/100 g	Palmitic Acid g/100 g	Stearic Acid g/100 g	Oleic Acid g/100 g	P/S
50	1.33 (± 0.05)	0.86 (± 0.01)	0.14 (± 0.03)	0.65 (± 0.02)

[a]—The standard deviations are presented in the brackets.

The results show that acid-catalysed methylation is a suitable derivatisation method with reproducible results for the samples studied in this work. The other aspects of the GC–MS method were the same as in the validated procedure presented by Tammekivi et al. [30]. Therefore, the contents of fatty acids in different pigment and oil mixtures and the parameters derived from them are reliable and can be compared to one another.

3.2.2. Absolute Quantification of the Pigment and Oil Mixtures

Table 3 represents the absolute quantification of palmitic, stearic, and oleic acid in the paint mock-ups. In fresh linseed oil, these values were P = 4.2 g, S = 2.9 g, and O = 15.5 g/100 g. As can be inferred from Table 3, one of the lowest values of these fatty acids were measured for the similar red ochre + linseed oil and hematite + kaolinite + linseed oil mixtures, which agree with the statement suggested above that these paint mock-ups are the most oxidised.

Table 3. Absolute quantification of palmitic (P), stearic (S), and oleic (O) acid in g per 100 g of dried oil paint mock-ups. The name of the pigment represents the studied pigment and linseed oil mixture.

Oil Concentration in g/100 g (ca. [a])	Chrome Oxide Green			Natural Cinnabar			Red Ochre			Prussian Blue			Hematite + Kaolinite			Yellow Ochre			Zinc White		
	P	S	O	P	S	O	P	S	O	P	S	O	P	S	O	P	S	O	P	S	O
10				0.1	0.1	0.03															
15	0.2	0.2	0.1																0.6	0.4	0.9
20				0.2	0.2	0.1															
25	0.3	0.4	0.1				0.1	0.1	0.0							0.4	0.3	0.0	1.1	0.8	2.1
30	0.5	0.5	0.1	0.3	0.4	0.2	0.1	0.1	0.0	0.3	0.4	0.0	0.1	0.1	0.1	0.7	0.5	0.02	1.4	0.9	2.5
35	0.5	0.5	0.2	0.5	0.4	0.2	0.2	0.2	0.0	0.4	0.6	0.1	0.1	0.1	0.1	0.9	0.6	0.03	1.6	1.1	3.0
39							0.2	0.2	0.0												
40	0.7	0.8	0.2	0.5	0.6	0.3				0.4	0.6	0.1	0.2	0.2	0.1	1.0	0.7	0.04	1.8	1.2	3.3
42							0.2	0.2	0.1												
45	0.6	0.8	0.2	0.5	0.6	0.3	0.2	0.3	0.1	0.5	0.7	0.1	0.2	0.3	0.1	1.2	0.9	0.1	2.0	1.4	3.8
50	0.6	0.9	0.2	0.9	1.0	0.7	0.2	0.3	0.1	0.5	0.7	0.1	0.2	0.2	0.1	1.3	0.9	0.1	2.3	1.6	4.3
55	0.9	1.1	0.3	0.9	1.1	0.7	0.2	0.2	0.1	0.5	0.8	0.2	0.3	0.3	0.2	1.5	1.0	0.1	2.3	1.6	4.4
60	0.9	1.1	0.3	1.2	1.4	1.4	0.4	0.6	0.2	0.7	1.0	0.2	0.3	0.3	0.2	1.8	1.2	0.2	2.6	1.8	5.1
62										0.4	0.6	0.1									
65	1.3	1.5	0.4	1.6	1.9	1.8	0.7	0.9	0.2	0.6	1.0	0.2	0.4	0.4	0.3	2.0	1.3	0.2	2.8	2.0	5.4
70	1.2	1.5	0.6	1.1	1.3	1.0	0.7	0.8	0.3	0.6	1.0	0.2	0.6	0.7	0.6	2.3	1.5	0.4	2.9	2.0	5.7
75	1.8	2.0	0.8	2.1	2.1	2.3	0.9	1.0	0.3	0.8	1.2	0.3	1.0	1.1	1.0	2.6	1.7	0.5	3.1	2.2	6.1
80	2.3	2.4	1.5				1.1	1.3	0.4	0.9	1.4	0.3				2.8	1.9	0.7	3.4	2.4	6.3
85				2.5	2.6	3.1				1.2	1.7	0.3				3.0	2.0	0.8	3.4	2.4	5.9
90	2.5	2.6	2.3				1.5	1.8	0.9	1.2	1.8	0.3				3.5	2.3	1.5	3.7	2.5	6.3
95	3.6	3.1	3.1	4.0	3.4	5.8	1.8	2.1	0.8	1.5	2.0	0.4				3.8	2.5	2.2	4.1	2.8	6.6

[a]—These values are rounded. The exact values are presented in the Supplementary Materials.

Again, most peculiar are the results of the zinc white-containing oil paint mixtures. In addition to the higher palmitic and stearic acid contents, compared to other pigments throughout the set, the oleic acid content is especially high. This probably contributed to the low A/P and $\sum$D values because less azelaic acid and other dicarboxylic acids were produced. The contents of palmitic and stearic acid are also the most similar to fresh linseed oil. Interestingly, although natural cinnabar + linseed oil samples (that had 50 g or more linseed oil per 100 g of sample) also had higher oleic acid content than the other paints, the A/P and $\sum$D values were not remarkably low.

In Figure 3, the correlations between stearic acid absolute concentration and the pigment concentration in the paint mock-ups are shown. Linseed oil mixtures with chrome oxide green, natural cinnabar, Prussian blue, red ochre, and hematite + kaolinite are presented in Figure 3A; zinc white and yellow ochre oil paint sets are shown in Figure 3B. The same graphs for palmitic acid are presented in the Supplementary Materials. The deviations from the correlation lines in the case of some pigment + linseed oil mixtures (e.g., chrome oxide green and natural cinnabar sets) are likely caused by the inhomogeneities of the samples.

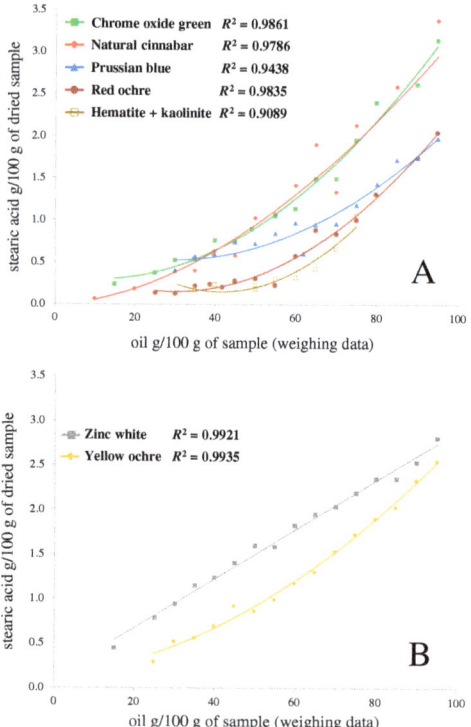

Figure 3. Correlations between stearic acid absolute quantity (g/100 g) vs. oil content (g/100 g) in the weighted sample. (**A**) chrome oxide green, natural cinnabar, Prussian blue, red ochre, and hematite + kaolinite mixtures with linseed oil. (**B**) zinc white and yellow ochre mixtures with linseed oil. The name of the pigment represents the studied pigment and linseed oil mixture.

In theory, if the concentration of a pigment does not influence the drying extent of linseed oil, then the correlation should be linear. The correlation in the case of chrome oxide green, natural cinnabar, red ochre, hematite + kaolinite, and Prussian blue are far from linear. This demonstrates that the higher the pigment concentration in these paint samples, the more stearic acid has evaporated from the paint. For zinc white and yellow ochre paints, the curves are the closest to a linear function. In the yellow ochre + linseed oil case, the reason for a curve closer to a linear function could be the fact that the samples were aged two months less, without artificially increased relative humidity, and the amount of the prepared paints was three times higher than for the paint samples presented in Figure 3A. For example, when looking at the absolute values in Table 3 and comparing the stearic acid quantities for 80 g of linseed oil per 100 g of sample to 40 g, it is evident that the decrease has been around 2.7 times, not 2. Therefore, it can be concluded that, also in the case of yellow ochre, the concentration of the pigment influences the drying processes.

Interestingly again, the zinc white + linseed oil paint set acts differently. When comparing the stearic acid quantities of 80 g of linseed oil per 100 g of sample to 40 g, the decrease is two times, which demonstrates that in the case of zinc white paints, the effect of the pigment concentration is the lowest (if present at all). Therefore, in zinc white paint mock-ups, it would be the simplest to suggest the pigment percentage in an unknown sample based on the stearic acid absolute content (if the same value for fresh oil is known).

4. Conclusions

In this study, seven artificially aged oil paint mock-ups in varying concentrations were analysed with GC–MS. In addition to the traditional relative quantification of fatty acids, the method developed in our workgroup was successfully used for the absolute quantification of fatty acids in aged paint samples.

The results show that the commonly used ratios to characterise and identify a drying oil (P/S, A/P, and $\sum$D) vary greatly because of both the different nature of the used pigments and different pigment concentrations. The P/S ratios of Prussian blue, chrome oxide green, natural cinnabar, hematite + kaolinite, and red ochre oil paint sets were generally below 1.0, which is uncommonly low for the used fresh linseed oil with a P/S value of 1.5. On the contrary, the P/S values for yellow ochre and zinc white-containing oil paints remained the same as in fresh oil (1.6 and 1.5, respectively). The high variance in the P/S ratio obtained in this study demonstrates that if this value is to be used to identify/confirm the type of drying oil, it should be performed highly judiciously. The pigment concentration effect could be detected by observing the A/P and $\sum$D ratios. In the case of chrome oxide green, natural cinnabar, yellow ochre, and red ochre paints, these values increased together with the increase of the pigment concentration. However, for Prussian blue and zinc white, the pigment concentration did not influence the A/P and $\sum$D values.

The examination of the absolute stearic acid content over the range of paint samples with varying pigment concentration showed that the correlation was far from linear for almost all the pigments (except for zinc white samples). This suggests that higher pigment concentration accelerates the evaporation of stearic acid from the paint samples. Therefore, only the absolute quantification shows that even when the ratios (P/S, A/P, $\sum$D) are stable or when no trend can be seen, the concentration of the pigment has an influence on the drying processes of oil-paint with almost all the pigments (except zinc white) studied in this work.

Supplementary Materials: The following are available online, **Table S1**: The exact pigment and linseed oil masses in the weighted mixtures [a]. **Table S2**: Oleic acid to stearic acid ratio (O/S), azelaic acid to suberic acid ratio (A/Su), and azelaic acid to sebacic acid ratio (A/Se) calculated from the GC–MS analyses. **Figure S1**: ATR–FT–IR spectrum of natural cinnabar and linseed oil aged mixture (50 g/100 g). **Figure S2**: ATR–FT–IR spectrum of Prussian blue and linseed oil aged mixture (50 g/100 g). **Figure S3**: ATR–FT–IR spectrum of hematite + kaolinite and linseed oil aged mixture (25 g of hematite + 25 g of kaolinite per 100 g of paint). **Figure S4**: ATR–FT–IR spectrum of chrome oxide green and linseed oil aged mixture (50 g/100 g). **Figure S5**: Correlations between palmitic acid absolute quantity (g/100 g) vs. oil content (g/100 g) in the weighted sample. (**A**) chrome oxide green, natural cinnabar, Prussian blue, red ochre, and hematite + kaolinite mixtures with linseed oil. (**B**) zinc white and yellow ochre mixtures with linseed oil. The name of the pigment represents the studied pigment and linseed oil mixture.

Author Contributions: Conceptualisation, E.T., S.V., and I.L.; methodology, E.T. and M.V.; validation, E.T.; investigation, E.T.; formal analysis, E.T.; resources, S.V. and M.V.; writing—original draft preparation, E.T.; writing—review and editing, S.V. and I.L.; visualisation, E.T.; funding acquisition, S.V. and I.L. All authors have read and agreed to the published version of the manuscript.

Funding: This research was funded by the ESTONIAN RESEARCH COUNCIL, grant numbers PUT1521 and PRG1198, as well as by the EU through the EUROPEAN REGIONAL DEVELOPMENT FUND, under project TK141 (2014–2020.4.01.15–0011).

Institutional Review Board Statement: Not applicable.

Informed Consent Statement: Not applicable.

Data Availability Statement: Not applicable.

Acknowledgments: This work was carried out using the instrumentation of the Estonian Centre of Analytical Chemistry (www.akki.ee, accessed on 15 September 2020). We would like to thank the Organic Chemistry department of the University of Tartu and Siim Salmar for enabling the usage of their GC–MS instrument.

Conflicts of Interest: The authors declare no conflict of interest.

Sample Availability: Small amounts of the paint mock-ups are available from Signe Vahur.

References

1. Sutherland, K. Gas chromatography/mass spectrometry techniques for the characterisation of organic materials in works of art. *Phys. Sci. Rev.* **2018**, *4*. [CrossRef]
2. Bonaduce, I.; Carlyle, L.; Colombini, M.P.; Duce, C.; Ferrari, C.; Ribechini, E.; Selleri, P.; Tinè, M.R. New Insights into the Ageing of Linseed Oil Paint Binder: A Qualitative and Quantitative Analytical Study. *PLOS ONE* **2012**, *7*, e49333. [CrossRef] [PubMed]
3. Colombini, M.P.; Modugno, F. *Organic Mass Spectrometry in Art and Archaeology*, 1st ed.; John Wiley & Sons, Ltd: Chichester, UK, 2009.
4. Castellá, F.; Pérez-Estebanez, M.; Mazurek, J.; Monkes, P.; Learner, T.; Niello, J.F.; Tascón, M.; Marte, F. A multi-analytical approach for the characterization of modern white paints used for Argentine concrete art paintings during 1940–1960. *Talanta* **2020**, *208*, 120472. [CrossRef]
5. Andersen, C.K.; Bonaduce, I.; Andreotti, A.; Van Lanschot, J.; Vila, A. Characterisation of preparation layers in nine Danish Golden Age canvas paintings by SEM–EDX, FTIR and GC–MS. *Heritage Sci.* **2017**, *5*, 34. [CrossRef]
6. Kasprzok, L.M.; Fabbri, D.; Rombolà, A.G.; Rovetta, T.; Malagodi, M. Identification of organic materials in historical stringed instruments by off-line analytical pyrolysis solid-phase microextraction with on-fiber silylation and gas chromatography-mass spectrometry. *J. Anal. Appl. Pyrolysis* **2020**, *145*, 104727. [CrossRef]
7. Winter, J.; Mills, J.S.; White, R. The Organic Chemistry of Museum Objects. *Stud. Conserv.* **1988**, *33*, 102. [CrossRef]
8. Mills, J.S. The Gas Chromatographic Examination of Paint Media. Part I. Fatty Acid Composition and Identification of Dried Oil Films. *Stud. Conserv.* **1966**, *11*, 92–107. [CrossRef]
9. Colombini, M.P.; Andreotti, A.; Bonaduce, I.; Modugno, F.; Ribechini, E. Analytical Strategies for Characterizing Organic Paint Media Using Gas Chromatography/Mass Spectrometry. *Accounts Chem. Res.* **2010**, *43*, 715–727. [CrossRef]
10. Keune, K.; Hoogland, F.; Boon, J.J.; Peggie, D.; Higgit, C. Comparative study of the effect of traditional pigments on artificially aged oil paint systems using complementary analytical techniques. In *Preprints of 15th Triennal Meeting of ICOM Committee for Conservation*; Allied Publishers Pvt. Ltd: New Delhi, India, 2008; Volume II, pp. 833–842.
11. Schilling, M.R.; Carson, D.M.; Khanjian, H.P. Evaporation of Fatty Acids and the Formation of Ghost Images by Framed Oil Paintings, WAAC Newsl. 1998. Available online: https://cool.culturalheritage.org/waac/wn/wn21/wn21-1/wn21-106.html (accessed on 9 February 2021).
12. Pitthard, V.; Staněk, S.; Griesser, M.; Muxeneder, T. Gas Chromatography—Mass Spectrometry of Binding Media from Early 20th Century Paint Samples from Arnold Schönberg's Palette. *Chromatographia* **2005**, *62*, 175–182. [CrossRef]
13. Kalinina, K.B.; Bonaduce, I.; Colombini, M.P.; Artemieva, I.S. An analytical investigation of the painting technique of Italian Renaissance master Lorenzo Lotto. *J. Cult. Heritage* **2012**, *13*, 259–274. [CrossRef]
14. Izzo, F.C. *20th Century Artists' Oil Paints: A Chemical-Physical Survey*; Ca' Foscari University of Venice: Venice, Italy, 2010.
15. Colombini, M.P.; Modugno, F.; Giacomelli, M.; Francesconi, S. Characterisation of proteinaceous binders and drying oils in wall painting samples by gas chromatography–mass spectrometry. *J. Chromatogr. A* **1999**, *846*, 113–124. [CrossRef]
16. Berg, J.V.D.; Berg, K.V.D.; Boon, J. Determination of the degree of hydrolysis of oil paint samples using a two-step derivatisation method and on-column GC/MS. *Prog. Org. Coatings* **2001**, *41*, 143–155. [CrossRef]
17. Berg, J.D.V.D.; Vermist, N.D.; Carlyle, L.; Holčapek, M.; Boon, J.J. Effects of traditional processing methods of linseed oil on the composition of its triacylglycerols. *J. Sep. Sci.* **2004**, *27*, 181–199. [CrossRef] [PubMed]

18. Bonaduce, I.; Ribechini, E.; Modugno, F.; Colombini, M.P. Analytical Approaches Based on Gas Chromatography Mass Spectrometry (GC/MS) to Study Organic Materials in Artworks and Archaeological Objects. *Top. Curr. Chem.* **2016**, *374*, 1–37. [CrossRef]
19. Colombini, M.P.; Modugno, F.; Fuoco, R.; Tognazzi, A. A GC-MS study on the deterioration of lipidic paint binders. *Microchem. J.* **2002**, *73*, 175–185. [CrossRef]
20. La Nasa, J.; Modugno, F.; Aloisi, M.; Lluveras-Tenorio, A.; Bonaduce, I. Development of a GC/MS method for the qualitative and quantitative analysis of mixtures of free fatty acids and metal soaps in paint samples. *Anal. Chim. Acta* **2018**, *1001*, 51–58. [CrossRef] [PubMed]
21. Llorent-Martínez, E.; Domínguez-Vidal, A.; Rubio-Domene, R.; Pascual-Reguera, M.; Ruiz-Medina, A.; Ayora-Cañada, M. Identification of lipidic binding media in plasterwork decorations from the Alhambra using GC–MS and chemometrics: Influence of pigments and aging. *Microchem. J.* **2014**, *115*, 11–18. [CrossRef]
22. Šefců, R.; Pitthard, V.; Chlumská, Š.; Turková, I. A multianalytical study of oil binding media and pigments on Bohemian Panel Paintings from the first half of the 14th century. *J. Cult. Heritage* **2017**, *23*, 77–86. [CrossRef]
23. Chiavari, G.; Fabbri, D.; Prati, S. Effect of pigments on the analysis of fatty acids in siccative oils by pyrolysis methylation and silylation. *J. Anal. Appl. Pyrolysis* **2005**, *74*, 39–44. [CrossRef]
24. Fuster-López, L.; Izzo, F.C.; Piovesan, M.; Yusá-Marco, D.J.; Sperni, L.; Zendri, E. Study of the chemical composition and the mechanical behaviour of 20th century commercial artists' oil paints containing manganese-based pigments. *Microchem. J.* **2016**, *124*, 962–973. [CrossRef]
25. Van Der Weerd, J.; Van Loon, A.; Boon, J.J. FTIR Studies of the Effects of Pigments on the Aging of Oil. *Stud. Conserv.* **2005**, *50*, 3–22. [CrossRef]
26. Gimeno-Adelantado, J.; Mateo-Castro, R.; Doménech-Carbó, M.; Bosch-Reig, F.; Doménech-Carbó, A.; Casas-Catalán, M.; Osete-Cortina, L. Identification of lipid binders in paintings by gas chromatography. *J. Chromatogr. A* **2001**, *922*, 385–390. [CrossRef]
27. Ioakimoglou, E.; Boyatzis, S.; Argitis, P.; Fostiridou, A.; Papapanagiotou, K.; Yannovits, N. Thin-Film Study on the Oxidation of Linseed Oil in the Presence of Selected Copper Pigments. *Chem. Mater.* **1999**, *11*, 2013–2022. [CrossRef]
28. Pitthard, V.; Griesser, M.; Stanek, S. Methodology and application of gc-ms to study altered organic binding media from objects of the Kunsthistorisches Museum, Vienna. *Ann. Chim.* **2006**, *96*, 561–573. [CrossRef]
29. Gautier, G.; Colombini, M.P. GC-MS identification of proteins in wall painting samples: A fast clean-up procedure to remove copper-based pigment interferences. *Talanta* **2007**, *73*, 95–102. [CrossRef] [PubMed]
30. Tammekivi, E.; Vahur, S.; Kekišev, O.; Van Der Werf, I.D.; Toom, L.; Herodes, K.; Leito, I. Comparison of derivatization methods for the quantitative gas chromatographic analysis of oils. *Anal. Methods* **2019**, *11*, 3514–3522. [CrossRef]
31. Doménech-Carbó, M.T. Novel analytical methods for characterising binding media and protective coatings in artworks. *Anal. Chim. Acta* **2008**, *621*, 109–139. [CrossRef] [PubMed]
32. Manzano, E.; Rodríguez-Simón, L.; Navas, N.; Checa-Moreno, R.; Romero-Gámez, M.; Capitán-Vallvey, L. Study of the GC–MS determination of the palmitic–stearic acid ratio for the characterisation of drying oil in painting: La Encarnación by Alonso Cano as a case study. *Talanta* **2011**, *84*, 1148–1154. [CrossRef]
33. Banti, D.; La Nasa, J.; Tenorio, A.L.; Modugno, F.; Berg, K.J.V.D.; Lee, J.; Ormsby, B.; Burnstock, A.; Bonaduce, I. A molecular study of modern oil paintings: Investigating the role of dicarboxylic acids in the water sensitivity of modern oil paints. *RSC Adv.* **2018**, *8*, 6001–6012. [CrossRef]
34. Modugno, F.; Di Gianvincenzo, F.; Degano, I.; Van Der Werf, I.D.; Bonaduce, I.; Berg, K.J.V.D. On the influence of relative humidity on the oxidation and hydrolysis of fresh and aged oil paints. *Sci. Rep.* **2019**, *9*, 1–16. [CrossRef]
35. Hermans, J.J.; Keune, K.; Van Loon, A.; Iedema, P.D. An infrared spectroscopic study of the nature of zinc carboxylates in oil paintings. *J. Anal. At. Spectrom.* **2015**, *30*, 1600–1608. [CrossRef]
36. Osmond, G. Zinc white: A review of zinc oxide pigment properties and implications for stability in oil-based paintings. *AICCM Bull.* **2012**, *33*, 20–29. [CrossRef]

Article

Two-Step Derivatization of Amino Acids for Stable-Isotope Dilution GC–MS Analysis: Long-Term Stability of Methyl Ester-Pentafluoropropionic Derivatives in Toluene Extracts

Svetlana Baskal, Alexander Bollenbach and Dimitrios Tsikas *

Institute of Toxicology, Core Unit Proteomics, Hannover Medical School, 30625 Hannover, Germany; baskal.svetlana@mh-hannover.de (S.B.); bollenbach.alex@gmail.com (A.B.)
* Correspondence: Tsikas.dimitros@mh-hannover.de

Abstract: Analysis of amino acids by gas chromatography-mass spectrometry (GC–MS) requires at least one derivatization step to enable solubility in GC–MS-compatible water-immiscible organic solvents such as toluene, to make them volatile to introduce into the gas chromatograph and thermally stable enough for separation in the GC column and introduction into the ion-source, and finally to increase their ionization by increasing their electronegativity using F-rich reagents. In this work we investigated the long-term stability of the methyl esters pentafluoropropionic (Me-PFP) derivatives of 21 urinary amino acids prepared by a two-step derivatization procedure and extraction by toluene. In situ prepared trideuteromethyl ester pentafluoropropionic derivatives were used as internal standards. GC–MS analysis (injection of 1 µL aliquots and quantification by selected-ion monitoring of specific mass fragments) was performed on days 1, 2, 8, and 15. Measured peak areas and calculated peak area ratios were used to evaluate the stability of the derivatives of endogenous amino acids and their internal standards, as well as the precision and the accuracy of the method. All analyses were performed under routine conditions. Me-PFP derivatives of endogenous amino acids and their stable-isotope labelled analogs were stable in toluene for 14 days. The peak area values of the derivatives of most amino acids and their internal standards were slightly higher on days 8 and 15 compared to days 1 and 2, yet the peak area ratio values of endogenous amino acids to their internal standards did not change. Our study indicates that Me-PFP derivatives of amino acids from human urine samples can easily be prepared, are stable at least for 14 days in the extraction solvent toluene, and allow for precise and accurate quantitative measurements by GC–MS using in situ prepared deuterium-labelled methyl ester as internal standard.

Keywords: amino acids; derivatization; esterification; GC–MS; pentafluoropropionic anhydride; stability; toluene

Citation: Baskal, S.; Bollenbach, A.; Tsikas, D. Two-Step Derivatization of Amino Acids for Stable-Isotope Dilution GC–MS Analysis: Long-Term Stability of Methyl Ester-Pentafluoropropionic Derivatives in Toluene Extracts. *Molecules* **2021**, *26*, 1726. https://doi.org/10.3390/molecules26061726

Academic Editors: Paraskevas D. Tzanavaras and Zhentian Lei

Received: 1 March 2021
Accepted: 16 March 2021
Published: 19 March 2021

Publisher's Note: MDPI stays neutral with regard to jurisdictional claims in published maps and institutional affiliations.

Copyright: © 2021 by the authors. Licensee MDPI, Basel, Switzerland. This article is an open access article distributed under the terms and conditions of the Creative Commons Attribution (CC BY) license (https://creativecommons.org/licenses/by/4.0/).

1. Introduction

Amino acids are carboxylic acids and contain at least one primary or secondary amine group and additional functionalities such as a hydroxyl (OH) group (Scheme 1). Amino acids are soluble in water and in water-miscible organic solvents such as methanol, but they are not soluble in water-immiscible organic solvents such as toluene. Amino acids and many of their metabolites are generally not accessible to gas chromatography (GC)-based analysis because they are not volatile and are thermally labile. Their injection in the gas chromatograph would lead to decomposition in the usually hot injector port to release small gases most likely including CO_2, NH_3 and H_2O. Thus, native amino acids are not compatible with GC-based techniques such as gas chromatography-mass spectrometry (GC–MS). This trouble can be solved by chemical reactions of the carboxylic (COOH) groups, amine (NH_2 and NH) groups, and OH groups with chemically reactive reagents to generate derivatives that are soluble in GC-compatible, water-immiscible organic, electroneutral, volatile and thermally stabile solvents [1–4]. Biological samples contain amino acids and their metabolites. GC–MS analysis of all amino acids in biological samples such as blood and urine usually requires a two-step derivatization procedure to protect the above mentioned functionalities. Amino acids can be converted to their corresponding methyl esters by heating the samples for instance with 2 M HCl in methanol (CH$_3$OH) or CD$_3$OD [3] (Scheme 2, upper panel). This reaction is specific for carboxylic groups of amino acids. Thus, additional functionalities must be reacted with a second derivatizing reagent such as an organic anhydride (Scheme 2, lower panel) [2,5]. The reaction order of the two derivatization steps is of particular importance. Anhydrides of organic acids can react with all functional groups including carboxylic groups, but such derivatives are not stable enough [6].

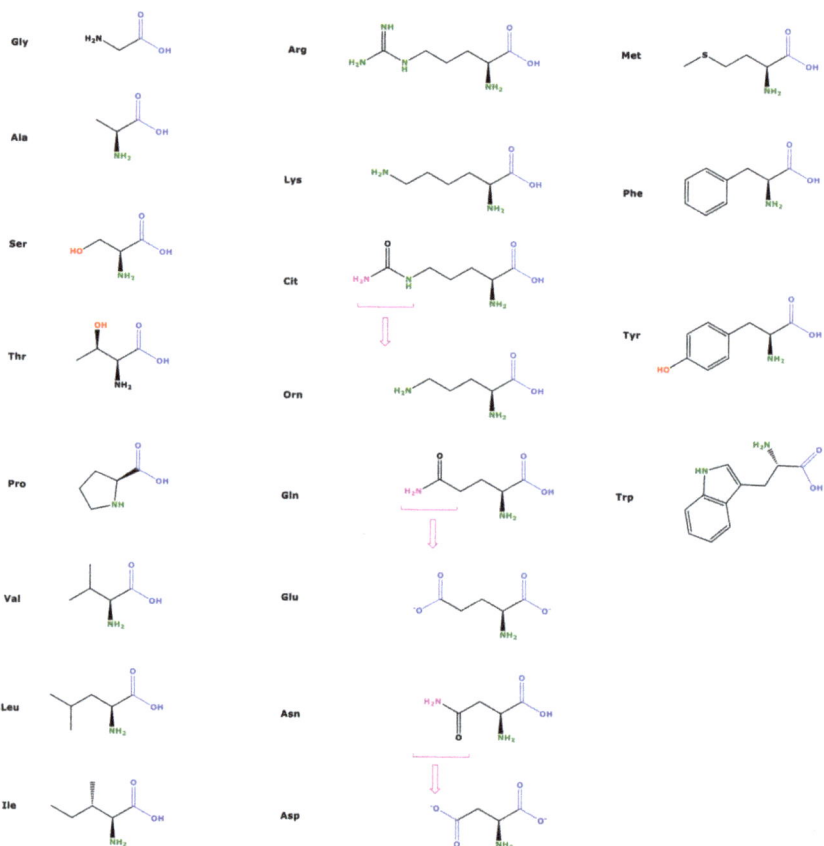

Scheme 1. Chemical structures of the amino acids investigated in the present study. The functionalities being accessible to derivatization are colored. Blue, carboxylic group; Green, amine; Red, OH; Magenta, ureido, carbamoyl. Cit, citrulline; Orn, ornithine. The chemical structure of homoarginine (hArg), the methylene homolog of arginine (Arg) is not shown.

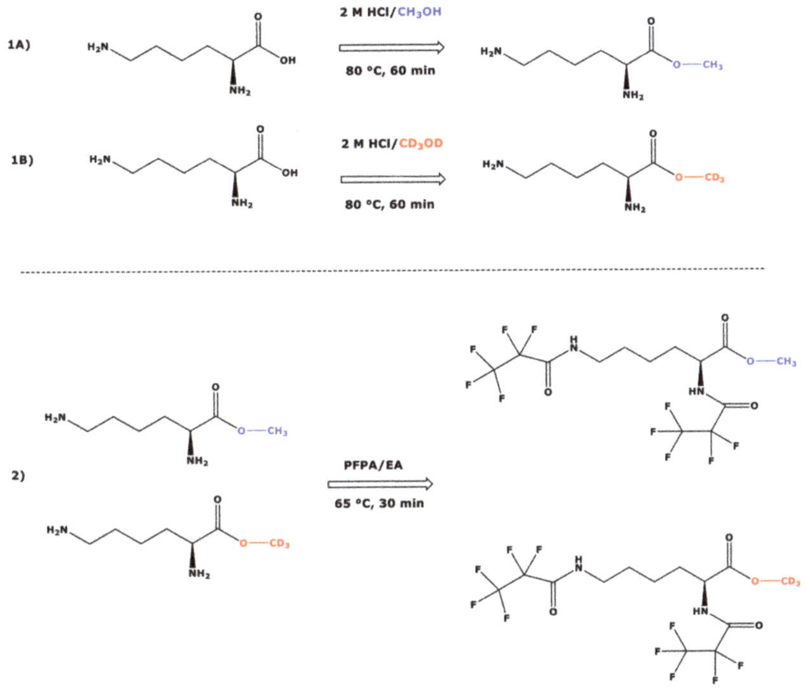

Scheme 2. Two-step derivatization of lysine representative of the chemical class of amino acids. Separate esterification with 2 M HCl in CH$_3$OH for biological amino acids (**1A**) and with 2 M HCl in CD$_3$OD for synthetic amino acids (**1B**) to serve as internal standards. (**2**) Combined acylation with pentafluoropropionic anhydride (PFPA) in ethyl acetate (EA; 1:4, v/v).

In previous work [7,8], we found that a two-step derivatization step (Scheme 2) is useful for the reliable simultaneous quantitative measurement of amino acids by GC–MS in different biological samples including plasma, serum and urine. A major advantage of this derivatization procedure is the generation of stable-isotope labeled amino acids analogs from synthetic amino acids for use as the respective internal standards. The latter are synthesized in situ by performing the first derivatization step (60 min, 80 °C) in a solution of 2 M HCl in commercially available, isotopically highly pure tetradeutero-methanol (CD$_3$OD) (Scheme 2) [5]. The fractions of unlabeled methyl esters (R–COOCH$_3$) of biological amino acids and their deuterium-labeled methyl esters (R–COOCD$_3$) are combined and subjected to the second common derivatization step (30 min, 65 °C) with pentafluoropropionic anhydride (PFPA) in ethyl acetate (1:4, v/v) (Scheme 2). Subsequently, organic solvents and excess PFPA are evaporated under a nitrogen stream, the residue is treated with borate buffer (200 µL) and immediately thereafter with toluene (200 µL). The sample is then vortex-mixed for 60 s to extract the amino acid derivatives into toluene, thereby keeping polar compounds such as pentafluoropropionic acid formed from hydrolyzed and reacted PFPA in the aqueous phase. A 150 µL aliquot of the upper toluene phase obtained by centrifugation (5 min, 800× g) is transferred into 0.2 mL microinserts which were placed in 1.5 mL autosampler glass vials, sealed tightly and subjected to GC–MS analysis [7,8].

Investigations of the stability of native analytes in their biological samples are essential in method validation. Stability studies of non-derivatized analytes in extracts such as in the mobile phase in liquid chromatography (LC)-based methods, for instance in autosampler vials during automated analysis, are widely performed and an indispensable part of method validation. Analogous stability studies in GC-based methods are relatively rare.

Yet, this may be a particular concern for certain analytes such as amino acids derivatized with PFPA [6] and OH-rich carbohydrates derivatized with silylating reagents such as N,O-bis(trimethylsilyl)trifluoroacetamide (BSTFA) [9].

The aim of the present study was to investigate the long-term stability (14 days) in toluene extracts of the methyl ester pentafluoropropionyl derivatives (Scheme 2) of several urinary amino acids (Scheme 1). As the GC–MS analysis of histidine (His) and cysteine (Cys) by this method is not reliable enough [8], these amino acids were not investigated in the present study.

2. Materials and Methods

2.1. Chemicals and Materials

All synthetic amino acids were of analytical grade and were commercially obtained from various manufacturers (Sigma-Aldrich, Merck, Germany). Tetradeuterated methanol (CD_3OD, 99% at ^2H) and pentafluoropropionic anhydride were supplied by Aldrich (Steinheim, Germany). Methanol was obtained from Chemsolute (Renningen, Germany). Hydrochloric acid (37 wt%) was purchased from Baker (Deventer, The Netherlands). Ethyl acetate was obtained from Merck (Darmstadt, Germany). Glass ware for GC–MS (1.5 mL autosampler glass vials and 0.2 mL microvials) and the fused-silica capillary column Optima 17 (15 m × 0.25 mm I.D., 0.25 µm film thickness) were purchased from Macherey-Nagel (Düren, Germany). Separate stock solutions of amino acids were prepared by dissolving accurately weighed amounts of commercially available unlabelled and stable-isotope labelled amino acids in deionized water. Stock solutions were diluted with deionized water as appropriate.

2.2. Derivatization Procedures for Amino Acids in Human Urine Samples

A quality control human urine sample (#29) and six 24-h collected urine samples from six renal transplant recipients (#364, #367, #377, #382, #388, #390) of a previously reported study [10] were used. Urine aliquots (10 µL) were taken and evaporated to dryness under a stream of nitrogen. Methyl esters of the amino acids and their metabolites analyzed in the present study were prepared as follows (see Scheme 2). The residues of the urine samples were reconstituted in 100 µL aliquots of a 2 M HCl/MeOH solution and the vials were tightly sealed. Esterification was performed by heating the samples for 60 min at 80 °C. After cooling to room temperature, urine extracts were spiked with 10 µL aliquots of the newly synthesized trideutero-methyl esters, i.e., the internal standard mixture, in order to reach relevant concentrations with respect to human urine, and the solvents were evaporated to dryness under a stream of nitrogen. Aliquots (100 µL) of a daily prepared PFPA solution in ethyl acetate (PFPA-EA, 1:4, v/v) were added, the glass vials were tightly sealed and heated for 30 min at 65 °C to prepare N-pentafluoropropionic amides of the methyl esters. After cooling to room temperature, solvents and reagents were evaporated to dryness under a stream of nitrogen. Subsequently, residues were treated first with 200 µL aliquots of 400 mM borate buffer, pH 8.5, and immediately thereafter with 200 µL aliquots of toluene, followed by vortex-mixing solvent extraction for 60 s and centrifugation (4000× g, 5 min, 18 °C). Aliquots (150 µL) of the upper organic phase were transferred into autosampler glass vials equipped with 200 µL microinserts, the samples were sealed and subjected to GC–MS analysis. After each GC–MS analysis (i.e., injection of 1 µL toluene aliquots, quantification by selected-ion monitoring of specific mass fragments), the samples were sealed again and stored at room temperature until the next analysis. This procedure was repeated on days 2, 8, and 15. The peak areas measured and the concentrations calculated at various storage times were used to evaluate the stability of the derivatives as well as the precision and the accuracy of the method. The concentrations of the internal standards were within expected concentrations of endogenous urinary amino acids (Table 1).

Table 1. Summary of the GC–MS conditions used for the simultaneous quantitative determination of the indicated amino acids (AA) in human urine using their stable-isotope labelled analogs as internal standards (IS).

Amino Acid	Derivative	AA/IS (m/z)	Retention Time (min)	Dwell Time (ms)	IS (μM)
Ala	Me-PFP	229/232	3.73/3.70	100	220
Thr	Me-(PFP)$_2$	259/262	4.07/4.05	50	165
Gly	Me-PFP	215/218	4.22/4.20	50	1100
Val	Me-PFP	257/260	4.44/4.42	50	33
Ser	Me-(PFP)$_2$	207/210	4.46/4.43	50	330
Leu/Ile	Me-PFP	271/274	5.09/5.07	100	88
Asn/Asp	(Me)$_2$-PFP	287/293	6.74/6.69	50	82.5
Pro	Me-PFP	255/258	7.18/7.16	100	22
Gln/Glu	(Me)$_2$-PFP	301/307	7.93/7.89	100	275
Met	Me-PFP	289/292	7.94/7.92	100	55
Orn/Cit	Me-(PFP)$_2$	418/421	8.60/8.58	50	27.5
Phe	Me-PFP	305/308	8.73/8.71	50	88
Tyr	Me-PFP	233/236	9.06/9.04	100	110
Lys	Me-(PFP)$_2$	432/425	9.51/9.49	50	110
Arg	Me-(PFP)$_3$	586/589	9.60/9.58	50	55
hArg	Me-(PFP)$_3$	600/603	10.39/10.37	100	5.5
Trp	Me-(PFP)$_2$	233/236	11.48/11.45	50	55

Abbreviations. AA, amino acid; IS, internal standard; m/z, mass-to-charge ratio; Me, methyl; PFP, pentafluoropropionyl. Orn, ornithine; Cit, citrulline; hArg, homoarginine.

2.3. Quantitative GC–MS Analyses of Amino Acids

All analyses were performed under routine conditions on a GC–MS apparatus consisting of a single-stage quadrupole mass spectrometer model ISQ, a Trace 1210 series gas chromatograph and an AS1310 autosampler from ThermoFisher (Dreieich, Germany) equipped with a 10 μL Hamilton needle. Toluene aliquots (1 μL) were injected in the splitless mode. The autosampler needle was cleaned automatically three times with toluene (5 μL) after each injection. Injector temperature was kept at 280 °C. Helium was used as the carrier gas at a constant flow rate of 1.0 mL/min. The oven temperature was held at 40 °C for 0.5 min and ramped to 210 °C at a rate of 15 °C/min and then to 320 °C at a rate of 35 °C/min. Interface and ion-source temperatures were set to 300 °C and 250 °C, respectively. Electron energy was 70 eV and electron current 50 μA. Methane was used as the reagent gas for negative-ion chemical ionization (NICI) at a constant flow rate of 2.4 mL/min. In quantitative analyses, the dwell time was 50 ms or 100 ms for each ion in the selected-ion monitoring (SIM) mode and the electron multiplier voltage was set to 1400 V.

The concentration of the amino in the human urine samples in quantitative analyses was determined in the SIM mode using one mass fragment (m/z) for the unlabelled amino acid and one ion for the corresponding mass fragment for the stable-isotope labelled amino acid serving as the internal standard (IS). The ions monitored are listed in Table 1. The peak area (PA) values of the urinary amino acids and of the respective internal standards were calculated automatically by the GC–MS software (Xcalibur and Quan Browser). The concentration of the amino acids was calculated by multiplying the peak area ratio of the endogenous urinary amino acid to the respective internal standard with the concentration of the respective internal standard. Statistical analyses and graphs were performed and prepared by GraphPad Prism 7 (San Diego, CA, USA).

3. Results and Discussion

The mean ratios of the PA values of the urinary amino acids to the PA values of the respective internal standards measured in the seven urine samples are summarized in Table 2. This Table indicates the inter-individual variation of the excretion of the analyzed amino acids. The concentrations of the amino acids in the 24 h collected urine samples of seven individuals varied from 18% to 20% for Met and from 89% to 93% hArg, indicating the biological variation of the urinary excretion of amino acids.

Table 2. Mean peak area ratio and its coefficient of variation (CV) of the indicated endogenous amino acids (d_0) in urine samples from seven volunteers to the respective internal standards (d_3) as measured by GC–MS analysis of 1 µL aliquots of the toluene extracts stored at room temperature for several days.

Amino Acid	Mean Peak Area Ratio (d_0/d_3)				CV (%)	Inter-Indivual Variation
	Day 1	Day 2	Day 8	Day 15	All Days	CV (%)
Ala	0.427	0.417	0.413	0.415	1.49	74–79
Thr	0.653	0.666	0.588	0.584	6.87	30–32
Gly	0.538	0.535	0.550	0.561	2.18	41–44
Val	0.966	0.960	0.813	0.805	10.1	29–36
Ser	0.724	0.721	0.767	0.782	4.10	36–37
Leu/Ile	0.580	0.575	0.564	0.554	2.04	49–50
Asn/Asp	2.204	2.216	2.506	2.571	8.07	32–37
Pro	0.081	0.081	0.079	0.077	2.41	21–25
Gln/Glu	2.754	2.754	3.505	3.372	12.9	36–39
Met	1.526	1.531	1.889	1.869	11.9	18–20
Orn/Cit	1.091	1.096	1.103	1.100	0.47	34–35
Phe	0.381	0.338	0.350	0.348	5.25	64–67
Tyr	0.812	0.813	0.818	0.804	0.71	47–49
Lys	0.818	0.814	0.855	0.845	2.41	67–70
Arg	0.437	0.430	0.438	0.434	0.83	52–55
hArg [a]	0.126	0.130	0.142	0.144	6.53	89–93
Trp	0.427	0.417	0.413	0.415	1.67	74–79

[a] hArg, homoarginine.

The coefficient of variation (CV) of the mean peak area ratio d_0/d_3 measured on days 1, 2, 8, 15 ranged between 0.47% for Orn/Cit and 12.9% for Gln/Glu. The higher CV values were due to higher values between days 1 and 2, and days 8 and 15 for some amino acids (Thr, Val) on the one hand, and lower values between days 1/2 and days 8/15 for other amino acids (Asn/Asp, Gln/Glu), one the other hand.

The mean peak area ratio values for the individual amino acids to their respective internal standards and the precision values (CV) in the toluene extracts of seven urine samples analyzed on days 1, 2, 8 and 15 are listed in Table S1. The CV values were below 10% for Ala, Thr, Gly, Ser, leu/Ile, Pro, Orn/Cit, Phe, Tyr, Lys, Arg, hArg and Trp. The CV values were in part between 10% and 28% for Val, Asn/Asp, Gln/Glu, Met. These data indicate that the Me-PFP derivatives of the analyzed amino acids are remarkably stable in the toluene extracts, with the stability apparently depending upon the amino acid and the biological matrix (i.e., urine).

In order to investigate the stability of the Me-PFP derivatives of the endogenous amino acids and of the internal standards we plotted separately their peak area values against the storage time of the toluene extracts. The results of this analysis are illustrated in the Supplementary Figure S1 for all amino acids and in Figure 1 exemplarily for lysine.

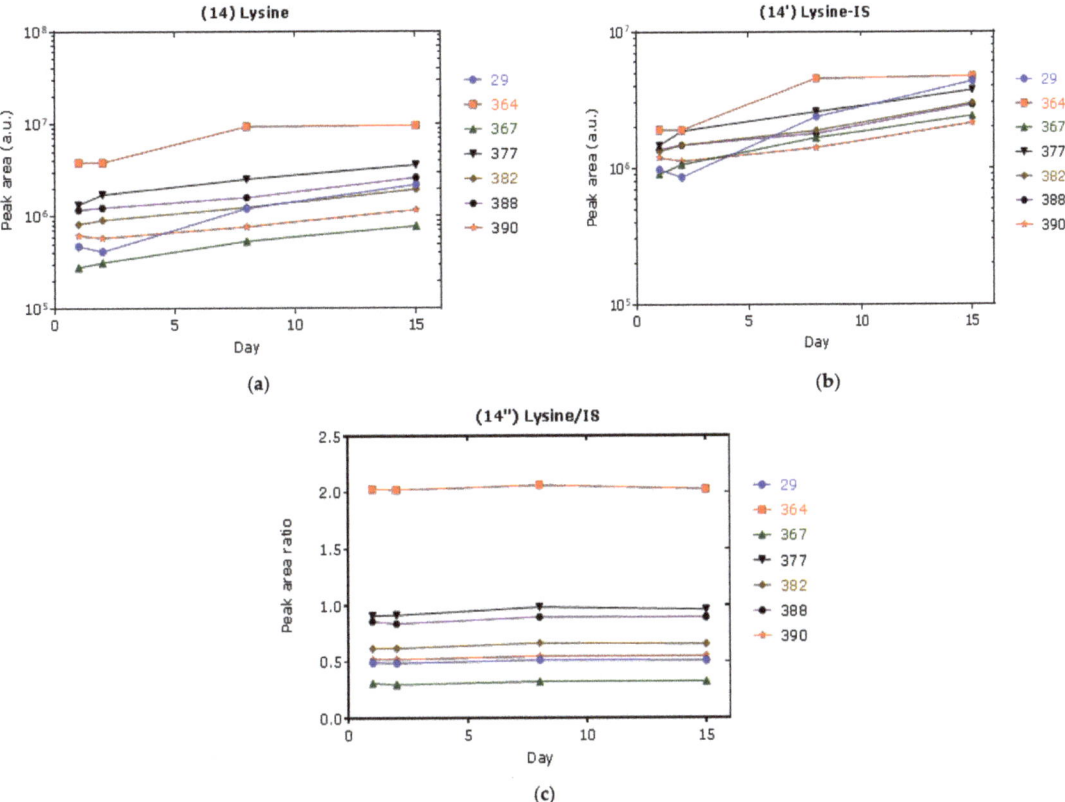

Figure 1. (a) Plots of the peak area of urinary lysine (upper left panel), (b) of the peak area of its internal standard (upper right panel), and (c) of the peak area ratio of lysine to its internal standard (lower panel) in seven 24 h collected urine samples against the storage time of the toluene extracts (day 1, 2, 8, 15). Note the decadic logarithm scale on the y axes in the left and middle panels. The plots of the other amino acids investigated in this study are found in the Supplementary Information. IS, internal standard; a.u., arbitrary units.

On day 1 and day 2 very similar peak areas were measured in the respective urine samples for all amino acids (Figure S1, left panels) and their internal standards (Figure S1, middle panels). On day 8 and day 15 also very similar peak areas were observed for the respective urine samples for all amino acids and their internal standards, except that the peak area values were generally higher (Figure S1, left and middle panels). The volume of toluene phase did not decrease in the microinserts during the storage of the sample. Enrichment of the derivatives is therefore unlikely to explain the higher peak area values of the amino acid derivatives. Because the toluene phases were not vortex-mixed or shaken between the analyses (for practical reasons), it cannot be excluded that exceeding PFPA ($\rho = 1.57$ g/mL) and Me-PFP derivatives "precipitated" in the toluene phase. Yet, the peak area values of the endogenous amino acids and their internal standards increased by almost the same factor, finally resulting in remarkably constant peak area ratios practically in all urine samples (Figure S1, right panels). These observations suggest that the methyl ester-pentafluoropropionyl derivatives of unlabeled amino acids of urinary origin and of the deuterium-labelled amino acids are stable for at least two weeks when stored in toluene at room temperature and analyzed by GC–MS in laboratory routine. These observations also provide convincing evidence that trideutero-methyl esters of amino acids undergo almost the same physicochemical reactions including chemical derivatization and perfectly correct for all possible changes during the whole analytical procedure including incomplete derivatization.

The derivatization conditions used in the present study, including reaction temperature and time, for the derivatization of the amino acids were found in previous studies to be optimum [7,8]. Reaction of amino acids with methanol in 2 M HCl at 80 °C requires about 60 min for a mean yield of 85% for endogenous and synthetic mono- and dicarboxylic amino acids. Ureido and carbamoyl amino acids are unstable under the esterification conditions (Scheme 1) [6,8]. Citrulline (Cit) is hydrolyzed to form ornithine (Orn); asparagine (Asn) and glutamine (Gln) are also hydrolyzed to form aspartate (Asp) and glutamate (Glu), respectively; Orn, Asp and Glu are further esterified to form their dimethyl esters [8]. OH-groups containing amino acids (Thr, Ser, Tyr) are not affected by the HCl-catalyzed esterification.

After esterification, reaction mixtures must be evaporated to complete dryness in order to avoid hydrolysis and to increase the yield of the second consecutive derivatization with PFPA in EA. This reaction requires lower reaction temperature and shorter reaction time given the high reactivity of PFPA. The reaction solutions (PFPA-EA, 1:4, v/v) should be freshly prepared for maximum yield. As PFPA is highly reactive and non-selective, it reacts with many functional groups including primary and secondary amines to form N-pentafluoropropionyl derivatives, hydroxyl groups to form O-pentafluoropropionic esters, and carboxylic groups to form mixed anhydrides. While pentafluoropropionamides are resistant towards water, mixed anhydrides are not stable even in virtually anhydrous toluene [6]. Thr and Ser contain each one OH group which also react with PFPA to finally produce the methyl ester-N,O-pentafluoropropionic derivatives. The results of the present study suggest that the methyl ester-N,O-pentafluoropropionic derivatives of Thr and Ser are very stable in toluene. Tyr contains an aromatic OH group. Under the derivatization conditions used in our study (Scheme 2), the acidic OH group of Tyr is likely to react with PFPA, yet with the O-pentafluoropropionic ester being very easily and quickly hydrolyzing during the borate buffer/toluene extraction step. This has been previously demonstrated by trimethylsilylating the non-derivatized OH group of the Me-PFP derivative of Tyr with BSTFA at room temperature [11].

The reaction mixture of the second derivatization step contains numerous compounds of urinary origin including derivatized and non-derivatized amino acids, huge excess of non-reacted PFPA, as well as pentafluoropropionic acid from hydrolyzed and reacted PFPA. For optimum analysis and minimum contamination of the GC–MS apparatus, extraction of the Me-PFP derivatives is highly recommended. Water-immiscible and GC-compatible organic solvents are best suited. We selected toluene for solvent extraction because it is practically not miscible with water (solubility: 29 µM water in 1 L toluene; i.e., about 160 times less soluble than water in ethyl acetate) and has a lower density than water (ρ = 0.87 g/mL). These features facilitate phase separation and easy transfer of the upper organic phase into autosampler glass vials, thereby effectively avoiding water-carryover. The results of the present study underline the superiority of toluene for the extraction and long-term storage of Me-PFP derivatives of amino acids. Because of the very high molar excess of PFPA over amino acids and other substances of urine origin, it is likely that remaining PFPA is also extracted into toluene and serves as an "in injector" derivatization reagent like ethyl acetate in the analysis of nitrous acid [12].

Trimethylsilyl (TMS) derivatives of many amino acids in pyridine-methoxyamine-N-trimethylsilyl-N-methyltrifluoroacetamide (MSTFA) were found to degrade during the storage in glass vials in the autosampler within a few hours [13]. The stability increased considerably when the derivative solutions were stored at 4 °C (for 12 h), with maximum stability (for 72 h) when stored at −20 °C [13]. These results suggest that TMS derivatives of amino acids are not best suitable for GC–MS analysis.

Alkyl chloroformate is widely used for the GC–MS analysis of amino acids [2]. This derivatization procedure was automated for high-throughput analysis of different classes of substances including amino acids [14]. Yet, no stable-isotope labelled analogs were used for quantitation, with the mode of quantitation being not clearly described in that work. Stability studies of chloroform extracts dried over anhydrous sodium sulfate and at stored at −80 °C. The precision (RSD) of the GC–MS method was used as a measure of the derivatives stability. With respect to the derivatives of amino acids from human urine the stability was reported to be <20% for 6 days storage for the majority of amino acids. Phenylalanine, glutamate, cysteine have been excluded from stability analyses due to lacking linearity.

4. Conclusions

Derivatization of the functional groups of amino acids, i.e., –COOH, –NH$_2$, –NH, –OH, –SH first with 2 M HCl in CH$_3$OH and then with PFPA in EA enables analysis of amino acids in biological samples such as human urine by GC–MS. Derivatization of the –COOH group(s) of synthetic amino acids with 2 M HCl in CD$_3$OD from commercial sources is a useful approach to prepare stable-isotope labelled analogs for use as internal standards in quantitative analyses. Toluene is excellently suitable for the instantaneous extraction and long-term storage of urinary amino acids and their internal standards as methyl ester-pentafluoropropionyl derivatives.

Supplementary Materials: The following are available online, Table S1 Mean peak area ratio (coefficient of variation, CV) of endogenous amino acids in the seven human urine samples to the respective internal standard as measured by GC–MS analysis of 1 µL aliquots of the toluene extracts stored and room temperature for several days. Figure S1. Plots of the peak area of urinary amino acids (upper panel), of the peak area of their internal standards (middle panel), and of the peak area ratio of the amino acids to their internal standards (lower panel) in seven 24 h collected urine samples against the storage time of the toluene extracts (day 1, 2, 8, 15). Note the decadic logarithm scale on the y axes in the left and middle panels.

Author Contributions: Conceptualization, D.T.; methodology, S.B.; software, S.B. and D.T.; validation, S.B., A.B. and D.T.; formal analysis, D.T.; investigation, S.B.; resources, D.T.; data curation, S.B. and D.T. writing—original draft preparation, D.T. and S.B.; writing, S.B. and D.T.; visualization, S.B. and D.T.; supervision, D.T.; project administration, D.T.; funding acquisition, D.T. All authors have read and agreed to the published version of the manuscript.

Funding: This research received no external funding. The APC was waived.

Institutional Review Board Statement: Ethical review and approval were waived for this study, due to the use of human urine samples originally collected in previously ethically approved study.

Informed Consent Statement: Patient consent was waived due to the use of human urine samples originally collected in previously ethically approved study.

Data Availability Statement: The study did not report any data.

Acknowledgments: We thank Bibiana Beckmann for administrative help.

Conflicts of Interest: The authors declare no conflict of interest.

Sample Availability: Not available.

References

1. Hušek, P.; Macek, K. Gas chromatography of amino acids. *J. Chromatogr. A* **1975**, *113*, 139–230. [CrossRef]
2. Hušek, P.; Švagera, Z.; Hanzlíková, D.; Řimnáčová, L.; Zahradníčková, H.; Opekarová, I.; Šimek, P. Profiling of urinary aminocarboxylic metabolites by in-situ heptafluorobutyl chloroformate mediated sample preparation and gas-mass spectrometry. *J. Chromatogr. A* **2016**, *1443*, 211–232. [CrossRef] [PubMed]
3. Ferré, S.; González-Ruiz, V.; Guillarme, D.; Rudaz, S. Analytical strategies for the determination of amino acids: Past, present and future trends. *J Chromatogr. B* **2019**, *1132*, 121819. [CrossRef] [PubMed]
4. Xu, W.; Zhong, C.; Zou, C.; Wang, B.; Zhang, N. Analytical methods for amino acid determination in organisms. *Amino Acids* **2020**, *52*, 1071–1088. [CrossRef] [PubMed]

5. Tsikas, D. De novo synthesis of trideuteromethyl esters of amino acids for use in GC-MS and GC-tandem MS exemplified for ADMA in human plasma and urine: Standardization, validation, comparison and proof of evidence for their aptitude as internal standards. *J. Chromatogr. B* **2009**, *877*, 2308–2320. [CrossRef] [PubMed]
6. Bollenbach, A.; Hanff, E.; Beckmann, B.; Kruger, R.; Tsikas, D. GC-MS quantification of urinary symmetric dimethylarginine (SDMA), a whole-body symmetric L-arginine methylation index. *Anal. Biochem.* **2018**, *556*, 40–44. [CrossRef] [PubMed]
7. Kayacelebi, A.A.; Beckmann, B.; Gutzki, F.M.; Jordan, J.; Tsikas, D. GC-MS and GC-MS/MS measurement of the cardiovascular risk factor homoarginine in biological samples. *Amino Acids* **2014**, *46*, 2205–2217. [CrossRef] [PubMed]
8. Hanff, E.; Ruben, S.; Kreuzer, M.; Bollenbach, A.; Kayacelebi, A.A.; Das, A.M.; von Versen-Höynck, F.; von Kaisenberg, C.; Haffner, D.; Ückert, S.; et al. Development and validation of GC-MS methods for the comprehensive analysis of amino acids in plasma and urine and applications to the HELLP syndrome and pediatric kidney transplantation: Evidence of altered methylation, transamidination, and arginase activity. *Amino Acids* **2019**, *51*, 529–547. [CrossRef] [PubMed]
9. Ruiz-Matute, A.I.; Hernández-Hernández, O.; Rodríguez-Sánchez, S.; Sanz, M.L.; Martínez-Castro, I. Derivatization of carbohydrates for GC and GC-MS analyses. *J. Chromatogr. B* **2011**, *879*, 1226–1240. [CrossRef] [PubMed]
10. Frenay, A.R.; Kayacelebi, A.A.; Beckmann, B.; Soedamah-Muhtu, S.S.; de Borst, M.H.; van den Berg, E.; van Goor, H.; Bakker, S.J.; Tsikas, D. High urinary homoarginine excretion is associated with low rates of all-cause mortality and graft failure in renal transplant recipients. *Amino Acids* **2015**, *47*, 1827–1836. [CrossRef] [PubMed]
11. Schwedhelm, E.; Tsikas, D.; Gutzki, F.M.; Frölich, J.C. Gas chromatographic-tandem mass spectrometric quantification of free 3-nitrotyrosine in human plasma at the basal state. *Anal. Biochem.* **1999**, *276*, 195–203. [CrossRef] [PubMed]
12. Tsikas, D.; Böhmer, A.; Mitschke, A. Gas chromatography-mass spectrometry analysis of nitrite in biological fluids without derivatization. *Anal. Chem.* **2010**, *82*, 5384–5390. [CrossRef] [PubMed]
13. Quéro, A.; Jousse, C.; Lequart-Pillon, M.; Gontier, E.; Guillot, X.; Courtois, B.; Courtois, J.; Pau-Roblot, C. Improved stability of TMS derivatives for the robust quantification of plant polar metabolites by gas chromatography-mass spectrometry. *J. Chromatogr. B* **2014**, *970*, 36–43. [CrossRef] [PubMed]
14. Zhao, L.; Ni, Y.; Su, M.; Li, H.; Dong, F.; Chen, W.; Wei, R.; Zhang, L.; Guiraud, S.P.; Martin, F.P.; et al. High Throughput and Quantitative Measurement of Microbial Metabolome by Gas Chromatography/Mass Spectrometry Using Automated Alkyl Chloroformate Derivatization. *Anal. Chem.* **2017**, *89*, 5565–5577. [CrossRef] [PubMed]

Article

Determination of Biogenic Amines in Different Parts of *Lycium barbarum* L. by HPLC with Precolumn Dansylation

Yun Ai [1,2], Yan Ni Sun [1], Li Liu [1], Fang Yuan Yao [1], Yan Zhang [1], Feng Yi Guo [1], Wen Jie Zhao [1], Jian Li Liu [1] and Ning Zhang [1,*]

1 Key Laboratory of Resource Biology and Biotechnology in Western China, Ministry of Education, School of Life Sience, Northwest University, Xi'an 710069, China; aiyun-vip@163.com (Y.A.); sunyanni@nwu.edu.cn (Y.N.S.); 18792597747@163.com (L.L.); yaoyao612525@163.com (F.Y.Y.); z18391092679@163.com (Y.Z.); guofengyii@163.com (F.Y.G.); 17835424667@163.com (W.J.Z.); jlliu@nwu.edu.cn (J.L.L.)
2 Xi'an Institute for Food and Drug Control, Xi'an 710054, China
* Correspondence: zhangning@nwu.edu.cn; Tel./Fax: +86-88302013

Abstract: The aim of this work was to characterize biogenic amines (BAs) in different parts of *Lycium barbarum* L. using HPLC with dansyl chloride derivatization, and jointly, to provide referential data for further exploration and utilization of *Lycium barbarum* L. The linear correlation coefficients for all BAs were above 0.9989. The limits of detection and quantification were 0.015–0.075 and 0.05–0.25 µg/mL, respectively. The relative standard deviations for the intra-day and inter-day precision were 0.66–2.69% and 0.91–4.38%. The described method has good repeatability and intermediate precision for the quantitative determination of BAs in different parts of *Lycium barbarum* L. Satisfactory recovery for all amines was obtained (79.3–110.3%). The result showed that there were four kinds of BAs. The highest putrescine content (20.9 ± 3.2 mg/kg) was found in the flower. The highest histamine content (102.7 ± 5.8 mg/kg) was detected in the bark, and the highest spermidine (13.3 ± 1.6 mg/kg) and spermine (23.7 ± 2.0 mg/kg) contents were detected in the young leaves. The high histamine (HIS) content in the bark may be one of the reasons why all of the parts of *Lycium barbarum* L., except the bark, are used for medicine or food in China. Meanwhile, the issue of the high concentration of HIS should be considered when exploiting or utilizing the bark of *Lycium barbarum* L.

Keywords: biogenic amines; *Lycium barbarum* L.; HPLC; derivatization

1. Introduction

Biogenic amines (BAs), such as tyramine (TYR), methylamine (MET), histamine (HIS), putrescine (PUT), cadaverine (CAD), tryptamine (TRY), phenylethylamine (PEA), spermine (SPM), and spermidine (SPD), are endogenous and indispensable components to living cells, playing important roles in cell proliferation and differentiation, regulation of nucleic acid function, protein synthesis, brain development, and nerve growth and regeneration [1]. Moreover, they have many biological activities, such as vasoconstriction [2], vasodilation [3], antioxidant [4], and promoting longevity [5]. For polyamines PUT, SPD, and SPM, their intracellular concentrations decline during human aging, and administration of SPD markedly extended the lifespans of yeast, flies, worms, and human immune cells. The SPD administration potently inhibited oxidative stress in aging mice. In aging yeast, spermidine treatment triggered epigenetic deacetylation of histone H_3 through the inhibition of histone acetyltransferases (HAT), suppressing oxidative stress and necrosis. Conversely, depletion of endogenous polyamines led to hyperacetylation, generation of reactive oxygen species, early necrotic death, and decreased lifespan [5]. In recent years, it was found that SPD and SPM potently inhibit oxidative stress in aging mice, which can increase the activity of antioxidant enzymes, reduce the accumulation of free radicals, improve the skeletal muscle cell membrane metabolism and anti-injury ability, and significantly delay the occurrence of mouse fatigue [6,7]. BAs in low concentrations are essential for many physiological functions; however, they can cause a variety of side effects in high concentrations, including rashes, headaches, nausea, hypo- or hyper-tension, cardiac palpitations, intracerebral hemorrhages, and anaphylactic shock, especially when alcohol or monoamine oxidase inhibitors are ingested at the same time [8]. Among the BAs, the toxicity of each amine is different, but HIS and TRY are the most toxic. HIS poisoning has a short incubation period, ranging from minutes to hours after ingestion. Symptoms include headache, facial flushing and sweating, rash and itching, nausea, vomiting, diarrhea, and heart palpitations [9]. TYR overdose may result in a hypertensive crisis, usually accompanied by a severe headache and sometimes intracranial hemorrhage and other neurological sequelae, cardiac failure, and pulmonary edema [10]. It was reported that 3 mg of PEA can cause migraine headaches in susceptible individuals [1]. PUT and CAD have not demonstrated direct adverse health effects, but they are known to enhance HIS toxicity by inhabiting HIS metabolizing enzymes [1]. The permissible limit of 50 mg/kg of HIS has been proposed by the US Food and Drug Administration (US FDA)—differently, 100 mg/kg was proposed by the European Union, South Africa, Canada, and Switzerland, and 200 mg/kg by Australia [11–13]. The acceptable level of TYR is 100–800 mg/kg and for PEA it is 30 mg/kg [1,4]. No recommendations have been suggested for PUT and CAD. Although estimating the total toxic dose of individual BAs is very difficult, Shalaby and Valsamaki et al. have stated that the safe sum of HIS, TRY, PUT, and CAD should not significantly exceed 900 mg/kg [14,15]. Therefore, the study of BAs is of great interest due to their activities and toxicological threats.

Lycium barbarum L. (*L. barbarum*) is a perennial bush produced mostly in Ningxia Hui Autonomous Region of China. The Compendium of Materia Medica in the Ming Dynasty recorded: the leaves of *L. barbarum* are called Tianjingcao; the flowers are called Changshengcao; the fruits are called Gouqizi (Lycii Fructus); the root bark is called Digupi (Cortex Lycii Radicis) [16]. In Traditional Chinese Medicine (TCM), the roots, stems, leaves, flowers, and fruits of *L. barbarum* are used as medicine [17]. The fruit of *L. barbarum*, known as the wolfberry, is a well-known product used for medicinal and food purposes in China [18,19]. It was first recorded in the book of "Shennong's Classic of Materia Medica" and described as making the body stronger, resistant to aging, and resistant to cold and heat, and enhancing physical fitness. It was listed in the Chinese Pharmacopoeia (2020). The dried fruit of *L. barbarum*, widely used as a very important commercial crop, has been developed and used in different kinds of products, including medlar wine, tea, drinks, health products, and cosmetics [20]. It is also used in food such as stewed soup and porridge. The fruit of *L. barbarum* also has been used for preventing and treating various diseases in East Asia for over 2000 years with excellent efficacy in nourishing the liver and kidney, brightening the eyes [21], and providing anti-aging and cytoprotective effects [22]. The leaves of *L. barbarum*, which are recorded in the book of Chinese Materia Medica, have effects of tonifying the kidney and providing a beneficial essence, removing heat and quenching thirst, nourishing the liver and improving eyesight, and promoting longevity [16]. Modern pharmacology shows that it has antioxidant, anti-aging, hypoglycemic, hypolipidemic, anti-tumor, antibacterial, anti-hypoxic, and anti-fatigue activities [23–25]. The leaves of *L. barbarum* also have been used as a daily health food, a health-enhancing herbal tea, and vegetables [26]. The leaves of *L. barbarum* and cactus are reportedly used to make a hypoglycemic health drink to control blood glucemin [27]. Stems and leaves of *L. barbarum* and garlic as mixed raw materials can be made as a composite function beverage [28]. The flower of *L. barbarum* has antioxidant, neuroprotective, and antibacterial effects [23]. The root of *L. barbarum* has anti-free radical activity, has immune regulation activity, lowers blood pressure, regulates blood lipids, is hypoglycemic, and is antibacterial [25]. Therefore, it is valuable to study the different parts of *L. barbarum* for its better exploitation and utilization.

For the study of the composition of *L. barbarum*, most studies have reported flavonoids, alkaloids, terpenoids, polysaccharides, peptides, phenolic amides, traceelements, and amino acids in the fruit or leaves of L. barbarum [20,29,30]. Ma Baolong et al. [31] successfully established a method for detecting chloride derivatives of SPM and SPD in L. barbarum leaves with high performance liquid chromatography (HPLC), but they did not provide examples, nor did they give specific contents. To the best of the authors' knowledge, there are no reports available about specific BA contents in the different parts of L. barbarum.

A variety of methods have been applied for the determination of BAs in foods, such as electrochemical biosensor methods [32], thin layer chromatography (TLC) [33], gas chromatography (GC) [34], ion chromatography (IC) [35], and reverse-phase HPLC (RP-HPLC) [36]. Since most BAs lack satisfactory absorption and significant fluorescence characteristics, chemical derivatization is usually performed to increase the sensitivity in determining BAs. O-phthalaldehyde (OPA) [37] and dansyl chloride (Dns-Cl) [38] are mostly used. The major disadvantage of OPA is that it only reacts with primary amines. However, Dns-Cl forms derivatives not only with primary but also with secondary amines. Their products are more stable than those formed by using OPA [39]. BAs were characterized in Chinese Tonic-Qi herbs, male silkworm moths, and rat plasma using HPLC with Dns-Cl derivatization in our previous studies [40–42]. This method has good repeatability and precision for the quantitative determination of BAs. This work aimed to characterize BAs in the different parts of L. barbarum using HPLC with Dns-Cl derivatization and to provide referential data for further exploration and utilization of L. barbarum. The identification of dansylated BAs was confirmed by HPLC-UV and ESI-Q-TOF-MS.

2. Results and Discussion

2.1. Preparation of Samples

In order to optimize the solution extracted, trichloroacetic acid, perchloric acid, and hydrochloric acid were employed to compare the extraction efficiency of BAs from L. barbarum. The best results came from the use of hydrochloric acid as the extraction reagent with an optimal concentration of 1 M. To determine low concentrations of BAs in the complex samples, it can be of great importance to remove interfering constituents in order to obtain the baseline separated peaks. Therefore, the amine was concentrated by liquid–liquid extraction and the interference components were removed. A total 10 mL of 1 M HCl extraction was collected, and the pH was adjusted to 12 to obtain free amines. The n-butyl alcohol-dichloromethane (1:1, v/v) was used to extract the free amines. One milliliter of 1 M HCl was added to the organic fractions before evaporation in order to obtain the stable amine hydrochloride.

BAs can be formed stable amine hydrochloride when extracted with 1 M HCl. An additional liquid–liquid extraction was employed to remove highly polar or ionic compounds. After organic solvent evaporation, the dry residue was dissolved in 0.1 M HCl (1 mL) for enrichment finally. The chemical constituents of L. barbarum are very complex, including anthocyanins, flavonoids, polysaccharides, fatty acids, amino acids, sugars, and some trace elements. Extraction of BAs from samples and pre-processing are important steps prior to the derivatization. Most of the methods available in the literature for BA determination involve an acidic extraction from a solid matrix; organic solvents are seldom used. The choice of acid has to be related to the characteristics of the matrix to be analyzed. Many compounds that might interfere can be eliminated in this step. In our previous study, an optimization of the extraction procedure was carried out by investigating the influences of extraction solvent (acid choice), extraction method, and extraction time on the outputs of the extraction [40–42]. To determine low concentrations of BAs in crude extracts from plant tissues, it can be of great importance to remove interfering compounds; purification is necessary. Therefore, an additional liquid–liquid extraction was employed to remove highly polar or ionic compounds from the plant extract. BAs were dissociated after adding NaOH, and extracted by an organic solvent which can remove water-soluble impurities (amino acids and sugar). After organic solvent evaporation, the dry residue was dissolved in 0.1M HCl (1 mL) for enrichment. This method has good repeatability, precision, and recovery for the quantitative determination of BAs.

2.2. Derivatization Procedure

To optimize the derivatization procedure, different amounts of Dns-Cl (between 3 and 10 mg/mL) and reaction times (from 30 to 60 min) were used to derivatize the 11 amines at different temperatures (between 25 and 80 °C). It was found that a 45 min derivatization at 60 °C by using 5 mg/mL Dns-Cl was sufficient to produce a quantitative conversion of the amines into stable Dns-Cl derivatives. As the dansylation reaction requires an alkaline state, the pH of the buffer is the main influence for dansylation on these 11 determined amines. Upon comparison of different pHs of the $NaHCO_3/Na_2CO_3$ buffer from 8.16 to 11.0, a suitable pH of 10.0 was determined. An addition of ammonia is necessary to quench an excess of Dns-Cl, and a reaction for 30 min is sufficient to stop the dansylation. Experiments have shown that the dansyl derivative is stable for more than 24 h at room temperature.

2.3. Identification of the BAs in L. barbarum

In the analysis of *L. barbarum* samples, the BAs peak identification was based on the comparison between the retention times of standard compounds and was confirmed by ESI-Q-TOF-MS. The chromatogram of a standard solution of dansylamines shown in Figure 1A was obtained using the gradient profile described in the Section 3.5. There was no overlap between each standard amine. The high content of acetonitrile (ACN) in the mobile phase within the twenty-fifth and thirty-fifth minutes is a necessity due to the slow movement of SPD and SPM in the column. After SPM elution, the composition of the mobile phase returned within 5 min to the initial composition. All of the standard amine peaks appeared within 35 min. The detection peaks of different parts of *L. barbarum* samples could be well separated with no endogenous interference observed at the retention times of both the analyte and the internal standard (IS). The sample of young leaves, as a representative one, is shown in Figure 1B–D. Major side-products of the dansyl reaction were eluted for 2–10 min and good separation was achieved from derivatized amines. The control group was detected using 10 mL of 1 M HCl to replace the 10 mL of sample extract derived with Dns-Cl and shown in Figure 1F.

Under the MS conditions, dansylated BAs produced stable and intense $[M + Na]^+$ ions seen in Figure 2 and Table 1. However, the product ion of SPM was not found, though the dansylated SPM was analyzed alone or the concentration of the analyte was increased. The dansylated SPM (MW: 1157) may be too large to detect under the MS conditions. SPM was identified by changing the mobile phase and the spiked standards [40].

Table 1. Characteristic mass fragments of dansylated BAs.

BAs	Dns-BAs	$[M + 23]^+$ (m/z)	$[M + 23]^+$ (m/z) found
PUT	Dns_2-PUT	577.1914	577.1921
HIS	Dns-HIS	367.1199	367.1215
SPD	Dns_3-SPD	867.3003	867.2959

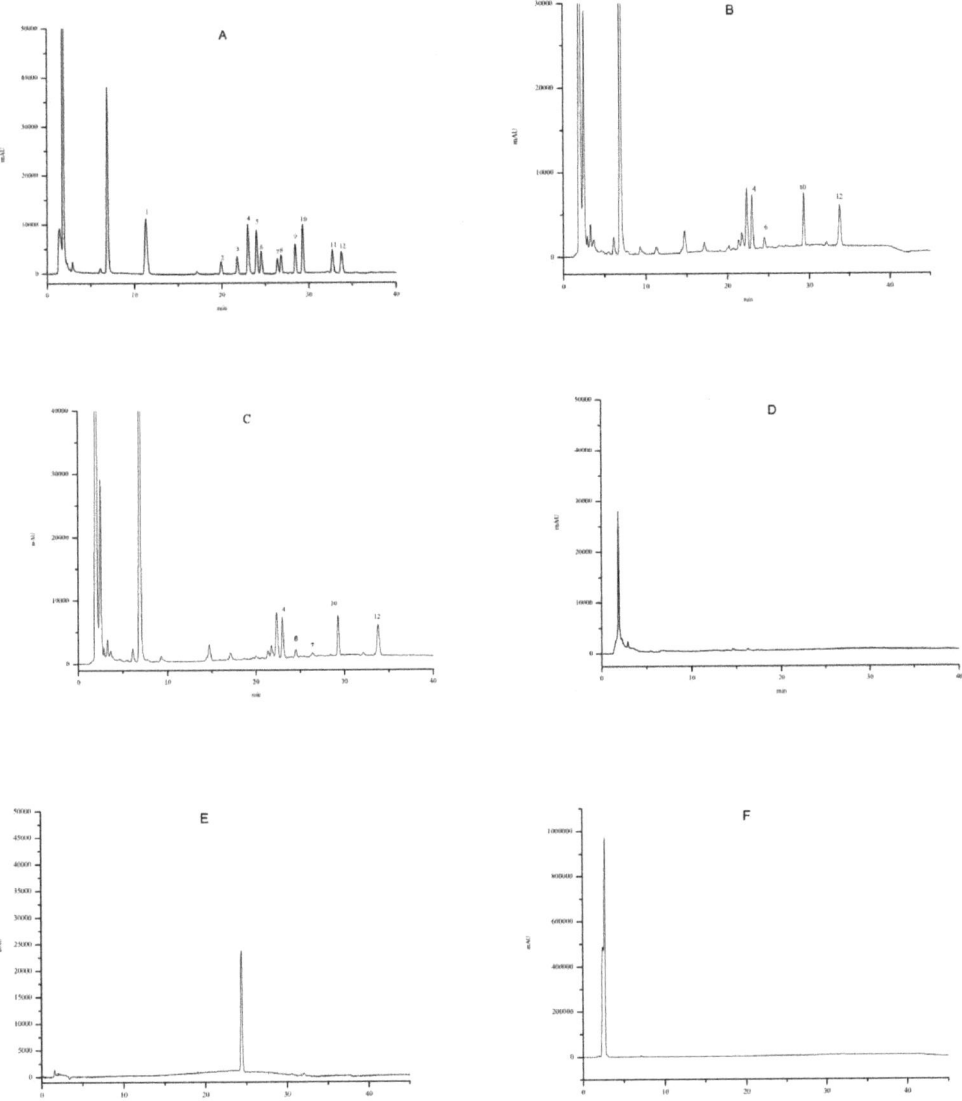

Figure 1. (**A**): Standard solution of BAs (10 μg/mL each); (**B**): *L. barbarum* leaves sample (IS unspiked); (**C**): *L. barbarum* leaves sample; (**D**): *L. barbarum* leaves sample (Dns-Cl unspiked); (**E**): Dns-Cl solution (20 μg/mL); (**F**): control group: 10 mL 1 M HCl replacing the 10 mL sample extract derived with Dns-Cl. Chromatographic peaks: (1) MET; (2) TRP; (3) PEA; (4) PUT; (5) CAD; (6) HIS; (7) DMP; (8) 5-HT; (9) TYR; (10) SPD; (11) DA; (12) SPM.

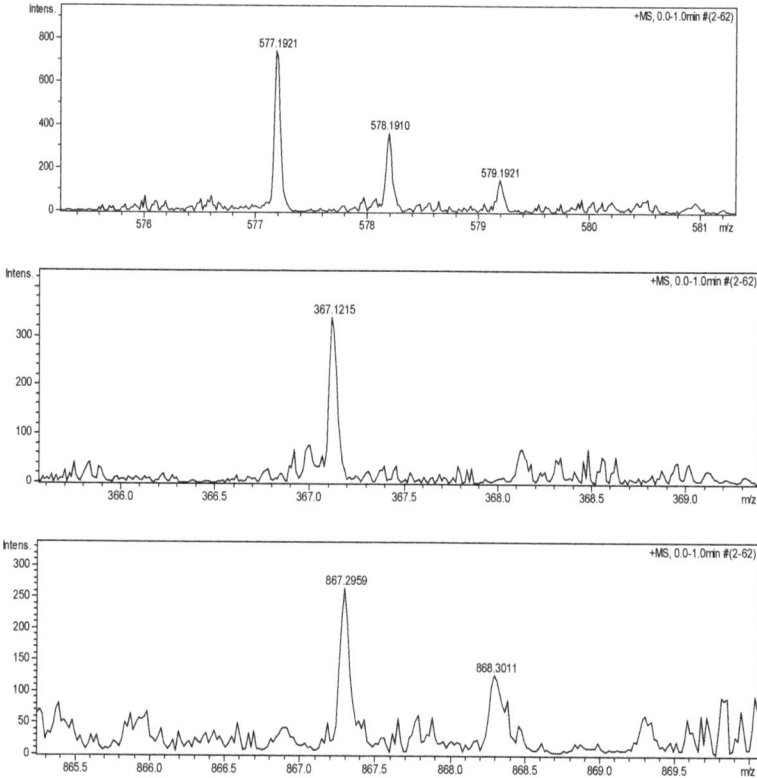

Figure 2. MS spectra of dansylated BAs.

2.4. The Matrix Effect

The matrix effect was assessed by the slope comparison method as performed by Flores and Jia groups [43,44]. The slope ratio of the matrix curve to the neat solution curve was calculated; the ratio value of 1.0 indicated no matrix effect. The results showed that most of the ratio values were close to 1.0, implying that there were no significant matrix effects in relatively complex plant matrices. Since no significant matrix effects were found for the relatively complex plant matrices, the proposed method might be applicable for the detection of BAs in other complex samples in the future.

2.5. Method Validation

The results are summarized in Table 2. Good fitting with the linear model for the response (R^2: 0.9989–0.9999) of each analyte was observed in the concentration ranges of 0.10–10, 0.25–10, and 0.10–50 µg/mL. The LOD and LOQ were in the ranges of 0.015–0.075 and 0.05–0.25 µg/mL, respectively. The relative standard deviations (R.S.Ds.) for the intra-day and inter-day precision were 0.66–2.69% and 0.91–4.38%, respectively. These results confirmed the good repeatability and intermediate precision of the described method. Satisfactory recovery for all amines was obtained (79.3–110.3%); young leaves sample, as a representative, is shown in Table 2.

Table 2. Sample characteristics obtained with HPLC analysis.

Biogenic Amines	Linear Range (µg/mL)	R^2	LOD (µg/mL)	LOQ (µg/mL)	Intra-Day Precision (%RSD) (n = 5)	Inter-Day Precision (%RSD) (n = 5)	Recovery (Young Leaves Samples) (% mean ± SD, n = 3)		
							1 µg/mL	5 µg/mL	10 µg/mL
MET	0.10–10	0.9999	0.015	0.05	0.98	2.54	89.1 ± 4.9	91.7 ± 1.3	86.5 ± 1.1
TRP	0.10–10	0.9990	0.030	0.10	0.81	1.53	79.3 ± 4.6	92.1 ± 1.2	91.7 ± 1.0
PEA	0.25–10	0.9997	0.075	0.25	1.01	1.73	92.6 ± 5.3	94.3 ± 1.6	90.5 ± 1.2
PUT	0.10–50	0.9996	0.015	0.05	2.69	4.38	88.2 ± 5.4	90.6 ± 1.6	88.6 ± 1.2
CAD	0.10–10	0.9998	0.030	0.10	0.93	1.67	86.3 ± 4.3	93.6 ± 1.6	95.5 ± 1.1
HIS	0.10–10	0.9993	0.030	0.10	1.27	2.39	89.5 ± 4.7	91.2 ± 1.2	87.9 ± 1.1
5-HT	0.25–10	0.9989	0.075	0.25	1.33	2.06	110.3 ± 5.1	88.1 ± 2.0	87.7 ± 1.7
TYR	0.25–10	0.9997	0.075	0.25	0.66	0.91	86.5 ± 4.1	88.2 ± 1.1	89.4 ± 1.3
SPD	0.10–10	0.9997	0.030	0.10	1.42	2.82	89.7 ± 5.7	92.6 ± 1.9	90.4 ± 1.4
DA	0.25–10	0.9989	0.075	0.25	1.09	2.43	93.1 ± 4.8	94.1 ± 2.5	88.5 ± 1.7
SPM	0.10–10	0.9996	0.030	0.10	2.07	3.31	96.5 ± 4.4	94.2 ± 1.4	90.1 ± 1.1

R^2: square of regression coefficient; LOD: limit of detection; LOQ: limit of quantification; RSD: relative standard deviation.

2.6. Distribution of the BAs in L. barbarum

Table 3 lists the BAs in different parts of *L. barbarum*.

Table 3. Content of BAs in different parts of *L. barbarum*. (n = 3).

Samples	Concentrations of BAs (mg/kg) (Mean ± SD)										
	MET	TRP	PEA	PUT	CAD	HIS	5-HT	TYR	SPD	DA	SPM
Young leaves	ND	ND	ND	12.94 ± 1.8 [ab]	ND	5.6 ± 0.9 [c]	ND	ND	13.3 ± 1.6 [a]	ND	23.7 ± 2.0 [a]
Mature leaves	ND	ND	ND	2.3 ± 0.4 [d]	ND	2.6 ± 0.3 [c]	ND	ND	10.7 ± 1.5 [abc]	ND	9.4 ± 1.1 [bc]
Young stem	ND	ND	ND	10.4 ± 1.4 [abc]	ND	10.2 ± 1.8 [c]	ND	ND	4.3 ± 0.8 [bc]	ND	13.3 ± 1.4 [b]
Mature stem	ND	ND	ND	5.2 ± 1.1 [acd]	ND	23.5 ± 2.2 [b]	ND	ND	3.6 ± 0.6 [bc]	ND	2.4 ± 0.4 [d]
Bark	ND	ND	ND	3.0 ± 0.5 [ad]	ND	102.7 ± 5.8 [a]	ND	ND	3.0 ± 0.5 [c]	ND	2.7 ± 0.4 [d]
Flower	ND	ND	ND	20.9 ± 3.2 [a]	ND	2.3 ± 0.4 [c]	ND	ND	10.7 ± 2.1 [abc]	ND	4.2 ± 0.7 [d]
Fruit	ND	ND	ND	1.3 ± 0.3 [d]	ND	ND	ND	ND	2.9 ± 0.1 [bc]	ND	6.3 ± 1.2 [cd]
Root	ND	ND	ND	5.2 ± 0.9 [abcd]	ND	17.0 ± 3.2 [bc]	ND	ND	3.0 ± 0.3 [bc]	ND	0.9 ± 0.1 [d]

ND: not detected. Different letters (such as a and b) indicate differences between period groups ($p < 0.05$), identical letters (including a and ab) indicate no differences between period groups, and no letters indicate no differences among all eight period groups.

As shown in Table 3, MET, TRP, PEA, CAD, 5-HT, TRY, and DA were not found in detectable concentrations in any of the samples. PUT, HIS, SPD, and SPM were found in all parts of *L. barbarum*, except that HIS was not found in detectable concentrations in the fruit.

The content of BAs was determined and compared in different parts of *L. barbarum*. The PUT content was the highest in the flower (20.9 ± 3.2 mg/kg), followed by young leaves, young stem, mature stem, root, bark, mature leaves, and fruit. The PUT content of the flower and young leaves, young stem, mature stem, root, and bark did not significantly differences among the six groups ($p > 0.05$); however, there were significant differences in PUT content among flower, mature leaves, and fruit ($p < 0.05$).

The highest HIS content was detected in the bark (102.7 ± 5.8 mg/kg), followed by mature stem (23.5 ± 2.2 mg/kg), root (17.0 ± 3.2 mg/kg), young stem (10.2 ± 1.8 mg/kg), young leaves (5.6 ± 0.9 mg/kg), mature leaves (2.6 ± 0.3 mg/kg), and flower (2.3 ± 0.4 mg/kg). HIS was not found in detectable concentrations in the fruit. The content of HIS in the bark was much higher than that in the other seven groups, and the difference was significant ($p < 0.05$). The HIS content in the mature stem was the second, with no significant difference from root, but there was a significant difference between the mature stem and young leaves, mature leaves, young stem, and flower ($p < 0.05$). The contents of HIS in mature leaves, young leaves, young stem, flower, and root were very low and were not significantly different ($p > 0.05$). The HIS content in the bark (102.7 ± 5.8 mg/kg) was higher than the limit standard (50 mg/kg by the US FDA, 100 mg/kg by the European Union, South Africa, Canada, and Switzerland). Considering the toxicity of HIS [1,9], it is necessary to pay attention to the high toxicity of HIS in the bark. The experimental result suggests that high HIS should be considered when exploiting or utilizing the bark of *L. Barbarum*. The HIS contents in young leaves, mature leaves, young stem, flower, and root were not significantly different, and were lower than regulated limits, so they can be used safely. The Compendium of Materia Medica recorded that the root, stem, leaves, flower, and fruit are used as medicine; only the bark is not. The bark is also not used for food in China. The high content of HIS in the bark may be one of the reasons.

Upon comparison with the SPD contents in different parts of *L. barbarum*, the highest amount of SPD was detected in the young leaves (13.4 ± 1.6 mg/kg), followed by mature leaves (10.7 ± 1.5 mg/kg) and flower (10.7 ± 2.1 mg/kg), all of which had significantly higher contents than the other parts. The difference between these three groups is not obvious ($p > 0.05$), but there was significant difference from other groups ($p < 0.05$). The highest SPM content was found in the young leaves, followed by young stem, mature leaves, fruit, flower, bark, mature stem, and root. Compared with the contents of SPM in different parts of *L. barbarum*, the contents of SPM in young leaves (23.7 ± 2.0 mg/kg), young stem (13.3 ± 1.4 mg/kg), mature leaves (9.4 ± 1.1 mg/kg), and fruit (6.3 ± 1.2 mg/kg) were significantly higher than those in the other parts, and the content of SPM in young leaves was the highest. A significant difference existed between the content of SPM in young leaves and the contents in the other seven groups ($p < 0.05$). The SPM contents in mature leaves and young stem were higher than in mature stem, bark, flower, fruit, and root, significantly ($p < 0.05$). No significant difference for SPM content was observed between mature leaves and young stem ($p > 0.05$).

3. Materials and Methods

3.1. Chemicals and Reagents

BAs, 2-phenylethylamine (PEA), methylamine (MET), putrescine dihydrochloride (PUT), cadaverine dihydrochloride (CAD), tryptamine (TRY), histamine dihydrochloride (HIS), serotonin hydrochloride (5-HT), tyramine hydrochloride (TYR), dopamine hydrochloride (DA), spermidine trihydrochloride (SPD), and spermine tetrahydrochloride (SPM) were obtained from Sigma-Aldrich (St. Louis, MO, USA). IS, 1,7-diaminoheptane (DMP) was supplied by Sigma-Aldrich (Steinheim, Germany). Dansyl chloride (Dns-Cl) from Sigma-Aldrich (USA) was used as a derivatization reagent. The HPLC grade acetonitrile (ACN) was purchased from Tedia (Fairfield, CT, USA). Deionized water was produced by a MILLPAK Reagent Water System (Millipore, MA, USA). All other reagents were of analytical grade.

3.2. Preparation of Standard Solutions

The mixed stock solution, containing 0.1 mg/mL of each individual amine, was prepared in 0.1 M HCl and stored at 4 °C. Standard solutions were prepared by diluting the stock solution and used to obtain calibration curves. A separate IS stock solution, containing 0.1 mg/mL DMP, was made in an analogous way. Dns-Cl standard solution was prepared in acetone and stored at 4 °C.

3.3. Preparation of Sample Solution

L. barbarum samples were collected from Xi'an Town, Zhongwei City, NingXia Hui Autonomous Region of China (31 August), and were categorized as young leaves, mature leaves, young stems, mature stems, main stem bark, flowers, fruit, and roots. Each sample was dried in the shade according to the traditional drying method of *L. barbarum* fruit (known as the wolfberry) [45].

The different parts were collected from multiple trees, then cut into small pieces (0.5 cm * 0.5 cm) and mixed, and then 3 samples (5 g each sample) were weighed randomly. Each sample was transferred into a 50 mL centrifuge tube, followed by the addition of 20 mL of 1 M HCl and 0.1 mL of IS working solution (0.1 mg/mL). The mixture was extracted in a constant-temperature shaker (SHA-C, China) for 60 min and then centrifuged (3600 rpm) for 10 min. The supernatant was collected with the residue extracted twice with the same volume of 1 M HCl. All supernatants were combined and filtered through a filter paper into a 50 mL brown volumetric flask. The final volume was adjusted to 50 mL with 1 M HCl. A 10 mL aliquot of sample extract was transferred to a 50 mL centrifuge tube and adjusted to pH 12 with 2 M NaOH. The mixture was extracted with 10 mL n-butyl alcohol-dichloromethane (1:1, *v/v*), vortex-mixed for 5 min and then centrifuged for 10 min at 3600 rpm. After separating the two phases, the lower organic phase was transferred to an evaporating dish. The extraction was repeated three times. The 1 mL of 1 M HCl was added to the combined organic fractions and evaporated to dryness under nitrogen in a water bath at 40 °C. The dried residue was dissolved in 1 mL of 0.1 M HCl, and 0.5 mL solution was collected for derivatization. The residue was used for the blank group. The control group was carried out with a 10 mL 1 M HCl replaced 10 mL extract, then followed the rest steps.

3.4. Derivatization Procedure

The 0.5 mL of dilute standard solution of the amines was mixed with 20 µL DMP (0.1 mg/mL), 0.5 mL of Dns-Cl (5 mg/mL in acetone), and 1 mL of Na_2CO_3-$NaHCO_3$ pH 10 buffer. The mixture was heated in a water bath at 60 °C for 45 min. The excess Dns-Cl was quenched by adding 100 µL of ammonia in the dark at room temperature. After 30 min, the mixture was adjusted to 5 mL with ACN. The solution was filtered through a 0.45 µm membrane filter and injected onto the chromatographic column for determination. For each of the extracted samples, a dansylation reaction was performed as described earlier for the standard solution, except that IS was not added.

3.5. Instrumentation and Conditions

The HPLC determination of dansyl derivatives of BAs was performed with a Shimadzu liquid chromatography system coupled with a system controller SCL-10A-Vp, two pumps (LC-10A-Vp), a DAD detector SPD-M-10A-Vp, a degasser GT-154, and a Rheodyne injector with a 20 µL loop (Tokyo, JPN). Separation was achieved using a Shimadzu shim-pack C18 analytical column (250 mm × 4.6 mm, Intenal Diameter, 5 µm) and protected with a Shimadzu extend C18 guard column (12.5 mm × 4.6 mm, Intenal Diameter, 5 µm) maintained at 30 °C. The mobile phase solution A was ultra-pure water and the mobile phase solution B was ACN. Gradient elution was from 45–50 % A for 0–7 min; 50–10% A for 8–25 min; 10% A for 26–35 min and 10–45 % A for 36–40 min at the flow rate of 0.8 mL/min. The detector was set at 254 nm; 20 µL of dansyl derivative was injected onto HPLC for analysis.

Mass spectrometry was carried out on a micrOTOF-Q II mass spectrometer instrument (Bruker Daltonics, Bremen, Germany). The conditions of MS analysis in the positive ion mode were as follows: dry gas; flow rate, 4.0 mL min^{-1}; dry heater temperature, 180 °C; scan range, 50–3000 *m/z*; capillary voltage, 4500 V; nebulizer press, 0.4 Bar.

3.6. Method Validation

The analytical method was validated in terms of its linearity, limit of detection (LOD), limit of quantification (LOQ), precision, and recovery. The calibration curves were generated by plotting the peak-area ratios of the analyte to the IS relative to the analyte concentrations against the seven concentrations of the standard mixtures. The LOD and LOQ were calculated based on the signal-to-noise ratios (S/N) of 3 and 10, respectively. Five replicates of the same sample were analyzed within one working day to determine the intra-day precision. The sample was continuously measured for three days, repeatedly, five times a day, to determine the inter-day precision. Recovery studies were carried out by adding 4 mL of a mixed standard solution at three different concentrations (1, 5, and 10 μg/mL) to three young leaves samples, the amine content of which had been predetermined. The standard solution was added to the sample in the first step (extracted with 1M HCl) [40].

3.7. Evaluation of the Matrix Effect

As it is very difficult to find blank plant matrix samples free of BAs, the slope comparison method was used instead to evaluate the matrix effect for this study. The leaf sample extracts, which were spiked with appropriate amounts of mixed standard solutions, as done for the apparent recovery measurement described in Section 3.6, were used to construct standard addition calibration curves. Then, the slopes of the calibration curves from the standard addition experiments were compared with the slopes obtained from the pure aqueous standards at the same concentration levels.

3.8. Statistical Analysis

The calculation of linearity relationship between peak-area ratios of analyte to the IS and concentrations of analyte were adopted by weighted-least-squares linear regression. SPSS 20.0 was used for all statistical tests. Data were presented as the mean ± standard deviation (SD). Single factor analysis of variance (One-way ANOVA) and Tukey's post hoc test were used for comparisons between groups. One-way ANOVA was used to determine group differences. The Tukey post hoc test was used for multiple comparisons among groups. A p-value < 0.05 was considered statistically significant.

4. Conclusions

In this work, a method involving HPLC with precolumn derivatization for determination of 11 BAs in *L. barbarum* was developed. This is the first report showing that different parts of *L. barbarum* contain PUT, HIS, SPD, and SPM. The PUT content was the highest in the flowers; the HIS content was the highest in the bark; the SPD and SPM contents were the highest in the young leaves. HIS was not found in the fruit and its content in the leaves was very low, which makes them safe when used in drugs and food. The high HIS content in the bark may be one of the reasons why all of the parts of *L. barbarum*, except the bark, are used for medicine or food in China. Meanwhile, the issue of high concentration of HIS should be considered when exploiting or utilizing the bark of *L. Barbarum*.

Author Contributions: Manuscript conception, N.Z.; writing, sample preparation, and original draft preparation, Y.A.; HPLC analysis Y.N.S.; editing, data analysis, and interpretation, L.L., F.Y.Y., Y.Z., F.Y.G., W.J.Z., and J.L.L. All authors have read and agreed to the published version of the manuscript.

Funding: This work was supported by the National Natural Science Foundation of China (20872118 and 30070905), the Natural Science Foundation of Shannxi Province, China (2020JM-419), and the Foundation of Shaanxi Administration of traditional Chinese Medicine, China (13-JC012 and 2019-ZZ-JC044).

Institutional Review Board Statement: Not applicable.

Informed Consent Statement: Not applicable.

Data Availability Statement: HPLC dates are available from the authors.

Conflicts of Interest: The authors declare no conflict of interest.

Sample Availability: Samples of *L. barbarum* are available from the authors.

Abbreviation

BAs: biogenic amines; MET, methylamine; PEA, 2-phenylethylamine; TYR, tyramine; HIS, histamine; PUT, putrescine; CAD, cadaverine dihydrochloride; TRY, tryptamine; PEA, phenylethylamine; 5-HT, serotonin; DA, dopamine; SPM, spermine; SPD, spermidine; IS,

Internal standard; DMP, 1,7-diaminoheptane; ACN, acetonitrile; TLC, thin layer chromatography; GC, gas chromatography; IC, ion chromatography; OPA, O-phthalaldehyde; Dns-Cl, dansyl chloride; LOD, limit of detection; LOQ, limit of quantification; S/N, signal-to-noise ratio; TCM, Traditional Chinese Medicine.

References

1. Silla Santos, M.H. Biogenic amines: Their importance in foods. *Int. J Food Microbiol.* **1996**, *29*, 213–231. [CrossRef]
2. Halász, A.; Baráth, Á.; Simon-Sarkadi, L.; Holzapfel, W. Biogenic amines and their production by microorganisms in food. *Trends Food Sci. Technol.* **1994**, *5*, 42–49. [CrossRef]
3. Önal, A. A review: Current analytical methods for the determination of biogenic amines in foods. *Food Chem.* **2007**, *103*, 1475–1486. [CrossRef]
4. Lovaas, E. Antioxidative effects of polyamines. *J. Am. Oil Chem.' Soc.* **1991**, *68*, 353–358. [CrossRef]
5. Eisenberg, T.; Knauer, H.; Schauer, A.; Büttner, S.; Ruckenstuhl, C.; Carmona-Gutierrez, D.; Ring, J.; Schroeder, S.; Magnes, C.; Antonacci, L. Induction of autophagy by spermidine promotes longevity. *Nat. Cell Biol.* **2009**, *11*, 1305–1314. [CrossRef]
6. He, E.P.; Tang, L.L.; Guo, Y.J. Effects of spermidine on free radical metabolism and anti-fatigue effects of skeletal muscle in mice. *Food Sci.* **2014**, *35*, 229–232.
7. Wei, C.; Wang, Y.; Li, M.; Li, H.; Lu, X.; Shao, H.; Xu, C. Spermine inhibits endoplasmic reticuum stress-induced apoptosis: A new strategy to prevent cardiomyo-cyte apoptosis. *Cell. Physiol. Biochem.* **2016**, *38*, 531–544. [CrossRef] [PubMed]
8. Lange, J.; Thomas, K.; Wittmann, C. Comparison of a capillary electrophoresis method with high-performance liquid chromatography for the determination of biogenic amines in various food samples. *J. Chromatogr. B. Analyt. Technol. Biomed. Life Sci.* **2002**, *779*, 229–239. [CrossRef]
9. Becker, K.; Southwick, K.; Reardon, J.; Berg, R.; MacCormack, J.N. Histamine poisoning associated with eating tuna burgers. *JAMA* **2001**, *285*, 1327–1330. [CrossRef] [PubMed]
10. Joosten, H.M.L.J. The biogenic amine contents of Dutch cheese and their toxicological significance. *Neth. Milk Dairy J.* **1988**, *42*, 25–42.
11. Tang, T.; Qian, K.; Shi, T.; Wang, F.; Li, J.; Cao, Y.; Hu, Q. Monitoring the contents of biogenic amines in sufu by HPLC with SPE and pre-column derivatization. *Food Control.* **2011**, *22*, 1203–1208. [CrossRef]
12. Saaid, M.; Saad, B.; Ali, A.S.M.; Saleh, M.I.; Basheer, C.; Lee, H.K. In situ derivatization hollow fibre liquid-phase microextraction for the determination of biogenic amines in food samples. *J. Chromatogr. A* **2009**, *1216*, 5165–5170. [CrossRef] [PubMed]
13. Saaid, M.; Saad, B.; Hashim, N.H.; Ali, A.S.M.; Saleh, M.I. Determination of biogenic amines in selected Malaysian food. *Food Chem.* **2009**, *113*, 1356–1362. [CrossRef]
14. Shalaby, A.R. Significance of biogenic amines to food safety and human health. *Food Res. Int.* **1996**, *29*, 675–690. [CrossRef]
15. Valsamaki, K.; Michaelidou, A.; Polychroniadou, A. Biogenic amine production in Feta cheese. *Food Chem.* **2000**, *71*, 259–266. [CrossRef]
16. People's Medical Publishing House. *Lishizhen Compendium of materia medica: On the university handbook, next volume*; People's Medical Publishing House: Beijing, China, 1982; pp. 2111–2117.
17. Zhao, J.; Jin, Y.; Yan, Y.; Qin, K.; Qian, D.; Zhang, W.; Peng, H.; Li, B.; Cai, Q.; Chen, Z.; et al. Herbal Textual Research on "Lycii Fructus" and "Lycii Cortex" in Chinese Classical Prescriptions. *Modern Chin. Med.* **2020**, *22*, 1269–1286.
18. Lu, A.M.; Wang, M.L. On the identification of the original plants in the modernization of Chinese herbal medicine - an example from the taxonomy and exploition of 'gouqi'. *Acta Bot. Boreal.-Occid. Sin.* **2003**, *23*, 1077–1083, (in Chinese English abstract).
19. Zhao, J.; Ge, L.Y.; Xiong, W.; Leong, F.; Huang, L.Q.; Li, S.P. Advanced development in phytochemicals analysis of medicine and food dual purposes plants used in China (2011–2014). *J. Chromatogr. A.* **2016**, *1428*, 39–54. [CrossRef] [PubMed]
20. Gao, K.; Ma, D.; Cheng, Y.; Tian, X.; Lu, Y.; Du, X.; Tang, H.; Chen, J. Three new dimers and two monomers of phenolic amides from the fruits of *Lycium barbarum* and their antioxidant activities. *Agric. Food Chem.* **2015**, *63*, 1067–1075. [CrossRef]
21. Amagase, H.; Farnsworth, N.R. A reviews of botanical characteristics phytochemistry, clinical relevance in efficacy and safety of *Lycium barbarum* fruit (Goji). *Food Res. Int.* **2011**, *44*, 1702–1717. [CrossRef]
22. Kulczyński, B.; Gramza-Michałowska, A. Goji Berry (*Lycium barbarum*): Composition and Health Effects-a Review. *Nephron Clin. Pract.* **2016**, *66*, 67–76.
23. Zhao, X. Study on Resource Chemistry of *Lycium Barbarum* Leaves and Flowers. Master Dissertation, Nanjing University of Chinese Medicine, Nanjing, China, 2020.
24. Chen, Y.; Tan, F.; Peng, Y. Research Progress in Leaves of *Lycium chinese* Mill and *Lycium barbarum* L. *Chin. Pharm. J.* **2017**, *52*, 358–361.
25. Guo, M.; Zhao, J.; Shi, W.; Xu, F.; Cong, W. Advances of Gouqi (Lycium) in Research of Anti-Aging. *Guid. J. Tradit. Chin. Med. Pharm.* **2019**, *25*, 124–128.
26. Yan, X.M.; Dong, J.Z.; Wang, Y. Comparision Studies of Main Active Compounds in young leaves of *L. barbarum*. *Ningxia Food Sci.* **2010**, *31*, 29–32.
27. Feng, W. Research Status and Trends of Cactus and *Lycium barbarum* Leaves Applied to Hypoglycemic Health Drink. *Food Sci. Technol. Res.* **2018**, *1*, 136.
28. Gu, J.Q. Preparation of Compound Healthy Beverage from Garlic and Medlar. *Food Sci. Technol.* **2002**, *10*, 46.

29. Gan, X.; Zhang, Y.X. Development of fermented Tea Fermented Beverage. *CCOC—China Chamber of Commerce Industry* **1998**, *10*, 12–13.
30. Li, H.Y.; Wu, D. Comparison of Trace Element Contents in Tea Leaves of Ningxia with Other Teas. *Stud. Trace Elem. Health* **2008**, *25*, 37–38.
31. Ma, B.L.; Qu, X.B. Determination of spermine and spermidine in wolfberry leaves by HPLC. *Ningxia Eng. Technol.* **2011**, *10*, 48–50.
32. Draisci, R.; Volpe, P.G.; Lucentini, O.L.; Cecilia, A.; Federico, R.; Palleschi, G. Determination of biogenic amines with an electrochemical biosensor and its application to salted anchovies. *Food Chem.* **1998**, *62*, 225–232. [CrossRef]
33. Shalaby, A.R. Multidetection semiquantitative method for determining biogenic amines in foods. *Food Chem.* **1995**, *52*, 367–372. [CrossRef]
34. Antoine, F.R.; Wei, C.I.; Otwell, W.S.; Sims, C.A.; Littell, R.C.; Hogle, A.D.; Marshall, M.R. Gas chromatographic analysis of histamine in mahi-mahi (*Coryohaena hippurus*). *J. Agric. Food Chem.* **2002**, *50*, 4754–4759. [CrossRef]
35. Draisci, R.; Giannetti, L.; Boria, P.; Lucentini, L.; Palleschi, L.; Cavalli, S. Improved ion chromatography-integrated pulsed amperometric detection method for the evaluation of biogenic amines in food of vegetable or animal origin and in fermented foods. *J. Chromatogr. A* **1998**, *798*, 109–116. [CrossRef]
36. Vinci, G.; Antonelli, M.L. Biogenic amine: Quality index of freshness in red and white meat. *Food Control.* **2002**, *13*, 519–524. [CrossRef]
37. De Mey, E.; Drabik-Markiewicz, G.; De Maere, H.; Peeters, M.C.; Derdelinckx, G.; Paelinck, H.; Kowalska, T. Dabsyl derivatisation as an alternative for dansylation in the detection of biogenic amines in fermented meat products by reversed phase high performance liquid chromatography. *Food Chem.* **2012**, *130*, 1017–1023. [CrossRef]
38. Buiatti, S.; Boschelle, O.; Mozzon, M.; Battistutta, F. Determination of biogenic amines in alcoholic and non-alcoholic beers by HPLC. *Food Chem.* **1995**, *52*, 199–202. [CrossRef]
39. Anlı, R.E.; Vural, N.; Yılmaz, S.; Vural, Y.H. The determination of biogenic amines in Turkish red wines. *J. Food Compos. Anal.* **2004**, *17*, 53–62.
40. Zhang, N.; Wang, Y.Q.; Wang, X.N.; Fan, M.Q.; Sun, Y.N.; Liu, Z.L.; Liu, J.L. Identification and determination of biogenic amines in nine kinds of Chinese Tonic-Qi herbs by RP-HPLC with pre-column derivatization. *Chin. J. Pharm. Anal.* **2017**, *37*, 1791–1798.
41. Fan, M.; Ai, Y.; Zhao, W.; Sun, Y.; Liu, J.; Zhang, N. Characterization of 10 biogenic amines in Male Silkworm Moth by HPLC with precolumn derivatization. *Curr. Pharm. Anal.* **2019**, *16*, 608–614. [CrossRef]
42. Wang, X.; Liang, Y.; Wang, Y.; Fan, M.; Sun, Y.; Liu, J.; Zhang, N. Simultaneous determination of 10 kinds of biogenic amines in rat plasma using high-performance liquid chromatography coupled with fluorescence detection. *Biomed. Chromatogr.* **2018**, *32*, e4211. [CrossRef] [PubMed]
43. Flores, M.I.A.; Moreno, J.L.F.; Frenich, A.G.; Vidal, J.L.M. Fast determination of myo-inositol in milk powder by ultra high performance liquid chromatography coupled to tandem mass spectrometry. *Food Chem.* **2011**, *129*, 1281–1286. [CrossRef] [PubMed]
44. Jia, S.; Kang, Y.P.; Park, J.H.; Lee, J.; Kwon, S.W. Simultaneous determination of 23 amino acids and 7 biogenic amines in fermented food samples by liquid chromatography/quadrupole time-of-flight mass spectrometry. *J. Chromatogr. A* **2011**, *1218*, 9174–9182. [CrossRef] [PubMed]
45. Liu, Z.; Chai, B.; Liu, J.; Jiang, P.; Zhao, D. Current Situation and Trend of Drying Way of Wolfberry. *Agric. Eng.* **2020**, *10*, 46–49.

Article

Experimental and Computational Evaluation of Chloranilic Acid as an Universal Chromogenic Reagent for the Development of a Novel 96-Microwell Spectrophotometric Assay for Tyrosine Kinase Inhibitors

Ibrahim A. Darwish [1,*], Hany W. Darwish [1,2], Nasr Y. Khalil [1] and Ahmed Y. A. Sayed [1]

[1] Department of Pharmaceutical Chemistry, College of Pharmacy, King Saud University, P.O. Box 2457, Riyadh 11451, Saudi Arabia; hdarwish@ksu.edu.sa (H.W.D.); nkhalil@ksu.edu.sa (N.Y.K.); ahmedyahia@ksu.edu.sa (A.Y.A.S.)
[2] Department of Analytical Chemistry, Faculty of Pharmacy, Cairo University, Kasr El-Aini St., Cairo 11562, Egypt
* Correspondence: idarwish@ksu.edu.sa

Abstract: The tyrosine kinase inhibitors (TKIs) are chemotherapeutic drugs used for the targeted therapy of various types of cancer. This work discusses the experimental and computational evaluation of chloranilic acid (CLA) as a universal chromogenic reagent for developing a novel 96-microwell spectrophotometric assay (MW-SPA) for TKIs. The reaction resulted in an instantaneous formation of intensely purple colored products with TKIs. Spectrophotometric results confirmed that the reactions proceeded via the formation of charge-transfer complexes (CTCs). The physical parameters were determined for the CTCs of all TKIs. Computational calculations and molecular modelling for the CTCs were conducted, and the site(s) of interaction on each TKI molecule were determined. Under the optimized conditions, Beer's law correlating the absorbances of the CTCs with the concentrations of TKIs were obeyed in the range of 10–500 µg/well with good correlation coefficients (0.9993–0.9998). The proposed MW-SPA fully validated and successfully applied for the determination of all TKIs in their bulk forms and pharmaceutical formulations (tablets). The proposed MW-SPA is the first assay that can analyze all the TKIs on a single assay system without modifications in the detection wavelength. The advantages of the proposed MW-SPA are simple, economic and, more importantly, have high throughput.

Keywords: tyrosine kinase inhibitors; chloranilic acid; charge-transfer reaction; 96-microwell spectrophotometric assay; high-throughput pharmaceutical analysis

Citation: Darwish, I.A.; Darwish, H.W.; Khalil, N.Y.; Sayed, A.Y.A. Experimental and Computational Evaluation of Chloranilic Acid as an Universal Chromogenic Reagent for the Development of a Novel 96-Microwell Spectrophotometric Assay for Tyrosine Kinase Inhibitors. *Molecules* **2021**, *26*, 744. https://doi.org/10.3390/molecules26030744

Academic Editor: Paraskevas D. Tzanavaras
Received: 9 January 2021
Accepted: 28 January 2021
Published: 31 January 2021

Publisher's Note: MDPI stays neutral with regard to jurisdictional claims in published maps and institutional affiliations.

Copyright: © 2021 by the authors. Licensee MDPI, Basel, Switzerland. This article is an open access article distributed under the terms and conditions of the Creative Commons Attribution (CC BY) license (https://creativecommons.org/licenses/by/4.0/).

1. Introduction

Cancer is the world's second major cause of death among males and females. This reflects ~9.6 million deaths in 2018 (~13% of all deaths). Cancer deaths worldwide are expected to increase significantly, with ~13.1 million deaths in 2030 [1]. According to many reported studies, cancer is a growing problem [2–4] and therefore it continues to spread worldwide, with a tremendous burden on individuals, families, communities and healthcare systems, both mentally and socially [5]. The majority of health care systems in developed countries are little prepared to deal with this challenge, and many cancer patients globally will not have the privilege of early treatment. In these countries, nearly two-thirds of cancer patients die [6]. There is evidence that shows that in countries where national health services are well-developed, strategies such as early detection, treatment and survivorship care are improving the survival rates of many types of cancers [1,6].

There are a number of treatment options for cancer such as surgery, radiation therapy and/or chemotherapy. For treating localized cancers, radiation therapy and surgery are preferred, while chemotherapy is considered safe for systemic cancers. Although chemotherapeutic agents exert their cytotoxic effect by way of interfering with the synthesis or the function of proteins and other needed cellular biomolecules, they also attack RNA, DNA and other vital proteins. Consequently, one must be careful when administering chemotherapy, as it has major side/toxic effects since it is capable of killing cells. A perfect recorded chemotherapy would be highly selective and specific to sick or cancerous tissues, while leaving healthy ones untouched. Unsurprisingly, the reality always differs from expectations, as most chemotherapies are heavily toxic, particularly for short-lived cells [7].

Recent developments in life science research in recent years have led to increased understanding of signaling transductions in tumor cells, apoptosis triggering, cellular interactions and other critical processes [8]. In addition, these chemotherapeutic drugs may be highly selective for DNA and other cellular targets that present in cancer cells as well as normal cells. The use of major intracellular signal transduction enzymes relevant to the differentiation and spread of cancer cells as drug testing objectives has become a hot area for medical research and development of antitumor drugs, and the continued improvement of effective, safe and specific new targeted anticancer drugs [9].

Tyrosine kinase (TK) is an enzyme that is responsible for transferring phosphate group from adenosine triphosphate (ATP) to the tyrosine residues of certain proteins inside the living cell [10]. An abnormality in TK could lead to a series of body diseases. Earlier investigations have demonstrated that there are TK activities for more than half of mutated proto-oncogenes and oncogenes, and the irregular expression of these genes can contribute to abnormal cell proliferation and eventually cancer [11]. Additionally, in conjunction with tumorigenesis, neovascularization and chemotherapy resistance, an abnormal expression of TK has been observed [12]. For that reason, targeting TK has attracted even more interest for the pharmaceutical industry for developing new chemotherapeutic drugs. Researchers and pharmaceutical organizations have been given high priority to developing tyrosine kinase inhibitors (TKIs) that may affect unique molecular pathways [13]. In 2001, the Food and Drug Administration (FDA) quickly approved the first targeted TKI drug imatinib and stimulated new energetic thought to cancer treatment [14–16]. To date, the FDA has approved over 20 TKI drugs [17]. These drugs are characterized by high effectiveness and low drawbacks and dominate the management of various types of cancer [10–17] compared to traditional cytotoxic antinineoplastics, some of which have become the foremost cancer treatment. The safety and efficacy of TKI medications is principally based on their corresponding pharmaceutical formulation quality, including drug content and uniformity.

For the quality control (QC) of TKIs, it is absolutely necessary to use proper analytical techniques. The analytical methods mentioned in the literature for QC of TKIs in their dosage forms are HPLC and HPTLC [18–27], voltammetry [28], spectrofluorometry [29–32] and spectrophotometry [33–44]. The most accessible methodology is spectrophotometry, which is quite easy, the least expensive and available in most quality control laboratories. Nevertheless, almost all of these spectrophotometric assays for TKIs are based on measurements of native UV absorption, which are not selective [33–39]. Few visible-spectrophotometric methodologies with varying chromogenic reagents have been developed for assaying TKIs [40–44]. Even worse, these methods incorporate laborious extraction steps and utilize large volumes of toxic organic solvents [45–48]. Furthermore, due to the differences in chemical structures of analyzed TKIs, these assays were individually developed. Besides, these assays have limitations regarding throughput, as they use traditional spectrophotometry.

From the presented information, establishing a universal spectrophotometric method for assaying any TKI irrespective of its chemical structure would be of great benefit and convenience. This research is directed to establish a novel 96-microwell-based spectrophotometric assay (MW-SPA) that could be applied in QC laboratories for reliable and accurate determination of any TKI. The proposed approach is based on formation of charge-transfer complexes (CTCs) between TKIs and chloranilic acid (CLA). The proposed procedure was established and validated for five TKIs. These TKIs were seliciclib (SEL), vandetanib (VAN), tozasertib (TOZ), dasatinib (DAS) and olaparib (OLA); their chemical structures are given in Figure 1. Their chemical names, molecular formulae and molecular weights are given in Table 1.

$$D + A \rightleftharpoons (D - A) \underset{\text{polar solvent}}{\rightleftharpoons} D^{·+} + A^{·-}$$
$$\text{complex} \quad \text{radical ions}$$

Chloranilic acid (CLA)   Seliciclib (SEL)   Vandetanib (VAN)

Tozasertib (TOZ)

Dasatinib (DAS)   Olaparib (OLA)

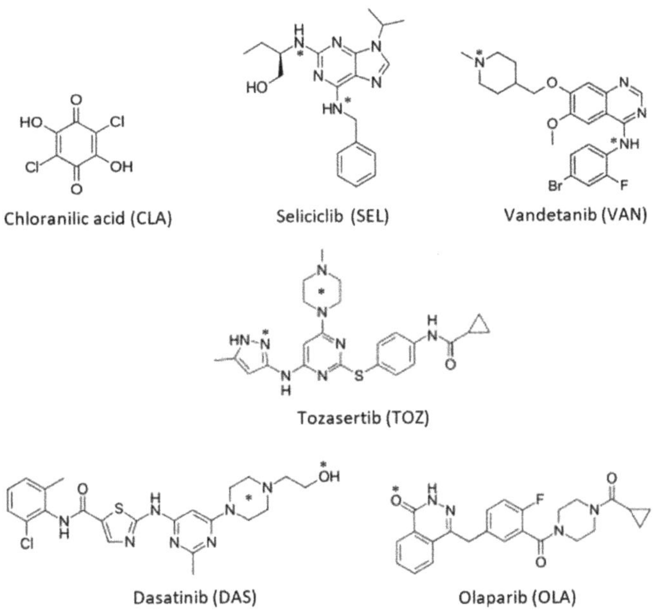

Figure 1. The chemical structures of chloranilic acid (CLA) and the investigated tyrosine kinase inhibitors (TKIs) with their abbreviations. Asterisks symbol (*) denotes the electron-donating sites of interactions of TKIs with CLA and the charge-transfer complexes (CTCs) formed.

Table 1. The investigated TKIs with names, abbreviations, IUPAC names, molecular formulae and molecular weights.

TKI Name	Abbreviation	IUPAC Name	Molecular formula	Molecular Weight
Seliciclib	SEL	(2R)-2-{[6-(benzylamino)-9-(propan-2-yl)-9H-purin-2-yl]amino}butan-1-ol	$C_{19}H_{26}N_6O$	354.46
Vandetanib	VAN	N-(4-bromo-2-fluorophenyl)-6-methoxy-7-[(1-methylpiperidin-4-yl)methoxy]quinazolin-4-amine	$C_{22}H_{24}BrFN_4O_2$	475.40
Tozasertib	TOZ	N-[4-[4-(4-methylpiperazin-1-yl)-6-[(5-methyl-1H-pyrazol-3-yl) amino] pyrimidin-2-yl] sulfanylphenyl]cyclopropanecarboxamide	$C_{23}H_{28}N_8OS$	464.59
Dasatinib	DAS	N-(2-chloro-6-methylphenyl)-2-[[6-[4-(2-hydroxyethyl) piperazin-1-yl]-2-methylpyrimidin-4-yl]amino]-1,3-thiazole-5-carboxamide	$C_{22}H_{26}ClN_7O_2S$	488.01
Olaparib	OLA	4-[[3-[4-(cyclopropanecarbonyl)piperazine-1-carbonyl]-4-fluorophenyl] methyl]-2H-phthalazin-1-one	$C_{24}H_{23}FN_4O_3$	434.47

2. Results and Discussion

2.1. Strategy and Design of the Study

Our selection for TKIs as target analytes is due to their therapeutic relevance and the need to develop a globalized analytical methodology for their quantification in their dosage forms regardless of their differences in chemical structure. By virtue of its simplicity and spreadability in almost all analytical laboratories, spectrophotometry was adopted for assaying TKIs in the current work. The CT reaction of the investigated TKIs were examined in this section on the basis of their expected high electron-donating capability due to the presence of multiple potentially electron-donating sites on the chemical structures of all TKIs (Figure 1); these sites may form CTCs with electron-acceptors. Previous research, which include CT reactions with several polyhalo-/polycyanoquinone electron acceptors, have shown that CLA is by far the most reactive acceptor because its reactions are instant at room temperature [49,50]. This is the reason why CLA was selected from other acceptors for the current work. Since traditional CT-based spectrophotometric methods have a restricted throughput and utilize large volumes of organic solvents that are costly and, more crucially, cause toxicity to analysts [45–48], this study was dedicated to developing a spectrophotometric assay for TKIs that is free from these demerits. Accomplishing this goal was achieved by performing a CT reaction between TKIs and CLA in 96-microwell assay plates and recording the color intensity by a microplate absorbance reader. This technique employs low volumes of organic solvents and offers a high-performance analysis that serves the needs of QC laboratories, as it allows analysts to rapidly perform huge numbers of samples and to obtain large datasets that would otherwise exhaust assets in terms of costs, effort and time [50,51].

2.2. UV–Visible Absorption Spectra and Band Gap Energy

UV–visible absorption spectra of various solutions of TKIs were measured in the range of 200–400 nm. The spectra exhibited various shapes, maxima and molar absorptivities. This is due to variability in their chemical structures (Figure 1). Nevertheless, none of the studied TKIs exhibited any reading above 360 nm (Figure 2). When TKI solutions were mixed individually with a yellow–orange color solution of CLA (its (λ_{max} of 444 nm), the solutions changed to purple color and the corresponding absorption spectra shifted toward larger wavelengths of both CLA and TKIs (Figure 2). Table 2 depicts all (λ_{max}) and the molar absorptivities (ε) for all TKIs. It was confirmed that the new absorption bands of TKI–CLA product were generated as a result of the reaction between CLA and TKIs, and the absorption intensities depended on the concentrations of TKIs. The TKI–CLA absorption spectra produced was shown to have the same form and pattern as the CLA radical anion formed as a result of the reduction procedure, as published previously [49,52]. The reaction was therefore suggested to be a CT interaction between TKI (electron donor (D)) and CLA (electron acceptor (A)) and the reaction initiated in methanol (polar solvent) to form the CTC (D–A). This complex was then disassociated by the ionizing power of methanol and formed the CLA radical anion, as shown below.

Table 2. Spectrophotometric parameters of the CT complex reaction of CLA with TKIs.

TKIs	λ_{max} (nm)	$\varepsilon_{max} \times 10^3$ (L mol^{-1} cm^{-1})	Band Gap Energy Value (eV)	Association Constant (L mol^{-1} $\times 10^8$)	DG^0 (J mol^{-1} $\times 10^4$)	Molar Ratio [TKI:CLA]
SEL	497	0.83	1.90	2.00	−4.74	1:2
VAN	517	1.54	1.90	2.18	−4.76	1:2
TOZ	512	1.47	1.90	1.13	−4.60	1:2
DAS	475	1.09	1.90	2.30	−4.77	1:2
OLA	494	0.73	2.14	1.97	−4.73	1:1

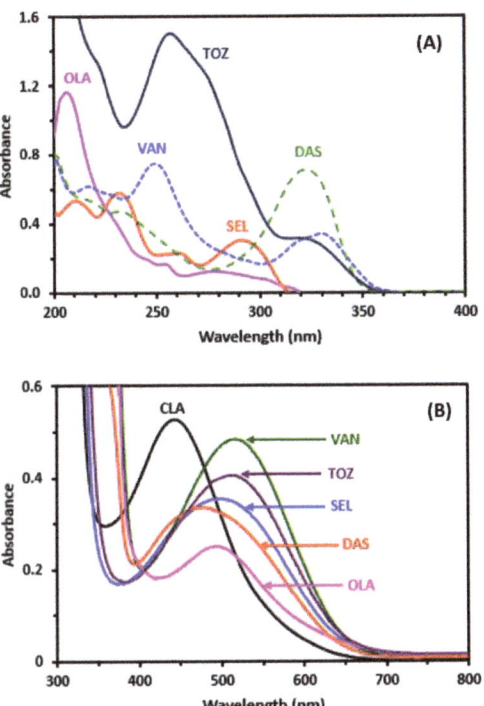

Figure 2. Panel (A): Absorption spectra of TKIs (10 μg/mL). Panel (B): absorption spectra of CLA (0.1%, *w/v*), and its reaction mixtures with TKIs (100 μg/mL). All solutions were in methanol.

It is well known that CLA has three different ionic forms depending on the pH value. The first form that occurs at low pH is the neutral yellow–orange form (H_2A), while the second form is the purple form (HA^-), which is stable at pH = 3, and lastly the third form is the pale violet form (A^{2-}), which is stable at high pH [53]. It was observed that the resulting products of all the investigated TKIs with CLA were purple; accordingly, the form HA^- was the concluded form of CLA that is involved in the current reaction. Additionally, there was other evidence for the suggested reaction, namely the disappearance of the formed purple color upon addition of mineral acids to the reaction mixtures. All of these findings support the development of the CTC between CAL and TKIs. The band gap energy (Eg) is the smallest required energy for excitation of an electron from the lower energy valence band into the higher energy band to participate in formation of a conduction band [54]. In order to compute Eg values, a Tauc graph was created from the absorption spectra of the TKI–CLA complexes by drawing energy values (hυ, in eV) against $(\alpha h\upsilon)^2$ (Figure 3). Eg values were attained by extrapolation of the linear segments of the plots to $(\alpha h\upsilon)^2 = 0$ [55]. The results showed that Eg values ranged from 1.90 to 1.92 eV for all the investigated TKIs (Table 2). These results illustrate the easiness of electron transfer from TKI to CLA and the formation of CTC absorption bands.

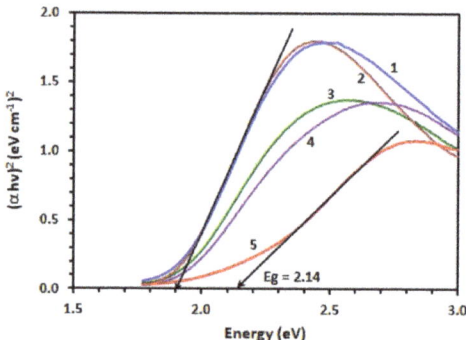

Figure 3. Tauc plots of energy (h) against (αh)2 for CT complex of CLA with TOZ (1), VAN (2), SEL (3), DAS (4) and OLA (5).

2.3. Optimizing Conditions for CT Reaction of CLA with TKIs

In order to pick the best solvent for color development optimum reaction conditions, various solvents with different dielectric constants [56] and polarity indexes [57] were tried and the absorption spectra reported. These solvents were acetonitrile, methanol, ethanol, acetone, propanol, butanol, dichloroethane, dichloromethane, chloroform, diethyl ether, benzene and dioxan. Small shifts were noticed in the λ_{max} values, as well as changes in molar absorptivity (ε) values. As anticipated, the interactions in more polar solvents that possess large dielectric constant values, such as methanol and acetonitrile, provided ε values when compared with less polar solvents, such as diethyl ether and chloroform. This was attributed to the complete transfer of electrons from the TKI molecule to CLA in polar solvents; hence, methanol was chosen throughout this work.

2.4. Association Constants and Free Energy Change for the CTC of TKI–CLA

The Benesi–Hildebrand method [58] was applied for calculation of the association constants (K_c) at room temperature (25 ± 2 °C) and at the λ_{max} of the formed TKI–CLA complexes. As shown in Figure 4 as a representative example, straight lines were obtained from which the association constants of the CTC were computed. The obtained values of the association constants ranged from 1.13×10^8 to 2.3×10^8 L mol^{-1}, as depicted in Table 2.

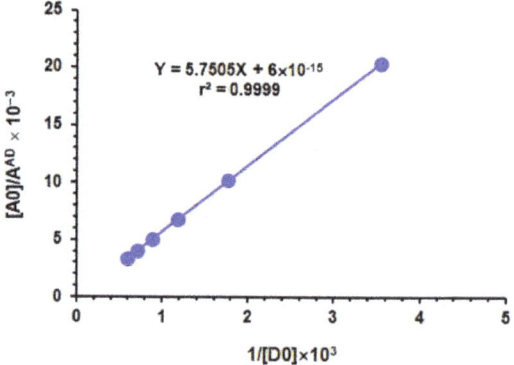

Figure 4. Benesi–Hildebrand plot of the CT complex of CLA with SEL, and the linear fitting equation with correlation coefficient (r). [A_0], A^{AD} and [D_0] are the molar concentration of CLA, absorbance of the complex reaction mixture and molar concentration of SEL, respectively.

The standard free energy change (ΔG^0) of the CTC were calculated using the following formula:

$$\Delta G^0 = -2.303\, RT \log K_c$$

where ΔG^0 is the standard free energy change of the complex (KJ mol^{-1}), R is the gas constant (8.314 KJ mol^{-1}), T is the absolute temperature in Kelvin (°C + 273) and K_c is the association constant of the complex (L mol^{-1}). ΔG^0 values were found to be comparable to all TKIs (~ 4.72 × 10^4 J mol^{-1}). These values proposed the easiness of the TKI interaction with CLA, as well as the stability of the formed CTCs [59].

2.5. Molar Ratio of the Reaction, Molecular Modelling of CTCs and Determination of the Sites of Interaction

The published spectrophotometric titration methodology [52] was adopted to find out the molar ratio of TKI to CLA, and it was found to be 1:1 for OLA and 1:2 for SEL, VAN, TOZ and DAS (Figure 5). The molecular modelling and energy minimization of the TKI molecules and the CTC were performed, and the electron density of each atom was computed for allocation of these sites from the multiple electron-donating sites that exist on TKI molecules (Figure 1).

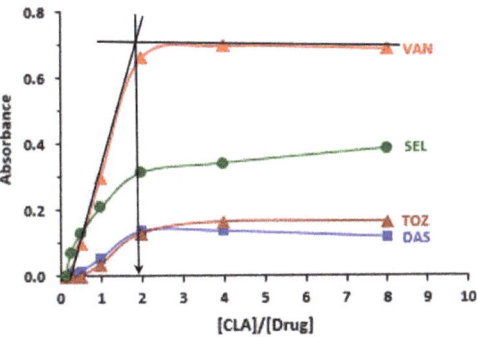

Figure 5. Plot of absorbance versus molar ratio of [CLA]/[Drug] obtained from reaction mixtures containing a fixed concentration of drug and varying concentrations of CLA. The mole ratio corresponds to the point of intersection of the tangents of straight-line portions of the plots (as shown for VAN). Measurements were carried out at 490 nm.

The molecular modelling was done with CS Chem3D Ultra, version 16.0, and execution took place by molecular orbital computations software (MOPAC) and molecular dynamics computations software (MM2 and MMFF94). The most likely sites for the interaction between CLA and TKI are found on the TKI, which have the highest electron density (Table 3). For verification of these sites' participation, energy minimization was performed for one molecule of TKI with the number of CLA molecules obtained from the molar ratios. The CLA molecule was observed adjacent to the suggested sites with the highest electron density (Figure 6). Exceptionally in the case of OLA, the CLA molecule was adjacent to the carbonyl oxygen atom (O11), although it had lower electron density than those on the amide nitrogen atoms (N22 and N25). This information confirmed that CLA–TKI interactions happen through $n \rightarrow \pi^*$ interactions. From the results of the molar ratio and computational molecular modelling, it was clear that these are the electron-donating sites on TKI molecules that are involved in generation of the produced CTCs with CLA.

Table 3. The molar ratios of the reaction of TKIs with CLA, types of atoms proposed as site(s) of interaction on TKIs molecules and charges on these atoms.

TKI	TKI:CLA Molar Ratio	Atom Type(s) Proposed as Site(s) of Interaction [a]	Charge [b]
SEL	1:2	(N10): Enamine or aniline nitrogen, delocalized lone pair of electrons	−0.8691
		(N21): Enamine or aniline nitrogen, delocalized lone pair of electrons	−0.8691
VAN	1:2	(N11): Enamine or aniline nitrogen, delocalized lone pair of electrons	−0.6
		(N23): Amine nitrogen	−0.81
TOZ	1:2	(N2): Aromatic 5-ring nitrogen	−0.7068
		(N14): Enamine or aniline nitrogen, delocalized lone pair of electrons	−0.8382
		(N17): Amine nitrogen	−0.81
DAS	1:2	(N28): Enamine or aniline nitrogen, delocalized lone pair of electrons	−0.8382
		(N29): Amine nitrogen, N-hydroxyethyl	−0.81
		(O30): Alcohol or ether oxygen	−0.68
OLA	1:1	(O11): Carbonyl oxygen in amide	−0.57
		(O21): Carbonyl oxygen in amide	−0.57
		(N22): Amide nitrogen	−0.6602
		(N25): Amide nitrogen	−0.6602

[a] These sites of interactions are denoted on the chemical structures of the TKIs (Figure 1). [b] The negative sign indicates negative electron density.

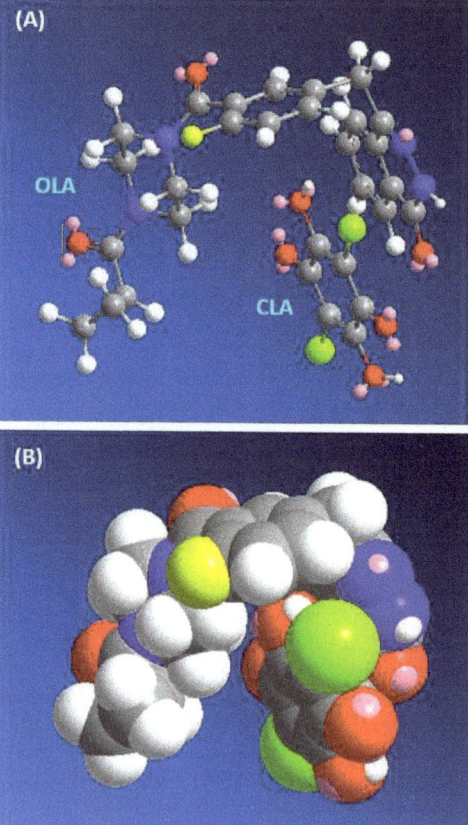

Figure 6. Energy-minimized CTC of CLA with OLA in the conformational (A) and 3D structures (B).

2.6. Optimization of MW-SPA Conditions

In order to have an excellent result in the 96-microwell assay plate, the experimental conditions were adjusted by adopting a "change one factor at a time" approach. Among the different tested solvents (Table 4), methanol was the optimum solvent, and it was used through the whole study, and measurements were recorded at 490 nm (the nearest filter to the λ_{max} of all investigated TKIs complexes of CLA). The observed results of changing CLA concentrations and the time of the reaction at room temperature (25 ± 2 °C) showed that the optimum CLA concentrations ranged from 0.2 to 0.8% (w/v), as shown in Figure 7. Similar experiments were conducted in order to optimize the reaction time, and it was discovered that the reaction was instantaneous; nevertheless, for obtaining the best reading precision, the measurements were carried out after 5 min from the starting point of the reaction. A summary of the condition ranges studied and the optimum value selected for the development of the proposed MW-SPA are included in Table 4.

Table 4. Optimization of experimental conditions for the 96-microwell spectrophotometric assay for TKIs based on their CT reaction with CLA.

Condition	Studied Range	Optimum Value [a]
CLA conc. (%, w/v)	0.01–0.8	0.4
Solvent	Different [b]	Methanol
Reaction time (min)	0–40	5
Temperature (°C)	25–60	25
Measuring wavelength (nm)	400–800	490

[a] Optimum values were used for all TKIs. [b] Solvents used were acetonitrile, methanol, ethanol, acetone, propanol, butanol, dichloroethane, dichloromethane, chloroform, diethyl ether, benzene and dioxan.

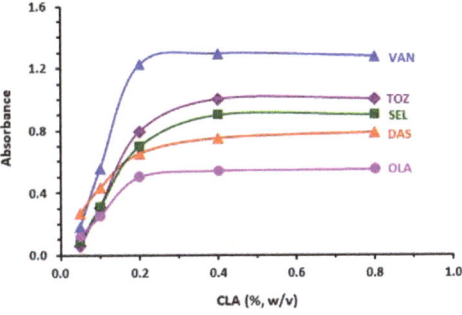

Figure 7. Effect of CLA concentration on its reaction with TKIs (200 µg/well). Measurements were carried out at 490 nm.

2.7. Validation of MW-SPA

2.7.1. Linear Range and Sensitivity

The calibration curves were constructed (Figure 8) according to optimum conditions of the MW-SPA (Table 4), and the least square method was used for linear data regression. Calibration curves in the range of 10–500 µg/well (100 µL) were linear with excellent correlation coefficients. The limits of detection (LOD) and quantitation (LOQ) were determined based on the International Conference on Harmonization (ICH) guidelines [60]. The LOD and LOQ levels lay at 3.78–8.16 and 11.36–24.46 µg/well, respectively. A summary for the calibration and validation parameters of the current MW-SPA is given in Table 5.

Table 5. Calibration parameters for the analysis of TKIs by the 96-microwell spectrophotometric assay based on their CT reaction with CLA.

TKIs	Linear Range [a]	Intercept	SDa [b]	Slope	SDb [b]	r [b]	LOD [a]	LOQ [a]
SEL	10–500	0.0034	0.52×10^{-2}	0.0041	1.5×10^{-3}	0.9995	4.12	12.38
VAN	15–300	0.0028	0.78×10^{-2}	0.0067	1.2×10^{-3}	0.9997	3.84	11.52
TOZ	10–300	0.0081	0.60×10^{-2}	0.0053	0.6×10^{-2}	0.9998	3.78	11.36
DAS	20–500	0.0056	0.68×10^{-2}	0.0033	0.1×10^{-2}	0.9993	6.78	20.32
OLA	25–500	0.0065	0.64×10^{-3}	0.0028	4.3×10^{-3}	0.9996	8.16	24.46

[a] Values are in µg/well. [b] SDa = standard deviation of the intercept, SDb = standard deviation of the slope, r = correlation coefficient.

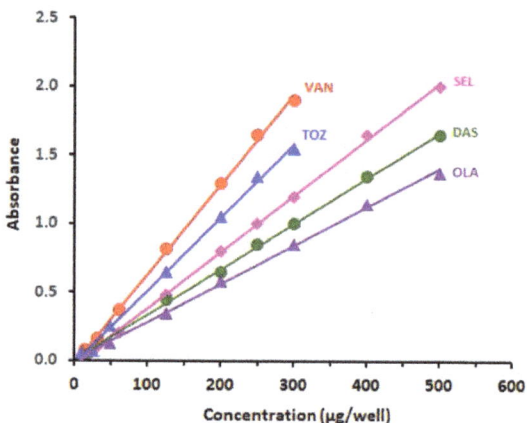

Figure 8. Calibration curves for determination of TKIs by the proposed 96-microwell-based spectrophotometric assay based on their reaction with CLA. Measurements were carried out at 490 nm.

2.7.2. Precision and Accuracy

The accuracy of the proposed MW-SPA was determined utilizing samples of TKI solutions at various concentration levels (Table 6). The values of relative standard deviation (RSD) were 1.24–2.24 and 1.51–2.87% for intra- and inter-assay accuracy, respectively. The high precision of the method was proved by these low RSD values. The accuracy of the proposed method was assessed by the recovery studies. The recovery values ranged from 97.2 to 102.4% (Table 6), reflecting the accuracy of the assay.

Table 6. Precision and accuracy of the proposed 96-microwell spectrophotometric assay for TKIs via their CT reactions with CLA.

TKIs	Relative Standard Deviation (%) [a]		Recovery (% ± SD) [a]
	Intra−Assay, n = 3	Inter−Assay, n = 3	
SEL	2.12	2.25	101.7 ± 2.3
VAN	1.24	1.51	102.3 ± 2.2
TOZ	1.53	2.11	97.6 ± 1.9
DAS	2.24	2.87	102.4 ± 2.3
OLA	2.01	2.54	97.2 ± 2.6

[a] Values are the means of three determinations.

2.7.3. Robustness and Ruggedness

The robustness of the method (effect of small changes in the variables on its performance) was assessed [60]. The results of the test were found to be not significantly affected by small variations in the studied variables; the recovery values ranged between 97.5 and 102.3 ± 1.76 and 2.49%, respectively. This confirmed that the proposed test was convenient for routine TKI analysis.

Additionally, the ruggedness was tested by performing the method by at least two different analysts on three different days [60]. The results were reproducible since RSD values never exceeded 2.8%.

2.7.4. Specificity and Interference

The advantage of the suggested MW-SPA is that measurements in the visible region are performed away from UV-absorbing interfering substances, which may be co-extracted from TKI-containing pharmaceutical formulations. Possible interference of additives in dosage forms was also studied. Mixing known quantity of TKI with different quantities of the familiar excipients was done to produce samples. These excipients included microcrystalline cellulose, magnesium stearate, sodium starch glycolate, colloidal silicon dioxide and anhydrous dibasic calcium phosphate. The results given in Table 7 showed that no interference was noted from any of the mentioned excipients with the suggested methods, as the recovery values ranged from 97.5 to 102.9%. The absence of interference with these excipients was caused by the organic solvent extraction of the TKI target from samples, where the excipients were not dissolved.

Table 7. Analysis of TKIs in the presence of the excipients in solid pharmaceutical tablets by the proposed 96-microwell spectrophotometric assay based on their CT reactions with CLA.

Excipient [b]	Recovery (% ± SD) [a]				
	SEL	VAN	TOZ	DAS	OLA
MCC (50) [c]	100.9 ± 0.8	101.3 ± 0.56	99.7 ± 1.1	99.5 ± 1.2	101.5 ± 1.2
CSD (10)	97.5 ± 1.5	98.6 ± 1.3	100.6 ± 0.9	100.9 ± 0.9	100.1 ± 1.8
ADCP (5)	101.3 ± 0.9	101.3 ± 1.2	98.8 ± 1.4	102.3 ± 1.7	98.8 ± 1.9
SSG (5)	101.2 ± 0.8	102.9 ± 0.8	99.2 ± 1.6	101.6 ± 1.3	99.2 ± 1.6
MS (5)	99.4 ± 1.2	101.8 ± 1.5	97.9 ± 1.9	100.4 ± 1.4	102.1 ± 1.4

[a] Values are means of three determinations. [b] Abbreviations: MCC = microcrystalline cellulose, CSD = colloidal silicone dioxide, ADCP = anhydrous dibasic calcium phosphate, MS = magnesium stearate. [c] Figures in parenthesis are the amounts in mg added per 50 mg of TKI.

2.8. Application of MW-SPA in the Analysis of TKIs in Pharmaceutical Formulations

The successfulness of the validation results demonstrated that the suggested procedure was suitable for routine QC analysis of the investigated TKIs. The MW-SPA was used to determine TKIs in various pharmaceutical formulations, and the results are depicted in Table 8. The acquired mean values of the marked amounts were in the range of 98.6% to 103.1%. The results proved that the proposed MW-SPA was appropriate for assaying the investigated TKIs in their tablets.

Table 8. Determination of TKIs in their pharmaceutical formulations by the proposed MW-SPA based on their CT reaction with CLA.

Taken Conc. (µg/well)	Recovery (% ± SD) [a]				
	Caprelsa Tablets (300 mg VAN)	Sprycel Tablets (70 mg DAS)	Lynparza Tablets (150 mg OLA)	LM TOZ Tablets [b] (100 mg TOZ)	LM SEL Tablets [b] (100 mg SEL)
50	102.4 ± 1.3	100.6 ± 1.3	97.9 ± 2.2	101.4 ± 1.8	99.1 ± 1.7
100	98.9 ± 1.2	101.9 ± 1.6	99.2 ± 1.4	103.1 ± 2.6	101.5 ± 1.4
150	101.8 ± 1.8	100.8 ± 1.1	101.3 ± 1.6	99.3 ± 1.2	98.6 ± 1.9
250	99.7 ± 2.1	99.5 ± 1.8	100.6 ± 1.5	101.3 ± 2.8	100.8 ± 1.6

[a] Values are mean of three determinations. [b] LM means laboratory made.

In the present study, the detailed spectrophotometric investigations and quantitative analysis were given for five TKIs; however, other TKIs (more than 10) were tested in our laboratory for their ability to form CT complexes with CLA, and their results were positive. Their universal ability to form CT complexes was attributed to the fact that all TKIs contain electron-donating atoms in their chemical structure; that is the main requirement for formation of charge transfer complexes. In addition, we confirmed that all TKIs, regardless their chemical structures, did not absorb above 400 nm (the UV cut off wavelength) because they all were not colored. Furthermore, we confirmed that, even if the TKI molecule absorbed at above 400 nm, it could be analyzed by its reaction with CLA as long as its absorption spectrum did not extend to the maximum absorption peak of the complex (475–517 nm).

3. Experimental

3.1. Apparatus

A double beam ultraviolet–visible spectrophotometer with matched 1-cm quartz cells (UV-1800, Shimadzu Co. Ltd., Kyoto, Japan) was operated for the scanning of all the generated UV–visible spectra. An absorbance microplate reader (ELx808: Bio-Tek Instruments Inc., Winooski, VT, USA) powered by KC Junior software provided with the instrument was used.

3.2. Chemicals and Materials

All the investigated TKIs were procured from LC Laboratories (Woburn, MA, USA) and Weihua Pharma Co. Limited (Hangzhou, Zhejiang, China) and utilized as provided. Their purity was >99% (as claimed by the providing companies), and their solutions remained stable for at least 7 days under refrigeration. CLA was bought from BDH Chemicals Co. (Langenfeld, Germany). Transparent 96-microwell plates were procured from Corning/Costar Inc (Cambridge, USA). AdjusTable 8-channel pipettes were obtained from Sigma-Aldrich Chemicals Co (St. Louis, Missouri, USA). BRAND® PP reagent tanks with lids for the pipettes were acquired from Merck KGaA (Darmstadt, Germany). The other reagents and solvents were of analytical grade (Fisher Scientific, California, CA, USA). The pharmaceutical formulations were caprelsa tablets (AstraZeneca, Cambridge, United Kingdom) labelled to contain 300 mg of VAN; sprycel tablets (Bristol Myers Squibb, New York, NY, USA) labelled to contain 50 mg DAS per tablet; and lynparza tablets (AstraZeneca, Cambridge, United Kingdom) labelled to contain 150 mg OLA per tablet. Laboratory-made tablets were prepared in the lab by combining individually accurate amounts (100 mg) of TOZ and SEL with 25 mg of each of starch, lactose monohydrate, microcrystalline cellulose and hydroxypropyl cellulose.

3.3. Preparation of TKIs Standard Solutions

Stock solutions of 5 mg/mL of SEL, VAN and OLA were obtained by dissolving 50 mg of the standard material in 10 mL of methanol, while 0.5 mg/mL of DAS and TOZ solutions were made by dissolving 5 mg of the standard material in 10 mL of methanol. These above-mentioned stock solutions were found to be stable over a period of 14 days when stored in the refrigerator.

3.4. Determination of Association Constants

A set of TKI solutions ranging from 1.79×10^{-4} to 1.85×10^{-3} M were swirled with a constant CLA concentration (4.8×10^{-3} M). The reaction was instant at room temperature (25 ± 2 °C). The absorbances of the colored solutions were recorded at their absorption peaks against exactly prepared reagent blanks. The measured absorbances were utilized to produce the plot of Benesi–Hildebrand [58] by plotting the values $[A_0]/A^{AD}$ versus $1/[D_0]$. Linear regression analysis was shown for the data using the following Benesi–Hildebrand equation [60]:

$$\frac{[A_0]}{A^{AD}} = \frac{1}{\varepsilon^{AD}} + \frac{1}{K_c^{AD} \cdot \varepsilon^{AD}} \times \frac{1}{[D_0]}$$

where $[A_0]$ represents the CLA molar concentration (the acceptor); $[D_0]$ represents TKI molar concentration (the donor); A^{AD} represents the absorbance of the CTC reaction mixture formed; ε^{AD} represents the molar absorptivity of the CTC; and Kc^{AD} represents the formation constant of the complex (L mol^{-1}). The intercept of the linear fitting equation was equivalent to $1/\varepsilon^{AD}$, and the formation constant was calculated from the derived value of ε^{AD} in addition to the slope of the equation.

3.5. Determination of CLA:TKI Molar Ratio

The spectrophotometric titration methods were applied in the current work [61]. Principal solutions of the investigated TKIs (2.5×10^{-3} M) and CLA (2×10^{-2} M) were readily prepared (i.e., molar concentration of CLA was 8 times bigger than that of TKI). Exceptionally, PEL concentration was (1×10^{-3} M) and that of CLA was 8×10^{-3} M. A set of master solutions of both the investigated TKIs and CLA were made to give molar ratios of TKI:CLA of 0.25:8. These solutions were always composed of constant TKI concentrations. The temperature was 25 ± 2 °C, and the corrected absorbances of the products were measured at 490 nm (after subtracting the readings of blanks that were treated similarly but using methanol instead of the sample) A graph was drawn by plotting the corrected measured absorbances versus the TKI:CLA molar ratio. From this graph, the molar ratio of the reaction was computed. The mole ratio corresponds to the point of intersection of the tangents of straight-line portions of the plots.

3.6. Preparation of TKI Tablets Solutions

The total amount of commercialized or synthetic tablets accounting for 50 mg of TKI (5 mg was used in case of DAS and TOZ) was placed in a 10-mL measuring flask, dissolved in a volume of 5 mL methanol, mixed thoroughly and sonicated for 5 min, completed to the mark with methanol, well shaken for 10 min and then filtered. The first portion of the filtrate was thrown away, and the exact volume of the filtrate was diluted with methanol. The final concentrations of TKIs ranged from 50 to 5000 µg mL^{-1}.

3.7. Procedure of MW-SPA

Aliquots (100 microliters) of the standard or tablet sample solutions were composed of varying concentrations of TKI, ranging from 5 to 500 µg, and were placed to 96-well plates in addition to 100 microliters of 0.4% *w/v* of CLA solution. The reaction was carried out at 25 ± 2 °C for 5 min. Absorbances of the product were measured by the plate reader at a selected wavelength (490 nm). The blank wells were treated exactly the same as the other wells, except for the addition of 100 µL of methanol to each of them instead of the TKI solutions. Then, the absorbances of the samples were corrected by subtracting those of the blanks.

4. Conclusions

The present study found the CLA reagent to be a universal chromogenic reagent for TKIs. The experiment proved how the reaction proceeded, through formation of colored CTCs between CLA and TKIs. The reaction was used to develop novel MW-SPA for the five investigated TKIs. The proposed assay outperformed all the established assays for TKIs, since it could be applied for assaying all TKIs irrespective of the differences in their chemical structures. In the presented work, five TKIs were tested; however, the universal applicability of the proposed assay was supported by another study that was carried out in our laboratory [62]. Extended advantages of the suggested MW-SPA are the easiness of the procedure (simplicity), the use of affordable analytical reagents (economic), the requirement of minimal volumes of reagent and solvent (eco sustainable "green" approach) and high throughput. All these advantages make the suggested MW-SPA an effective universal TKI assay for routine QC laboratory use.

Author Contributions: I.A.D.: conceptualization, methodology, validation, formal analysis, resources, writing—review and editing, supervision. H.W.D.: investigation, methodology, formal analysis, writing—review and editing, data curation. N.Y.K.: investigation, visualization, data curation. A.Y.A.S.: investigation, formal analysis. All authors have read and agreed to the published version of the manuscript.

Funding: The Deanship of Scientific Research at King Saud University through the research group No. RGP-225.

Institutional Review Board Statement: Not applicable.

Informed Consent Statement: Not applicable.

Data Availability Statement: Data is contained within the article.

Acknowledgments: The authors would like to extend their appreciation to the Deanship of Scientific Research at King Saud University for its funding of this research through the research group No. RGP-225.

Conflicts of Interest: The authors declare that they have no known competing financial or other conflict of interest.

Sample Availability: Samples of the compounds are not available from the authors.

References

1. World Health Organization. Geneva, Report 12 September 2018. Available online: https://www.who.int/news-room/fact-sheets/detail/cancer (accessed on 24 November 2020).
2. Milaat, W.A. Knowledge of secondary-school female students on breast cancer and breast self-examination in Jeddah, Saudi Arabia. *East. Mediterr. Health J.* **2000**, *6*, 338–343. [PubMed]
3. Ezzat, A.A.; Ibrahim, E.M.; Raja, M.A.; Al-Sobhi, S.; Rostom, A.; Stuart, R.K. Locally advanced breast cancer in Saudi Arabia: High frequency of stage III in a young population. *Med. Oncol.* **1999**, *16*, 95–103. [CrossRef] [PubMed]
4. Hashim, T.J. Adolescents and cancer: A survey of knowledge and attitudes about cancer in eastern province of Saudi Arabia. *J. Fam. Community Med.* **2000**, *7*, 29–35.
5. Prager, G.W.; Braga, S.; Bystricky, B.; Qvortrup, C.; Criscitiello, C.; Esin, E.; Sonke, G.S.; Martínez, G.A.; Frenel, J.-S.; Karamouzis, M.; et al. Global cancer control: Responding to the growing burden, rising costs and inequalities in access. *ESMO Open* **2018**, *3*, e000285. [CrossRef]
6. Nwagbara, U.I.; Ginindza, T.G.; Hlongwana, K.W. Health systems influence on the pathways of care for lung cancer in low- and middle-income countries: A scoping review. *Global. Health* **2020**, *16*, 23. [CrossRef]
7. Bertino, J.R.; Hait, W. Principles of cancer therapy. In *Textbook of Medicine*, 22nd ed.; Golden, L., Ausiello, D., Eds.; WB Saunders: Philadelphia, PA, USA, 2004; pp. 1137–1150.
8. Whittaker, S.; Marais, R.; Zhu, A.X. The role of signaling pathways in the development and treatment of hepatocellular carcinoma. *Oncogene* **2010**, *29*, 4989–5005. [CrossRef]
9. Agarwal, E.; Brattain, M.G.; Chowdhury, S. Cell survival and metastasis regulation by Akt signaling in colorectal cancer. *Cell Signal.* **2013**, *25*, 1711–1719. [CrossRef]
10. Wang, Z.; Cole, P.A. Catalytic mechanisms and regulation of protein kinases. *Methods Enzymol.* **2014**, *548*, 1–21.
11. Drake, J.M.; Lee, J.K.; Witte, O.N. Clinical targeting of mutated and wild-type protein tyrosine kinases in cancer. *Mol. Cell Biol.* **2014**, *34*, 1722–1732. [CrossRef]
12. Knosel, T.; Kampmann, E.; Kirchner, T.; Altendorf-Hofmann, A. tyrosine kinases in soft tissue tumors. *Pathologe* **2014**, *35* (Suppl. S2), 198–201.
13. Winkler, G.C.; Barle, E.L.; Galati, G.; Kluwe, W.M. Functional differentiation of cytotoxic cancer drugs and targeted cancer therapeutics. *Regul. Toxicol. Pharmacol.* **2014**, *70*, 46–53. [CrossRef] [PubMed]
14. Martin, H.C.; Grant, W.; John, R.J.; John, D.; Jogarao, G.; Atiqur, R.; Kimberly, B.; John, L.; Sung, K.K.; Rebecca, W.; et al. Approval summary for imatinib mesylate capsules in the treatment of chronic myelogenous leukemia. *Clin. Cancer Res.* **2002**, *8*, 935–942.
15. Druker, B.; Guilhot, F.; O'Brien, S.G.; Gathmann, I.; Kantarjian, H.; Gattermann, N.; Deininger, M.W.N.; Silver, R.T.; Goldman, J.M.; Stone, R.M.; et al. Five-year follow-up of patients receiving imatinib for chronic myeloid leukemia. *N. Engl. J. Med.* **2006**, *355*, 2408–2417. [CrossRef] [PubMed]
16. Christopher, F. Targeted chronic myeloid leukemia therapy: Seeking a cure. *J. Mang. Care Pharm.* **2007**, *13*, S8–S12.
17. Jiao, Q.; Bi, L.; Ren, Y.; Song, S.; Wang, Q.; Wang, Y.S. Advances in studies of tyrosine kinase inhibitors and their acquired resistance. *Mol. Cancer* **2018**, *17*, 36. [CrossRef]
18. Darwish, I.A.; Khalil, N.Y.; AlZeer, M. ICH/FDA Guidelines-Compliant Validated Stability-Indicating HPLC-UV Method for the Determination of Axitinib in Bulk and Dosage Forms. *Curr. Anal. Chem.* **2020**, *16*, 1106–1112. [CrossRef]
19. Khandare, B.; Musle, A.C.; Arole, S.S.; Popalghat, P.V. Analytical method development and validation of olmutinib bulk drug as per ICH Q2 guidelines by using RP-HPLC Method. *J. Drug Deliv. Ther.* **2019**, *9*, 608–611.
20. Khalil, N.Y.; Darwish, I.A.; Alshammari, M.F.; Wani, T.A. ICH Guidelines-compliant HPLC-UV Method for Pharmaceutical Quality Control and Therapeutic Drug Monitoring of the Multi-targeted Tyrosine Kinase Inhibitor Pazopanib. *S. Afr. J. Chem.* **2017**, *70*, 60–66. [CrossRef]
21. Latha, S.T.; Thangadurai, S.A.; Jambulingam, M.; Sereya, K.; Kamalakannan, D.; Anilkumar, M. Development and validation of RP-HPLC method for the estimation of Erlotinib in pharmaceutical formulation. *Arab. J. Chem.* **2017**, *10*, S1138–S1144. [CrossRef]
22. Ashok, G.; Mondal, S.; Ganapaty, S.; Bandla, J. Development and validation of stability indicating method for the estimation of pazopanib hydrochloride in pharmaceutical dosage forms by RP-HPLC. *Pharm. Lett.* **2015**, *7*, 234–241.
23. Bende, G.; Kollipara, S.; Kolachina, V.; Saha, R. Development and validation of a stability indicating RP-LC method for determination of imatinib mesylate. *Chromatographia* **2007**, *66*, 859–866. [CrossRef]

24. Hajmalek, M.; Goudarzi, M.; Ghaffari, S.; Attar, H.; Mazlaghan, M.G. Development and validation of a HPTLC method for analysis of sunitinib malate. *Braz. J. Pharm. Sci.* **2016**, *52*, 595–601. [CrossRef]
25. Dutta, D.; Das, S.; Ghosn, M. Validated HPTLC method for the determination of nintedanib in bulk drug. *Proceedings* **2019**, *9*, 22. [CrossRef]
26. Vadera, N.; Subramanian, G.; Musmade, P. Stability-indicating HPTLC determination of imatinib mesylate in bulk drug and pharmaceutical dosage form. *J. Pharm. Biomed. Anal.* **2007**, *43*, 722–726. [CrossRef] [PubMed]
27. Mhaske, D.V.; Dhaneshwar, S.R. Stability indicating HPTLC and LC determination of dasatinib in pharmaceutical dosage form. *Chromatographia* **2007**, *66*, 95–102. [CrossRef]
28. Reddy, C.N.; Prasad, P.; Sreedhar, N.Y. Voltammetric behavior of gefitinib and its adsorptive stripping voltammetric determination in pharmaceutical formulations and urine samples. *Int. J. Pharm. Pharm. Sci.* **2011**, *3*, 141–145.
29. Rajesh, V.; Jagathi, V.; Sindhuri, K.; Devala Rao, G. Spectrofluorimetric method for the estimation of Erlotinib hydrochloride in pure and pharmaceutical formulations. *E-J. Chem.* **2011**, *8*, S304–S308. [CrossRef]
30. Mandal, B.; Balabathula, P.; Mittal, N.; Wood, G.C. Himanshu Bhattacharjee. Development and validation of a spectrofluorimetric method for the determination of Erlotinib in spiked human plasma. *J. Fluoresc.* **2012**, *22*, 1425–1429. [CrossRef]
31. Zawaneh, A.H.; Khalil, N.N.; Ibrahim, S.A.; Al-Dafiri, W.N.; Maher, H.M. Micelle-enhanced direct spectrofluorimetric method for the determination of linifanib: Application to stability studies. *Luminescence* **2017**, *32*, 1162–1168. [CrossRef]
32. Maher, H.M.; Alzoman, N.Z.; Shehata, S.M. An eco-friendly direct spectrofluorimetric method for the determination of irreversible tyrosine kinase inhibitors, neratinib and pelitinib: Application to stability studies. *Luminescence* **2017**, *32*, 149–158. [CrossRef]
33. Padmalatha, H.; Vidyasagar, G. Development and validation of UV spectrophotometric method for the determination of erlotinib in tablet formulation. *Imp. J. Med. Org. Chem.* **2011**, *1*, 26–30.
34. Sankar, F.G.; Latha, P.V.; Krishna, M.V. UV-spectrophotometric determination of imatinib mesylate. *Asian J. Chem.* **2006**, *18*, 1543–1544.
35. Sankar, D.G.; Rajeswari, A.; Babu, A.N.; Krishna, M.V. UV-spectrophotometric determination of dasatinib in pharmaceutical dosage forms. *Asian J. Chem.* **2009**, *21*, 5777–5779.
36. Annapurna, M.M.; Venkatesh, B.; Chaitanya, R.K. Analytical techniques for the determination of erlotinib hcl in pharmaceutical dosage forms by spectrophotometry. *Chem. Sci. Trans.* **2014**, *3*, 840–846.
37. Khandare, B.; Dudhe, P.B.; Upasani, S.; Dhoke, M. Spectrophotometric determination of vandetanib in bulk by area under curve and first order derivative methods. *Int. J. PharmTech Res.* **2019**, *12*, 103–110. [CrossRef]
38. Annapurna, M.M.; Venkatesh, B.; Chaitanya, R.K. New derivative spectrophotometric methods for the determination of erlotinib hydrochloride (a tyrosine kinase inhibitor). *Indo Am. J. Pharm. Res.* **2013**, *3*, 9270–9276.
39. Khandare, B.; Musle, A.C.; Arole, S.S.; Pravin, V.; Popalghat, V. Spectrophotometric determination of olmutinib in bulk by area under curve and first order derivative methods and its validation as per ICH guidelines. *J. Drug Deliv. Ther.* **2019**, *9*, 349–354. [CrossRef]
40. Souria, E.; Amoon, E.; Ravarib, N.S.; Keyghobadia, F.; Tehrania, M.B. Spectrophotometric methods for determination of sunitinib in pharmaceutical dosage forms based on ion-pair complex formation. *Iran. J. Pharm. Res.* **2020**, *19*, 103–109.
41. Rani, G.U.; Chandrasekhar, B.; Devanna, N. Extractive colorimetric method development and validation for erlotinib in bulk and tablet dosage form. *J. Appl. Pharm. Sci.* **2011**, *1*, 176–179.
42. Balaram, V.M.; Rao, J.V.; Khan, M.M.; Sharma, J.V.; Anupama, K. Visible spectrophotometric determination of imatinib mesylate in bulk drug and pharmaceutical formulations. *Asian J. Chem.* **2009**, *21*, 5241–5244.
43. Wani, T.A.; Ibrahim, I.A. A novel 96-microwell-based high-throughput spectrophotometric assay for pharmaceutical quality control of crizotinib, a novel potent drug for the treatment of non-small cell lung cancer. *Braz. J. Pharm. Sci.* **2015**, *51*, 439–447. [CrossRef]
44. Alzoman, N.Z.; Alshehri, J.M.; Darwish, I.A.; Khalil, N.Y.; Abdel-Rahman, H.M. Charge–transfer reaction of 2,3-dichloro-1,4-naphthoquinone with crizotinib: Spectrophotometric study, computational molecular modeling and use in development of microwell assay for crizotinib. *Saudi Pharm. J.* **2015**, *23*, 75–84. [CrossRef] [PubMed]
45. Wennborg, H.; Bonde, J.P.; Stenbeck, M.; Olsen, J. Adverse reproduction outcomes among employee in biomedical research laboratories. *Scand. J. Work Environ. Health* **2002**, *28*, 5–11. [CrossRef] [PubMed]
46. Lindbohm, M.L.; Taskinen, H.T.; Sallman, M.; Hemminki, K. Spontaneous abortions among women exposed to organic solvents. *Am. J. Ind. Med.* **2007**, *17*, 449–463. [CrossRef] [PubMed]
47. Wennborg, H.; Lennart, B.; Harri, V.; Gösta, A. Pregnancy outcome of personnel in Swedish biomedical research laboratories. *J. Occup. Environ. Med.* **2000**, *42*, 38–446. [CrossRef]
48. Kristensen, P.; Hilt, B.; Svendsen, K.; Grimsrud, T.K. Incidence of lymphohaematopoietic cancer at university laboratory: A cluster investigation. *Eur. J. Epidemiol.* **2008**, *23*, 11–15. [CrossRef]
49. Darwish, I.A. Analytical study for the charge-transfer complexes of losartan potassium. *Anal. Chim. Acta* **2005**, *549*, 212–220. [CrossRef]
50. Sittampalam, G.S.; Kahl, S.D.; Janzen, W.P. High-throughput screening: Advances in assay technologies. *Curr. Opin. Chem. Biol.* **1997**, *1*, 384–391. [CrossRef]
51. Burbaum, J.J. Miniaturization technologies in HTS: How fast, how small, how soon? *Drug Discov. Today* **1998**, *3*, 313–322. [CrossRef]

52. Alzoman, N.Z.; Sultan, M.A.; Maher, H.M.; Alshehri, M.M.; Wani, T.A.; Darwish, I.A. Analytical study for the charge-transfer complexes of rosuvastatin calcium with π-acceptors. *Molecules* **2013**, *18*, 7711–7725. [CrossRef]
53. Gohar, G.A.; Habeeb, M.M. Proton transfer equilibria, temperature and substituent effects on hydrogen bonded complexes between chloranilic acid and anilines. *Spectroscopy* **2000**, *14*, 99–107. [CrossRef]
54. Karipcin, F.; Dede, B.; Caglar, Y.; Hur, D.; Ilican, S.; Caglar, M.; Sahin, Y. A new dioxime ligand and its trinuclear copper(II) complex: Synthesis, characterization and optical properties. *Opt. Commun.* **2007**, *272*, 131–137. [CrossRef]
55. Makuła, P.; Pacia, M.; Macyk, W. How to correctly determine the band gap energy of modified semiconductor photocatalysts based on UV–Vis spectra. *J. Phys. Chem. Lett.* **2018**, *9*, 6814–6817. [CrossRef] [PubMed]
56. Vogel, A.I.; Tatchell, A.R.; Furnis, B.S.; Hannaford, A.J.; Smith, P.G. *Vogel's Textbook of Practical Organic Chemistry*, 5th ed.; Longman Group UK Ltd.: London, UK, 1989.
57. Polarity Index. Available online: http://macro.lsu.edu/howto/solvents/polarity%20index.htm (accessed on 24 November 2020).
58. Benesi, H.A.; Hildebrand, J. *Physical Pharmacy*, 4th ed.; Lea & Febiger: Philadelphia, PA, USA, 1993; p. 266.
59. Pandeeswaran, M.; Elango, K.P. Solvent effect on the charge transfer complex of oxatomide with 2,3-dichloro-5,6-dicyanobenzoquinone. *Spectrochim. Acta A* **2006**, *65*, 1148–1153. [CrossRef] [PubMed]
60. ICH Guideline, Q2(R1). *Validation of Analytical Procedures: Text and Methodology*; The International Conference on Harmonization: London, UK, 2005.
61. Skoog, D.A. *Principle of Instrumental Analysis*, 3rd ed.; Saunders: New York, NY, USA, 1985.
62. Darwish, I.A.; Khalil, N.Y.; Darwish, H.W.; Alzoma, N.Z.; Al-Hossaini, A.M. Spectrophotometric and computational investigations of charge transfer complexes of chloranilic acid with tyrosine kinase inhibitors and application to development of novel universal 96-microwell assay for their determination in pharmaceutical formulations. *Spectrochim. Acta A* **2021**, in press. [CrossRef]

MDPI
St. Alban-Anlage 66
4052 Basel
Switzerland
Tel. +41 61 683 77 34
Fax +41 61 302 89 18
www.mdpi.com

Molecules Editorial Office
E-mail: molecules@mdpi.com
www.mdpi.com/journal/molecules

www.ingramcontent.com/pod-product-compliance
Lightning Source LLC
LaVergne TN
LVHW070737100526
838202LV00013B/1253